Care, Power, Information

This book is a critique and provincialization of Western social science and Global Northern academia, by the author of *The Digital Coloniality of Power.* It exposes shared colonial and extractive rationalities and histories of research, higher education, digitalization, and bioeconomy while proposing in the idea of *BluesCollarship*, a sketch for an alternative culture of worlding and commoning knowledge work and for making care matter in research and higher education.

In a discourse analysis and provincialization of research and higher education, a tradition of elitist White-Collaredness in academia and in the social sciences, in particular, is criticized, and an alternative attitude towards the production, transfer, and use of knowledge – *BluesCollarship* – is proposed. The latter is rooted in a different idea of what "infrastructure" is, and in practices of decoloniality. Noting the current political climate of propaganda and populism, the persistence of social inequalities as well as of racism and misogyny, it is proposed that *how* people give warrant for knowledge claims should be reviewed under different terms. A coherent theme is that there is a genealogical root for current neo-extractive and neo-colonial rationalities in the Athenian idea of *oikos*, which conflates family, household, and property.

In taking a distinctly *writerly* approach – rather than giving ready-made answers – the book aims at permanently provoking readers at every turn to think further, as well as *before-and-beyond* what is written, but to do so in *thinking together with Others.* Thus the book addresses scholars and students from across the social sciences who seek challenges to established ways of thinking in academia without simply replacing one canon for another. This book is for those who think of themselves as knowledge and culture laborers in this age of precarization, who seek to replace the university and cognitive capitalism with a pluriversity and an infrastructure built on knowledge and culture as fundamental values.

Alexander I. Stingl is a WIRL-CoFund Fellow at the Institute for Advanced Studies and Centre for Interdisciplinary Methodologies at the University of Warwick. He is the author of *The Digital Coloniality of Power.*

Routledge Advances in Sociology

268 Urban Environments for Healthy Ageing
A Global Perspective
Edited by Anna P Lane

269 Conflict and the Social Bond
Peace in Modern Societies
Michalis Lianos

270 Boundaries of European Social Citizenship
EU Citizens' Transnational Social Security in Regulations, Discourses and Experiences
Edited by Anna Amelina, Emma Carmel, Ann Runfors and Elisabeth Scheibelhofer

271 Cultures, Citizenship and Human Rights
Edited by Rosemarie Buikema, Antoine Buyse and Antonius Robben

272 Regional Cultures, Economies, and Creativity
Innovating Through Place in Australia and Beyond
Edited by Ariella Van Luyn and Eduardo de la Fuente

273 Care, Power, Information
For the Love of BluesCollarship in the Age of Digital Culture, Bioeconomy, and (Post-)Trumpism
Alexander I. Stingl

274 The Field of Water Policy
Power and Scarcity in the American Southwest
Franck Poupeau with Brian F. O'Neill, Joan Cortinas Muñoz, Murielle Coeurdray and Eliza Benites-Gambirazio

For more information about this series, please visit: www.routledge.com/Routledge-Advances-in-Sociology/book-series/SE0511

Care, Power, Information

For the Love of BluesCollarship in the Age of Digital Culture, Bioeconomy, and (Post-)Trumpism

Alexander I. Stingl

Routledge
Taylor & Francis Group
NEW YORK AND LONDON

First published 2020
by Routledge
605 Third Avenue, New York, NY 10017

and by Routledge
2 Park Square, Milton Park, Abingdon, Oxon OX14 4RN

First issued in paperback 2021

Routledge is an imprint of the Taylor & Francis Group, an informa business

Publisher's Note
The publisher has gone to great lengths to ensure the quality of this reprint but points out that some imperfections in the original copies may be apparent.

British Library Cataloguing-in-Publication Data
A catalogue record for this book is available from the British Library

Library of Congress Cataloging-in-Publication Data
Names: Stingl, Alexander I., author.
Title: Care, power, information : for the love of bluescollarship in the age of digital culture, bioeconomy, and (post-)Trumpism / Alexander I. Stingl.
Description: Milton Park, Abingdon, Oxon ; New York, NY : Routledge, 2020. |
Series: Routledge advances in sociology | Includes bibliographical references and index.
Identifiers: LCCN 2019039341 (print) | LCCN 2019039342 (ebook) | ISBN 9781138100954 (hardback) | ISBN 9781315657301 (ebook)
Subjects: LCSH: Social sciences—Philosophy. | Social sciences—Western influences. | Equality.
Classification: LCC H61 .S86 2020 (print) | LCC H61 (ebook) | DDC 300.1—dc23
LC record available at https://lccn.loc.gov/2019039341
LC ebook record available at https://lccn.loc.gov/2019039342

ISBN 13: 978-1-03-208271-4 (pbk)
ISBN 13: 978-1-138-10095-4 (hbk)

Typeset in Bembo
by codeMantra

In memory
of
Donald N. Levine
(1931–2015)

Contents

Foreword

Stories

It is from a place and time, from a position that one can feel a current of rampant anti-intellectualism on the one hand, to which a response seems lacking by a certain class of "oligarchic elite" in "the academy/university" that, on the other hand, comes from and ushers in privilege while also embracing neoliberalization sometimes openly, sometimes under the covers. It is this combination, which must be opposed, resisted, and thoroughly taken from behind from a transdisciplinary view with an appreciation for the ([Settler] colonial) histories of the knowledges involved, and with an appreciation for all kinds of "crafts(wo)manship". In this regard, in a recent essay, Achile Mbembe (2015) describes it pretty well:

> In these times of urgency, when weak and lazy minds would like us to oppose "thought" to "direct action"; and when, precisely because of this propensity for "thoughtless action", everything is framed in the nihilistic terms of power for the sake of power – in such times what follows might mistakenly be construed as contemptuous.
>
> And yet, as new struggles unfold, hard questions have to be asked. They have to be asked if, in an infernal cycle of repetition but no difference, one form of damaged life is not simply to be replaced by another.

A Sensei from Afar

This book is dedicated to Donald N. Levine, or rather, now, it is dedicated in memory of and for his legacy. A dedication for Don was already intended to be part of this book I first drew up concrete plans for the manuscript, but at that time Don was still with us. I would have liked nothing better than to give him a copy in person, expressing once again my gratitude to him. I last met Don in person during the American Sociological Association meeting in 2013 in New York, if I recall correctly, where he gave a speech in memory of the recently passed Robert Bellah, a speech that was both kind and humble yet managed to celebrate the grandiose intellect, academic rigor, and political courage of Bellah. In the days afterward, we shared some anecdotes and news. I told him of my recent adventures we shared in my (at the time still existent, now expired) personal luck and my struggles with and disappointments in academia today. On his part, Don let me in on his publication plans and his concerns over whether or not he would still be able to realize them, having found out that he was seriously ill (which he did not make a secret, so I can mention it here), which – he was perhaps even more concerned about – might also prevent him from traveling once again to Northern Africa (these, too, are not confidential details, I have to add, but aspects Don made widely known). It was the last time that we met in person, though shortly after he sent me one of the few desk-copies that he had produced of a book, collecting his writings on Aikido. It contained the kindest and most encouraging dedication scribbled on the front page that I have ever received from any author friend. But even with all this told, it does not do justice to the kind of scholar Don was, which is the example of what a *Gentleman Scholar* **should** be. And, please, let me emphasize that I believe – and Don would be the first to agree – that this should not be related to any terms of male privilege or paternalism. Quite on the contrary, there must be room in dialogue for scholars (and other interested participants) of all kinds (and ethnicities, and religions, and sexes, and genders, and so forth). I like the idea of the *gentleman scholar* in the same way I like the idea of *feminist theorists, modest cyborgs, daft punks,* and *mad critics,* because those who take these ideas and what they stand for seriously are allies; they can be allies with one another as well as allies in countering what I call *White Collar Sociologists.* Don was, in my view, *a* very possible anti-thesis and anti-dote to the *White Collar Sociologist* even if such a person, like Don, can still be located in the discourse of the theoretical mainstream and cite and study the canonical figures.

Perhaps, to make this clearer, I should tell the story how Don and I met in the context of my previous academic experiences:

I was working on my dissertation from 2005 to 2008 at a German University, which was an interesting and eye-opening time for numerous reasons. But certainly, I can say, that during this time, it became evidently clearer and clearer that concept of "care" is not considered to be part of the relationship between an advisor and a PhD candidate in Germany. I had on occasion read about this as a problem and more often heard about this, before beginning my own PhD adventure, but did not necessarily believe this to be true as a general rule. But my experiences confirmed the rumors, and over time I met many PhDs and PhD candidates whose experiences compare to and, sadly, often rival mine. Some are willing to write about this, but the majority is too afraid of negative consequences for their careers if they were to speak up (fearing to be identified even if they were to write under a pseudonym). Contemporary academia in general, but specifically with the bureaucratic path-dependencies in Germany, punishes any non-allowed form of dissent – yes, there are some "formulaic" and "conventionalized" forms of permissible dissent, which is then, however, not dissent anymore; Michel Foucault, among others, had already written about these mechanisms. And it should be said that, with the tightening of labor markets, the supposed over-production of graduates, and the encouragement of a hyperinflation of applications ("Apply for every job that you might even remotely be a fit for!"), while many jobs are realistically handed down within a network of allies and politics, have led to a situation where the academic-workforce is large enough that any single act of non-conformation is punishable by career-death. Again, Foucault (and others) has written extensively about the emergence of this situation in the 18th and 19th century in modern practices of labor force control, although it leads to a *re-prebendalization* and/or *re-feudalization* of the worlds of work in general and in academia in particular. As David Graeber attests to this in a recent interview (Graeber/Jeffries 2015), culminating in the judgment that

> [a]cademia was, he [Graeber] muses, once a haven for oddballs – it was one of the reasons he went into it. "It was a place of refuge. Not any more. Now, if you can't act a little like a professional executive, you can kiss goodbye to the idea of an academic career."

But within sociology, this is, I think, also a problem of (many sociologists' unreasonable "fear of equivocation [not mention abstraction, and speculation]" and their subsequent overemphasized attention to) translation. Don, instead, managed to take up the conflicting or violent energy and turn them into opportunities for dialog. He struck up a dialog with me, among many other people. What I learned from him was to try and see the conflict within academia and the structural violence in it, against dissenters and against different voices and ideas, and try and turn those into opportunities – sometimes I succeed in trying (whether my efforts succeed depends on further factors) sometimes, I, too, have bad days, but from Don I also learned that one must be persistent in effort but that doesn't mean one can always succeed in mastering it. But we must create opportunities from within the problems as they present themselves.

Accordingly, in 2007, I received a tiny amount of funding to travel to the United States and do some archive research and conduct some interviews for my dissertation. My research (and dissertation) focus then was on the history of the concepts of the social sciences, ending with the Social Relations movement and its central figure Talcott Parsons. My interest in this topic lay in a certain distrust that I developed for sociology's origin myths (specifically those focusing on economic ideas) that sociological textbooks and most of my previous teachers offered. These myths seemed like nicely adorned, comfortable and complacent, and suspiciously neat just-so stories; but something about them and the narrative behind them never made sense to me whenever I read Simmel, Weber, Parsons, Foucault, and other sociological and affiliated authors, specifically when I read them in light of Kant and in light of my knowledge about biology (and biomedicine) and its history. So, eventually I did a lot of archive work, especially on fatigue research and on social relations research at the Harvard University Archives, including an extensive investigation into the papers of Talcott Parsons, Chester Barnard, Elton Mayo, and Lawrence Henderson. During this time, I got in touch with Victor Lidz, a former student of Parsons, who invited me to come to Philadelphia and have a talk with him. I booked a flight and an acquaintance from Boston managed to organize a room for me with an old friend of his. Victor, who was also amazingly kind and generous to me, organized two further conversations: For one, he put me in touch with another former assistant of Parsons, Mark Gould at Haverford, with whom I met for another interview over a late lunch outside of Philadelphia; and for the second, Victor drove us to Harold Bershady's house, picking up Renée Fox on the way. So I ended up, luck had it, spending an afternoon with three amazing and impressive sociologists, who I already greatly admired. I think that few people who know her or of her will be surprised by the lasting impression that Renee Fox left on me. She is one of the most courageous, smart, and strong persons I have met in my life, however brief our few exchanges were, and she remains an incredible inspiration.

The kindness of these people, the time for and, above all, "care" they took of me, at the time a mere PhD student, was something I was not used to from my native German academic milieu.

Whatever their level of prestige and merit, you will believe me when I say that it is little wonder that I was already impressed with the warm reception by this small group of scholars in the United States around Don – and then some: a warm reception by professors of great international reputation. I also have to emphasize here that many (though not all) of the German professors in question are male professors – there still is, after all, a large gender division in academia in Germany as well. I have to emphasize this, because I do have some great and wonderful experiences with at least a handful of German professors, but almost all of them are women.[1] In particular, I am very grateful to have had the opportunity to study with Elisabeth Beck-Gernsheim, who has shaped the way I teach today a lot through her example of being an excellent teacher as well as being an outstanding researcher (typically, one does often hear her male colleagues misrepresent the work she and her late husband did together, emphasizing in utter misconstrual his contribution over hers), and who was also instrumental in my decision to pursue an academic career and with a focus around sociology by encouraging such a choice.[2] In light of working on my dissertation, I also had the, however brief, opportunity or other to talk with Uta Gerhardt, both on the phone and at conferences, and she gave me a lot of insights and advise, specifically about archive research in general and the Parsons archives in particular. In her work, one can find an example of what good research should look like, since she is one of the most meticulous and rigorous researchers I have ever read (or met). I found in both, Elisabeth Beck-Gernsheim and Uta Gerhardt, two amazing, intelligent, meticulous, patient, and kind scholars. Now, one is left to wonder, why is it that in Germany, many male professors of sociology I encountered during my time as a graduate student, during my dissertation (my committee included), and as a post-doc kept "warning" me on a variety of occasions about certain professors - without exception this warning was issued about a professor who was a woman each - by calling them "difficult" (German: *schwierig*)? Now I don't have to answer that, do I? One can easily see the misogynist attitude in play, I guess (see Manne 2018).

I may point out, too, that this may have at least partially changed in the last decade, since I was a PhD student – but to what degree it's really changed, well, that's another question. Misogynistic attitudes and practices can, of course, be found in the entirety of Western academia – and other kinds of discrimination should be mentioned, too. In Germany, however, it is conflated with the occasional return to "mandarin" culture (precisely in correlation with the rise of the administrative class in higher education), which gives it a characteristic flavor, while my first experiences with US-American scholars were so radically different (although this, too, is not necessarily representative for present US-American academia at large, where certainly all is not well). And Don, well, Don was an above-and-beyond exception even herein in the interactions not only I had with him but also those I witnessed and was told about by others.

Shortly, before my trip to Philadelphia to meet Victor, I wrote an email to Don. It was basically a cold approach, I didn't know Don at the time, I only had read about his work on Simmel, and I was interested in the context of the missing Simmel chapters in Parsons's early work – Simmel's debt to Kant and Lotze, as well as his early work on yodeling and knowledge of physiology and ethnology seemed to me an interesting problem in light of the history of ideas. I never thought I'd meet Don in person, who was a professor in Chicago, since a trip to Chicago certainly wasn't possible for time and budget reasons. When he learned that I was visiting Victor, he was ecstatic, pointing out that he'd be in Philadelphia soon, visiting family. Via email, we compared itineraries and it turned out that he'd be leaving exactly one day before I arrived in Philadelphia. I proposed a phone interview after I returned from Philadelphia. Shortly after, I received an email from Don. He postponed his flight for a day, so we could meet for lunch at the airport, crossing lines of flight, quite literally. I want you to imagine this: Here you have a German PhD candidate who is used to not being taken seriously in a system where you are told that you are only allowed a voice once you have become a full, tenured professor. And then here you have this German nobody student meet with Don, who postpones his flight for a day so this student had a chance for a conversation with him. This is how Don and I met, and this is the Donald Levine I will carry with me for the rest of my life.

But there is more to Don, the sociologist, the researcher, the educator, the college dean, the diplomat, the expert on Ethiopia, the Aikido sensei, among many other things that he seemed to bring together and fill with life with an unbelievable ease. I could unfold his many involvements, ideas, and his work in further detail, but I won't. If you are interested, it's more fun to find out yourself. The one thing I will point out, though, is the organization Aiki-extensions that Don helped bring to life. It is an organization that uses Aikido to resolve conflicts and to promote dialog and understanding. It's a great idea, and I understand the reason behind Don's involvement. They have created a memorial fund in Don's name, which I personally think is worthy of support.

Therefore, the following book is, among other things and like the book before it, an attempt at an aikido move in sociology: to identify antagonistic flows and turn them into opportunities for new directions.

Narratives: How to Place This Text

Care, Power, Information is, basically, the second in a trilogy of books – although it could be said that the as of yet unpublished studies in the manuscript titled *Anthropos's Scaffoldings* (circulated as a sort of gray book for now but not for long) serve as a sort of prequel to this trilogy.

Following in spirit if not entirely in method a line of inquiry first proposed by Hans Blumenberg as the question of the "legitimacy of the modern age", the three books embark on a shared and successive line of flight (which makes them into a trilogy of sorts) that engages the inherent contractions, structures of violence, but also the opportunities in inquiring after the *legitimacy of the digital age.*

In the first book, I have proposed that, in a sense, whatever digital culture is or assumes to be, it's not really anything "new under the sun". In claiming that we do have an opportunity to create decolonial options for the social sciences and a sociology of opportunity, I deployed – not a novel, but what I hope to be – a productive and writerly (Barthes) approach across (rather than from within) the sociological imagination, and created a small number of (exemplary) horizontal and, thus, topical platforms for future conversations with sociology in an integrative interdisciplinary register. The main claim of this book is that Digital Culture is merely an extension of the *coloniality of power and Being*, inherently silencing Radical Otherness and constructing its Other(s) as *disposable.*

With this present book – *Care, Power, Information*, named after a university course I designed and taught – I deploy (to remain in the spatial metaphoric realm) *mycelial platforms* to complement the *hyberbolic geometry* of the prior book. My aim is to *think in* the relations between these three concepts that allow us to *think with* these relations across different topics and unearth interdependencies, cross-connections, and analogies. In the first book, specifically its first chapter, as well as in the future third book, I pay a lot more attention to decolonial thinkers (Mignolo, Lugones, Dussel) and postcolonial sociologists (Go, Decoteau) than I do in this second book. This isn't due to an ignorance of some kind, but due to the fact that I try to elucidate – for the bigger project – what Western/Global Northern Society can know about itself in its own terms, since it can hardly pretend to speak in any other terms than its own without once again illegitimately appropriating and colonizing the terms of Others. Instead, I attempt a provincialization.

It could be said that the present book is not a "good" book (with Dussel, I question the "goodness" claim, and with different writing styles in each chapter – some more hermetic and convoluted than others, some annoyingly vulgar – I challenge the norms of academic writing), nor a nice one, not an easy one, for sure. It's what I have come to call my "a little bit crazy book", for it's a book written in, as, and for a transition. It is also an "angry" book. And while many of its parts appear as though they can exist individually, it's meant to be read as one. It is "one and many", and cannot be otherwise or either. The themes run through it as a mycelium. It is a cognitive exercise, which must be conducted whole. It's not a journey that can be interrupted or experienced in chosen pieces. It's the mad adventure of a daft punk – my, am I not pompous, obnoxious? [While reading, keep "tongue-in-cheek".]

In the eventual and yet-to-be-written third and final book (the most serious one, it will be [almost] fun-free, I pinky-swear), I will propose an approach towards a theory of justice, which argue that we cannot have *a* theory of this kind, but we can have insights into the (political) ontologies of production of how justice can become expressed/articulated in various forms; and I also converse with decolonial thinkers, i.e. what I tentatively call a theory of generative values for relational justice.

Each book is written in a different style and has various "moving parts", wherein the interlocking mechanism between the parts differs within each book: If *Anthropos's Scaffoldings* is truly the prequel to this trilogy, then it serves well that it is a collection of largely independent studies, held together by a few loose questions that were leftovers from my dissertation research and grad years. The trilogy's first book, *The Digital Coloniality of Power*, was (and still is) intended to be read from cover to cover and it does have a – however not always immediately obvious – connecting thread running through it (namely how Global Northern processes of classifications are producing both a form of *inclusive exclusion* and of *silencing radical Other[ness]*, which have been a constitutive process of colonization [since the 16th century], of Western social science [since mid-19th century], and of digitialization [which many see beginning after 1945, but which is deeply rooted in the late 19th century]) – Lisa Jo van den Scott wrote a very kind but also fair review in *Symbolic Interaction* (fair, since she does point out the one or two, in retrospect, obvious weaknesses of the book), which also serves as a good overview for the book. This present book, the second in the trilogy, presents as a different animal. Its various parts are highly interdependent, but they can each be read separately also (there are different *ways of reading* this book possible). Their goal is to deploy the complex relational systematicity (not: system!) of *care*, *power*, and *information* as a tool to think with. This present book is, also, not linear in its argument, style, temporalities, and direction of thought. Where in *The Digital Coloniality of Power* these dimensions were imagined in terms of (crocheted) coral reefs, *Care Power Information's* imaginary is that of mycelia and myxomycetes (or slime molds). Finally, the third book, yet to be written, will be the most "traditional" academic book, although it is largely a book in political philosophy and not originally rooted in sociology, in that it develops, step by step and preceded by "state of discussion/literature", an argument or theory.

Each style was chosen to fit the topic(s) presented in the book, though, in a way that could help elucidate how to actually "think" these topics best, without presenting merely as *readerly* texts. In the spirit of Roland Barthes, among others, I think of *readerly texts* as points of closure, which the reader is presented with to accept as "the knowledge they *are*". *Writerly* texts, on the other, are open texts that the reader can, if you like, continue to write or "do something with". I hope to have achieved this goal, of producing such texts throughout the project. I understand that *writerly* texts are less liked by certain readers, specifically students are often not encouraged to take up such texts by sociology professors (for reasons that should become clear throughout this book), but also by reviewers and critics, and, subsequently, they are often shunned by editors who have to consider (university library) markets. Nonetheless, I claim that both universities (and their libraries) and books (to even deserve that name), in order to be places of opportunity – and by "opportunity", I precisely do not mean this perverted noun from the discourse of Neoliberal®Speak™ – and potential (*potentia* [Dussel]) must retain substantial niches for *writerly* interventions and adventures. The case in point is that readerly texts and efforts also only occupy niches, and that the general ecology is not a universal space but comprised of a manifold of niches.

In *The Digital Coloniality of Power*, I also made similar but maybe less potent remarks on the need I see for developing writing style(s). I am not a Caribbean thinker, nor am I French, but I feel a kindred spirit with what Lewis Gordon told me in a conversation we had on writing/writers in Paris in spring 2018, and which he carved out so wonderfully in his seminal essay on Sylvia Wynter (Gordon n.d.-b):

> I speak, of course, of the designation *writer*. An unusual feature of the Caribbean writer, as with the French writer, is the inconceivability of limited scope. For such an individual, there may be initial interest – say, philosophical, fictional, poetic, historical, or natural scientific – but

> in the end, the concern is more about what needs to be said than on the credentialing or locating of disciplinary identities of who says it. Thus, the sociologist, historian, philosopher, economist, psychiatrist, or dramatist becomes, ironically, more temporary clothing for the salient body of *thought*. I mention this because of the difficulty and inappropriateness of determining exactly what Sylvia Wynter *is* and what she *does*. She has, in effect, transcended the collapse of means-and-ends argumentation by literally making disciplinary formation a mere *tool* for her greater set of projects. True, she has a degree in Medieval Spanish literature, and it is also true that she has worked as a dancer, an actress, and she has written plays and a novel, and it is also true that she is Professor Emerita of Spanish and Portuguese Studies and Black Studies at Stanford University. These designations are for her opportunities with which she had to work through her struggle to comprehend things greater than herself, which are, in effect, greater than all of us. Heavily rooted in the currents of social life, she is attuned to the value of transcendence as what Karl Jaspers would call a *cipher* of where we are. Put differently, we must stay attuned to a "there" in order to understand what it means to be "here". In the course of such attunement, Wynter has sought many guides. Hers is a world of multiple thinkers from multiple perspectives. In some, she finds a kindred spirit.

I am also thinking of what the philosopher Gilles Deleuze in reference to a little book on Kant (which is more often cited in this context than it is actually read) sketched – not as a project but – as attitude, which is often quoted and often translated into other languages to sound even more provocative by making it seem more vulgar. While there is an ambiguity towards a reference of anal sex,[3] Deleuze wanting to "take Kant (and other authors) from behind" to make them "give birth to monstrous children" – alluding to an act of *immaculate conception*, the fertility of which he had speculated on in other works. Nonsurprisingly, this only spurred on the imagination of more attention-seeking authors such as Slavoj Žižek.

It is an essential idea of this project to conduct an act of taking from behind sociology itself by illustrating how such an act of taking from behind can be conducted and how it can be a productive endeavor. Precisely, I insist on the importance of doing so in order to identify, unmask, and resist today's caste of *White Collar Sociologists* and make their *project* (their "throwing forward"), upholding a comfortable and complacent status quo, impossible to maintain. This book is, to riff on something Anna Tsing said in talking about her book in a seminar at CSI Mines ParisTech in June 2018, maybe not an obsessively well-tended but much cared-for "garden of patches".

But I think NK Jemisin in conversation with Neil Gaiman recently put it quite well, what I am looking for with this book, which is also in a sense the second in a trilogy of *orthoquels*:

> There's making a statement, and then there's the engagement with people. People, politics, polity, all of these things—that is what politics is ultimately about. If you are showing, like you said, just the people you have in your life, if you are showing reality as it exists, people as they exist—that's political as fuck.[4]

Notes

1 If there were professors who ascribe to a different gender-type, such as transgender, then this was not public knowledge and I was not aware of it.
2 I regret that circumstances at the time prevented me from choosing her as my PhD advisor.
3 The German translator for Berlin-based publisher Merve Verlag did, as far as I know, indeed translate it as *Arschfick* ("assfuck").
4 March 2018 at https://lithub.com/on-writing-the-comics-and-queer-characters-we-need/.

Acknowledgments

In *The Digital Coloniality of Power*, I have included in the Foreword a brief account of how this yet-to-be-concluded book trilogy itself came to be and how some parts of that first book were originally orphaned writings. This present and second book in this trilogy holds some sections that owe their existence to similar occasions as well as presents completely new and improved versions of ideas that I have introduced in blogs, smaller papers, and chapters in recent years. Nonetheless, the textual material in this book itself should be considered original, in terms of me – to the best of my knowledge – not having published it previously in printed form – save occasional early drafts here and there as sneak-peak previews for the book on my blog. This book originally featured what I hoped would be the comprehensive account of what I call *semantic agency theory* (SAT) at the start of the book; unfortunately once that particular part of the material grew to a 30,000-word-long chapter of its own, which still was missing a few aspects I thought it should cover, I decided to take it out of this book and it will be published in a soon to be forthcoming book called *Wild Doings*. I also continually toyed with long and explicit sections on each Grexit, Brexit, and the so-called refugee crisis – and then, well, Trump. While, eventually, I had to let go of most of these sections, their influence can still be felt in many places where I adapted ideas and points of view that should give this book both the boon of timeliness and the asset of transcending the present moment. Some materials were originally produced for talks and presentations, and even though I wrote them out, I never published them. Among these, there is an (in)famous *Guerilla paper* given around 4S in Copenhagen – a conference I had refused to attend officially, protesting conference fees I deemed unreasonable and, above all, the lack of transparency. Several talks that were first inklings of what would become parts of this book in improved versions were given at meetings of American Sociological Association (ASA), Eastern Sociological Society (ESS), Society for the Study of Social Problems (SSSP), the "Changing Societies" workshop at the Social Science Center Berlin (WZB), Society for the Study of Symbolic Interaction (SSSI), and two so-called *Klausurwochen* – one in Tübingen and one in Bochum – where I was invited as a sort of "promising international junior researcher" financed by the German Federal Ministry of Research and Education (BMBF). I included ideas first presented in talks and guest lectures I gave at various opportunities at Leuphana University and at Rochester Institute of Technology. I included ideas, which I used at various opportunities to provoke the shit out of academics who I just found decadent – in other words, I take Foucault's notion of "permanent provocation" quite seriously. Most notably, this book is a direct result of courses I created and held at Leuphana University under the title *Care, Power, Information*. This book took too long to write. It was finally almost completely aborted in late 2018, after I lost first my backup drive and then my computer broke in quick succession – it was only with some luck that the manuscript itself survived, but the bibliography, contained in a Zotero file, was lost, and had to be painstakingly reconstructed as far as possible but many references and citations, I had to delete in the process, because I could not reconfirm their original source: Respectively, there was a significant delay; and, more importantly, I have digested several times (and originally referenced in text) the amount of literature eventually included in the bibliography – to cite as example the work of Mark Bevir or Sighard Neckel of whom you find one or two works cited each, where originally there were about a dozen texts referenced each (the advantage is, however, that with less references in the text, readability has increased). My sincere apologies to the publisher and the editing team for the delay in handing this manuscript. While the craziness and hardship of these past years are not an excuse, at least it is an explanation.

The work on this book, the actual and eventual manuscript was written between February 2015 and August 2016, between January and May 2017, finalized between December 2018 and March 2019, and it covered a time, which was – and in all likelihood continues to be – very difficult: Between 2014 and 2017, I lost a number of friends and acquaintances to cancer, most of them in their 30s and early 40s (thus, about my age). 2015 began as a year that was very promising, personally and professionally. Within months, professional perspectives and opportunities that seemed as good as guaranteed evaporated in the way that a lot of perspectives in academia vanish today. Eventually, in 2016, my personal life was equally affected… again. I have to thank the kindness of some old and some new friends as well as curious and engaged students, that it was possible for me to finish a book at all in this difficult time, to keep up my work ethic – however crazy and erratic it is to begin with – and to continue forward with what I deem one of the most important issues of our time: Namely, saving what can be called many things and have many flavors, but for which "critical thinking" is a good (if currently somewhat occupied) name, and creating new spaces for "learning together", of which "the academy" (if it takes seriously that which I call *BluesCollarship* and overcomes the worship of privilege it assigns to *White Collar Academia*) can be one such place once it understands that it must not be a uni- but a pluriversity.

The years since *The Digital Coloniality of Power* were difficult for me, and continue to be so: Trying to work in research while being critical of academia is an uphill battle. But these were and are not positive political times: From the Eurocrisis, to the refugee crisis, the rise of right-wing extremism, the election of Donald Trump, to the *gilets jaunes*, etc.... this book's sensitivity to its moment in time made it problematic to finish.

I have lost a lot in the years this manuscript was in production and I can identify each moment of a loss in the respective section when re-reading the manuscript. But in every problem, every obstacle, every loss, there is an opportunity to develop something new. That was the insight I began and ended my book *The Digital Coloniality of Power* with: There are opportunities in every paragraph of this book for the readers to use creatively. That I focus on creating opportunities, rather than on the problems and losses, when writing this book, is owed to the many people I met along the way. I am sure, I will not manage to do justice here to all of them with these few expressions of gratitude.

First and foremost, my gratitude is to the many students who attended my courses, guest talks, and lectures. I am also thankful for the constructive criticism that I received from the audiences for the talks. I wish to thank the wonderful presiders, panelists, and members of the audience of the nine-panel mini-conference "Decolonial Options for the Social Sciences" that I organized at the Eastern Sociological Society Meeting in Boston in March 2016 – it was worth every headache.

I want to thank the powers-that-are at the Society for the Study of Symbolic Interaction (SSSI), especially Patrick McGinty, for making the inclusion of an entire decoloniality panel into the program for the Seattle meeting in 2016 possible, and for being enthusiastic about the conversation to be established between SI and decolonial options; and I wish to thank the wonderful panelists.

I am grateful to the organizers and participants of the "Cold Bodies, Warm Machines" conference in Düsseldorf in fall of 2016, where I was invited to give one of my performance lectures, probably confusing the crap out of everybody. I am told there is a video-recording of that somewhere on the web. I had some fun conversations there with Laura Hille and Erich Hörl (both of whom I also know from Lüneburg), Felix Hüttemann, Katherine Kramer, Klaas Werner, and Fabian Saavedra-Lara.

Between 2011 and 2016, Sabrina M. Weiss enlightened me, lifted me up, and made my life better. I hope I did a little bit of the same in return.

A shout-out to the team of Hotel Central Nürnberg. Parts of this book were drafted on good old paper at the *Thirsty Baker* and the *Irish Castle* in Nürnberg, as well as the *Maelzer* in Lüneburg, a nod to the great waiting and bar staffs in these locations, especially Paul, Nicole, Mario and Jens. I am deeply indebted to Corinna Dartenne, and her family, who is a brilliant historian, educator, conversationalist, and warrior-queen in the fight against the neoliberalization of higher education, who also runs a small bed&breakfast in Lüneburg and who more than one time made the impossible possible to find a place for me to stay on short notice (even when already fully booked).

I only ever met her once very briefly in passing and we have only had very short exchanges over Facebook, but Donna Haraway has been such an inspiration over the past few years, and from watching talks of hers, I felt encouraged to include even more fun and laughter into my talks and my teaching.

2016 and 2017 also brought surprises and, perhaps, avenues I had not considered – I feel immense gratitude to the WZB in Berlin for the conference they organized in Berlin and Paris to create and facilitate French-German collaborations, and for inviting me. In particular Susanne Beer and Friederike Molitor were especially kind and are two amazing scholars as well as organizers in their own right. A big and heartfelt thanks also to Lisa Crinon. As a consequence, attending the International Panel on Social Progress (IPSP) meeting in Lisbon, Portugal and having many insightful conversations were a great boon, which profited this book in the end; especially conversations with Prabhat Patnaik, Akeel Bilgrami, Saskia Sassen, Ama De Graft-Aikins, and Vivian Lin stand out, but many others, especially the many student helpers, should be mentioned here, too. Most importantly, though, these events led to a spring/summer fellowship in Paris in 2017 with the *Fondation Maison des sciences de l'homme* (FMSH) and a research association with the program on *Économie du bien-être et de la justice sociale* at the *Collège d'études mondiales*, where I could work on my own stuff while also functioning as a moderator for the final editing phase of a three-volume publication with Cambridge UP under the umbrella of the IPSP. The fellowship, which included my staying at FMSH's Maison Suger in Paris would not have been possible without some financial support of the WZB (and funding they had available from the BMBF) and a large junk of financial support by both FMSH and the Collège, nor would it have been possible without the help of Gwenaëlle Lieppe, Sara Guidani, Alison Sanders, Thomas Huet, Charlotte Huber, Victoire Ekambi, and Nadia Cheniour. Above all, Olivier Bouin and Marc Fleurbaey extended the invitation, had many conversations with me, which created interesting directions to explore, and have been kind to me in very many ways that I have very little idea how to repay in kind.

The final words of this manuscript, certainly not the least substantial ones, were penned in 2018/19, during my year-long Independent Scholar Fellowship awarded in the ISF4 competition by the Independent Social Research Foundation (ISRF), which allowed me to return to Paris and hosted at FMSH and the *Collège*, and staying, again, at Maison Suger. My heartfelt thanks to Louise Braddock, Lars Cornelissen, Stuart Wilson, and Rachael Kiddey at ISRF for organizing everything so wonderfully and in a collaborative spirit from their end.

There are countless debts to a former collaborator of mine, Sabrina M. Weiss – whatever she is doing right now, since telling me she would leave academia and, thus, our generative partnership behind, she will excel at it.

I am indebted to Eben Kirksey for taking me up on my invitation to come to Paris and give an amazing talk, wherein we were introduced to chemo-ethnography.

I feel very grateful for the welcome I have received in Paris by Annick Sardeing, Marie Guerin, Stephanie Alexander, Ernesto Sferrazza, Stéphane Dufoix, Françoise Vergès, Thomas Huet, Olivier Bouin, and Marc Fleurbaey at the *Collège d'études mondiales* with which I remain affiliated, as well as, of course, the conversations with people one just "runs into", such as Wilhelm Heitmeyer, Elena Esposito, Charles T. Wolfe, Giuseppe Bianco, Glenda Andrade, Maura Benegiamo, Matteo Vagelli, Nurul Huda Mohd Razif, Geoffrey Pleyers, Marianne Rosa-Diab, Emilie Rebourg, Annalisa Pelizza, Amy Colin, and Bastienne Schulz. For wonderful spring and summer adventures with wonderful conversations, sharing (always too much of) wonderful food and drink in Paris, and helping me discover unknown parts of the city, I want to express my deepest thanks and appreciation to journalist *extraordinaire* Delphine de Freitas, Wonder Woman Manca Slana, and, of course, the amazing Sara Neumane (who is my favorite neuroscientist in the world).

I am deeply grateful for both the good times and letting me have calm times in the corner for reading at both The Bombardier – thanks to Ertug (R2), Vittorio, Danny, Craig, Breff, Marie Charlotte, Danielle, Aless, and Emily – and WOS Bar – thanks to Kahina, Dan, Pierre, and Iina, and all the regulars with whom I had so many amazing chats.

Regarding Lüneburg, besides many fantastic students (and here I want to mention, among the many wonderful courses I taught, especially the students attending my 2018 block-seminar "the Bioeconomy"), a nod to the Maelzer is needed where we held our discussion evenings, tons of thanks must be expressed towards Cristina Blohm (and her army of amazing tutors), Andreas Jürgens, Lena Foelsche, the various students who have been TAs, and Andre Bader, Leon Kaiser, Lukas Beckenbauer, Trang Vo, Dieter Friedrichs, Tom Schmidt, Bastian Hagmeier, Marlene Anton, Julian Klosik, Anne-Sophie Wellna, Isabel von Wiltke, Liselotte Hermes da Fonseca, the Lovely Ladies (Maria M., Maryann H., and Sabrina V.), and, of course, to the human/non-human duo that became a home for too short a time and friends in continuity: Ouzo Foerster and his human Yvonne.

I certainly need to thank tons of individual people beyond that, but I focus here and wish to thank above all, for being in various versions of good to me in a very difficult time of my life, the following people (and I am sure I have forgotten to mention some, sorry, still scatterbrain me): Vilna Bashi Treitler, Manuela Boatcă, Melissa Weiner, Ulrike Boehmer, Beth Royce Small, Rachel Guo, Tracey Corley, Fritz Breithaupt, Walter

Mignolo, Jane and Lewis Gordon (who made the impossible possible in Boston 2016), Tom DeGloma, Erin Johnson, Anne-Marie Champagne, Jyoti Puri, Julian Go, Justin Groslight, Nicholas Rowland, Sarah Craig, Dirk vom Lehn, Maria Puig della Bellacasa, Noortje Marres, Clemens Driessen, Tamar Blickstein, Michael Weinman, Gareth Edel, Dorian Gittleman, Amit Prasad, Hans Bakker, Carina Ruppert, Alexandra Feuchtinger, Mohammad Golam Nabi Mozumder, Karol Kurnicki, David Selim Sayers, Evrim Selim-Sayers, Julien Seroussi, Franck Leibovici, Gaetan Robillard, Stef Fishel, Wendel J. Burkhardt, Petra Maitz, Tomo Masic, Benno Heidkamp, Sal Restivo, Kathrin Schwalbe ("und die Frauen" at Hotel Central), David Lindner, Tito Huq, Armin Monsorno, Bernhard Tilg, Caroline Gerbaud, and, very important: a huge thanks for everything and the company to my good friend Gilles Lhuilier – Gilles has, certainly, become my sounding board for some of my wilder ideas, a teacher on things transnational jurisprudence, and, when most needed, a mentor. And, of course, Sepp and Mu.

This is punk................

...... or slime mold.

Part A

(As in *a priori*) – Of Mythonauts and Mythophants

Introduction (or, rather: *hors d'oeuvre hors-texte*)

Against "White Collar" Sociology

The text, the book that follows, suffers: Among the many things and in the many ways it suffers, the ways and matters wherein it is imprecise, ambiguous, equivocal, and prone to misunderstandings, among those things there are also the most effervescent ones: Time and times.

Killers of Time (or: "Can There Be a Sociology of Empathy without Empathy for Sociology?" and Other Whiggish Questions)

But that *is* a problem, that I, as the (dare I say?) *author*, find myself in generally good company: People who are, however, minutely noteworthy by "the public" of some kind in some context or other – be they celebrities interviewed for a tabloid or a recently hired junior professor probed by a student newsletter – these people are often presented with some kind of standardized questionnaire, which asks for some biographical facts, likes and dislikes, and so on. The two questions about one's greatest strength and its opposite, the greatest failure, flaw, or vice, are pretty much staple ones for these kinds of questionnaires in newspapers or magazines. The philosopher Hans Blumenberg, whose *Work on Myth* (1996 [orig. 1979; engl. 1988]) serves as one of the main reflective (and thereby reflexive) surfaces that this little book is cast at (or upon, or maybe against, perhaps even along with), responded to the latter question, the question for his greatest flaw, that it was that he *is not able to say precise enough what it is that he sees.*[1]

But this kind of suffering (a flaw) doesn't end therewith; more to the point: the suffering begins with this. Suffering is, among others, related to *empathy* (a concept that does have a bit of a part to play in my theoretical work [see Weiss et al. 2014]), because of the *pathos* of the "*-pathy*" in *empathy. Pathos* evokes the affective dimensions of experience and expression, and is thus often translated from classic Greek to mean passion or "suffering". So I am writing this little *Preface* very *pathetically*, i.e. appealing to your emotions, which can be considered an interesting inquiry into the distinction between myth and dogma, which is specifically important in Blumenberg's book on myth. Also, if you are reading this now, you have to suffer me, the author of this little venture. In my native German, this kind of suffering can mean many things – notably, pointing out differences between languages is a little trick I ask my readers to suffer occasionally, for it helps clarify meanings quite productively (*empathy/Empathie* is a difficult case, for it is a word used in German just as it is in English, but it is equally ambiguous in both languages, and it is often confused with *sympathy/Mitgefühl/Mitleid[en]* or *feeling with/Mitfühlen*, while its philosophical history as *Einfühlen* is often told *whiggishly* without sufficient knowledge of its actual genealogy).

Now if you have to suffer me, your experience is either one of *erleiden* or of *ertragen*: You either are actually suffering in terms of a painful experience (*erleiden*) or, even if it's a burden, then it is one that you are carrying or have to carry (*ertragen*: the verb *tragen* means "carry"; *ertragen* is a passive form of carrying or standing something but making explicit that it is *at a cost* – be that pain, time, uncomfortability, resources, etc.). The latter is, in my view, a very important idea, for it (at least figuratively, at best literally) *carries further* than the notion of *tolerance/Toleranz*. "To tolerate" tends to contain a moment of violence or victimization (Forst 2013; Brown/Forst/et al. 2014), because it evokes a notion of distance and alienation that cannot be reconciled (except through assimilation), and in its enactment, its temporary inclusion of the Other can thus be easily revoked and the consequences of this revocation are – at least rhetorically – less explicit. In other words: In saying "I tolerate you", I am saying nothing other than "I recognize that you are different, in a way that there is something wrong with you, and I merely have decided not to act on this difference (yet), largely because it costs me nothing to tolerate" (we learn from Rainer Forst). In saying *Ich ertrage dich*, one can of course casually mean *tolerance*, but one actually states that one is carrying/suffering the Other. *Ertragen*, more than mere tolerance ever does, seems to demand something *of* me. In this regard, I prefer *ertragen/suffering* over *tolerance/Toleranz*, because it makes explicit an *effort* that "an Other(ness) so carried" elicits.[2] Some people may find this untenable, because it is politically more fashionable to be tolerant, and toleration's tactic is precisely to keep implicit and hidden that "standing someone's Otherness/difference/strangeness" is costly. But that is exactly why *tolerance* is, in a way, cheap. It costs nothing or little. *Suffering someone*, on the other hand and after a fashion, is *the enigma of a (very particular) gift*. And the enigma is that we are fully aware that we will pay a price, but not actually count the cost: Bravado? *Suffering someone Other*, therefore, does **not** allow to construct this Other through the *coloniality of power and Being* in the same way tolerance retains this coloniality as a constant and looming possibility, namely, tolerance constructs the Other as someone for whom support (i.e. *that which carries*) can be easily revoked, in other words an Other who can be constructed as "disposable". These are two different narratives, and if you are willing to *suffer me*, even though my way of *doing narrative* may be somewhat Other than what you are used to, it should become evidently clear that asking *how* and the *why* of the constructions of the relations between *information, power,* and *care* as *narratives* are, first of all, lines of inquiry that can serve as important (ethical) platforms for future discussions and opportunities in social justice and equality; and second, that it is important to accept that we must not simply tolerate but suffer Otherness.

In the end, it is now, perhaps, already presumably clear: I am interested in narratives. Narratives are not fictions or stories[3] but something more profound. They are not foundational in the terms of a perennial philosophy, but post-foundational – this has to do with how I understand and deploy the concept of *narrative empathy,* wherein I conjoin the ideas proposed by Fritz Breithaupt and Suzann Keene and view them through the lens of Barbara Tversky (see Stingl in Weiss/Restivo/Stingl 2014). *Narrative empathy* – we should, to reiterate, not forget – hides the *pathos* in *empathy,* and *pathos* is a form of suffering. There are many reasons why I am interested in narrative rather than story/fiction – narrative lies "before *and* beyond" stories and fictions – but one reason is that my work is about life not death, since

> [s]tories are told in order to kill something. In the most harmless, but not the least important case: to kill time. In another and more serious case: to kill fear.
>
> (Blumenberg 1988: 34)

What, one may ask on that note, is the story in and of the social sciences today? It's a story of self-impoverishment.

Signs of Poverty

In 2014, at one of the larger sociology conferences held in the United States around the year, I attended a panel that focused on the role that morality plays in the context of sociology. There emerged a number of things in the discourse of this panel, which evinced a new chapter in my reflections on the status of sociology at large.

Perhaps, first, a word on situating the context of my own views: In a famous declaration, Ian Hacking reminded his interlocutors that while his works span the history of probability, multiple personality disorder, and other seemingly unrelated topics, he was, to begin with and would remain throughout all his works, an analytic philosopher, but also one who would write about subject matters that an analytic philosopher is not expected to write about. In the same vein, I cannot but confess that, aside from my unusually broad multi-disciplinary training, the rigor and imagination that inform my research scholarship and the demands I make of my students to learn and deploy is first and foremost of a sociological cuisine (perhaps often using a rich broth of biology). This rigor and imagination inform my scholarship, even if I have been referred to by colleagues as an empirical philosopher, politologist, critical realist theorist, political philosopher, broadband intellectual (I have been called nasty things, too), and I am presently working on (too) many timely as well as untimely issues of *bio-digitality*, *bioeconomy*, and the question "What is a technological object?", on digital cultural capital and inequality, on decolonial political economies of science, on the history of the Western body concept, among others. At the same time, I strongly believe in integrative interdisciplinarity (or, if you prefer, what Lewis Gordon calls transdisciplinarity) and conversation, and here I always begin from a platform of biology. With regard to the social sciences, I have always been delighted that Talcott Parsons had in-depth bio-chemical knowledge (he originally wanted to become a physician and had taken courses, including at the Woods Hole Oceanographic Institution), and like Kant and Jaspers, he too had considered to study to work in medicine. I am equally fond of the story of Norbert Elias who brought a model of the human brain to his introduction to sociology lectures, of his comments that he considered the relation between biology and sociology as a decisive problem for sociology, and of his criticism of the idea that sociologists could learn the most from other sociologists whereas he favored to learn from other disciplines.

In the same vein, it was a only strange and accidental turn of fate that prevented my original plans to study biology and get into genetic and epigenetic research. But neither efforts in honing my basic biological knowledge nor my interest in that field has ever run dry.

Exposure to decolonial perspectives has given me life-long suspicion that Western knowledge is particularly "Western", ways to articulate this provincializing position explicitly in more interesting and more effective ways, and to try my best to learn from it and incorporate these. It is an, however, impossible and probably very arrogant attempt at (self-)provincialization of my knowledge and scholarship – as if I have ever done anything else. Perhaps it is a kind of negative anthropology that begins with, at the very least, taking apart (and if not even down) both the *anthropos* and the *logos* in "anthropology". It is entirely possible that the recent "(re) turn to *Ethnographic Theory*" in cultural anthropology indicates that ethnographic theory should be the new interdisciplinary name for (cultural) anthropology – remember that "names have power".[4]

With this contextualization of my own position, let us now return to 2014 and the panel on morality and sociology. What immediately stood out to me was the fact that each of the panelists – and, as it turns out, the majority of the audience – seemed to equate morality with religion. I found this ultimately puzzling. Perhaps given the presence of one scholar on the panel who self-describes as gradually related to a Durkheimian perspective, one may speculate that this is a typical American rendering, based on a particular reading of the sociological classics – consider, for example, that Max Weber's oeuvre is usually accounted for by many American sociologists through the lens of the Protestant Ethic (2002 [1905[5]]) rather than Weber's writings on rural-urban legal history or on the philosophy and interdisciplinarity of science (which Marianne Weber, posthumously, collected under the Fichtean notion of *Wissenschaftslehre* [1988]).[6] Within American sociology, there is a prominent stream (of scholarly collective consciousness, if you prefer) – or let me use the term *thought-scape*[7] to denote precisely this aspect of discourses – that can problematize morality and moral issues[8] almost exclusively in religious terms and contexts. I think it is highly problematic if the question of the relevance of morality for/in sociology (resp. as a sociological problem) is posed and then reduced to religion. This tendency can be seen even in such a seminal social constructionist study of obesity as Abigail Saguy's *What's Wrong with Fat*? (2012). In this book, one of the frameworks that are analyzed as accounting for the various obesity discourses is morality, which Saguy, however, deflates to mean religion – a maneuver that is also never justified (not in her nor in any other case I know of). The value judgments deriving from morality and the foundations of moralities are not only religious. And even if the equation were true for US-American society as an isolated case – and it certainly is not – the fact that Saguy uses as a point of comparison empirical materials from France, where it can be easily argued that morality cannot be identified with religion, renders this part of her (otherwise seminal) study problematic.[9]

Not that morality can't involve religion at all; quite on the contrary, it absolutely can, but that should not be the exclusive subject of discussion nor should morality be reduced to religion – explicitly or implicitly. This type of reductionist move, I propose, is also typically found in what I will come to call *White Collar sociology*.

In addition, the aforementioned conference panel was composed of two men and two women, who were, as you might guess, white, Ivy-League-educated, well-dressed (expensive suits, hair and face made-up and everyone well-manicured), and so on. I mention this because it gave the whole affair something of a clichéd touch. But to be honest, up to that point, I couldn't have cared less about this panel per se. Everyone presented their little bit, then one speaker – who explicitly claimed to tie her perspective to an evolutionary population biological approach – said something that I found completely horrifying, and the fact that no one else on this panel or hardly anyone of the admittedly smaller audience shared my sentiment[10] speaks volume:

"Having many children is a sign of poverty!"

I was floored and enraged, because this statement was presented as something like an anthropological universal truth. I was floored and enraged, because this statement is highly arrogant and utterly colonial and imperial. It represents Western/Global Northern social science at its worst.[11] Specifically in the context of the speaker's preceding talk, bringing in a supposedly biological perspective, the reintroduction of a religiously laced, neoliberal-technocratically founded social Darwinism became eerily present, made worse by the "I want the best for these people" attitude while pretending to be "positivistically objective".

My point here is not to generalize this anecdote beyond its particular place and time and context. My point is to inquire, as I often do, how it could become possible that this statement could become *assembled* and enunciated in the first place (and time). I am particularly interested in the epistemic authority it wields, the epistemic injustices it produces and is produced by, and about the apparent lack of epistemic responsibility and response-ability of the person who decided to utter these words. Behind it lies, among other things, a very problematic (somewhat Schmittian) understanding of "modernization", typical for the brand of social science that I call *White Collar.* This notion of "modernization" draws together and *assembles* a relation between *care, power,* and *information* in a very particular way, which reduces in particular the concept and dimension of care and renders it ultimately dependent on the other two dimensions.

Cold Care

In a paper on Hannah Arendt's work spanning the tension between *sacralization* and *secularization* (a conceptual tension I pondered in my previous book *The Digital Coloniality of Power* [2015]), Sybille de la Rosa (2014: 174) refocuses Arendt's position as one that manages to develop an alternative for a problematic dichotomy within political theory; this dichotomy, which de la Rosa identifies (following research of Markus Holzinger [2007]) is a problem arising out of modernity with regard to morality, since the increase of autonomy that follows with modernity has been brought on, in this account, with the *destruction of traditional and religious certainties. Compensation is provided by political theory* [and I would add within sociology also] *through either (arbitrary) forms of moral coding or through an orientation focus on contingency and "decisionism".*[12] I would add here that the latter diagnosis is not entirely wrong, but it should be *geopolitically* qualified (Mignolo) that "modernity" is a Western phenomenon that has been known to (epistemically) colonize the world *from* the Global North. I would, however, also argue that in the sociological frame of reference, the first option does not have to exclude the second with necessity, precisely as it sets up antagonistic structures. Also and related, while Left wing sociologists, like the various generations of the Frankfurt School, have held on to and promoted the values of a Leftist (and often petty) Bourgeoisie (which includes Adorno's hatred of Jazz same as it does the reification of male privilege in Habermas's public sphere), many American sociologists promote the nostalgic view of equating (a moral core of) religion with moral certainty. I think either version is largely problematic in terms that set their relation to the social worlds they study in an antagonistic and *decisionist* (or longing for *decisionist*) form.

While I believe that Arendt, by her account (see Benhabib 2003) of the life of Rahel Varnhagen (and contrast to Aristotle and Augustine), her critical adoption of and engagement with Karl Jaspers (Murai 2012), and, subsequently, her account of Kant's political philosophy – which promotes the notion of political imagination and judgment, which surprisingly few philosophers were willing to do with such rigor, though it *is* an important aspect in Kant (Arendt 1990) – managed to develop theoretical tools that may help us understand the problems and opportunities that lie (geopolitically) within the situation of Western modernity, I do not think that she necessarily provides us with a tangible, resilient alternative.[13] In other terms, Arendt's work is an excellent tool to have available for anamnesis and diagnosis, which is necessary for good therapy, but she is not providing us with a (philosophy-) therapeutic regime that could work on its own.

But decisionism, on the other hand, is also an important aspect in the emergence of *White Collar sociology*. Detachment from their subject of study is always vertical (and upward), never horizontal or diagonal/queer. More importantly, it always pre-supposes conceptual, empirical, and methodological frameworks that are inherently constructed – whether explicitly or implicitly – in terms of a vertical hierarchy that knows a below and an above. In terms of the idea of *care*, this kind of social science can only ever *care about*, it has already difficulties with the notion of *caring for*, but it is absolutely devoid of any notion of either *caring with* or *enabling care*.

About Schmitt: On Deciding on Being White Collared

When I talk about "white collar", this will evoke associations of varying kinds in different readers. Some will immediately associate the notion of the "white coat" as a marker of a paternalistic authority: The moment a medical doctor puts on the white coat, his power to diagnose and to prescribe can be performed with effect. This effect, its foundation, and its power have been studied and discussed, of course. It is such a common place that in many a Hollywood action film or TV-episode it is used for a chilling or a comedic effect, when a character without medical training dons a white coat in a hospital and is immediately requested to act the part performed by some unsuspecting side character.

In the 1930s, the term white collar, referring to white shirts such as worn by office workers, emerged to be used to denote non-agrarian and non-industrial work – who could be referred to as blue collar workers, while later on service work was also distinguished as pink collar. With systematic sociological work on issues of labor and class, such as C. Wright Mills's *White Collar* (2002 [orig. 1951]), the relation between the fable of social upward mobility, middle class, and attaining a *white collar* job was revealed as sociological topic, following Edwin Sutherlands preceding interest in white collar crime in the 1940s – crime committed by people who were otherwise associated with a high social status and respected occupation. The stereotypical Wall Street–Main Street distinction that we find ceaselessly reiterated in popular, propagandistic politics drinks from this same well.

The white coat and the white shirt, of course, do not lend themselves to situations where they run the danger of procuring any stains.

Finally, we often find associated a notion of privilege, and thus *white privilege* (the privilege enjoyed by persons who are considered ethnically white and often are male identifying as heterosexual).

Talking about *White Collar academia* in general, and *White Collar sociology* in particular, is not something I can, nor wish to, define *too* sharply: As a concept, it is one of those notions that one can identify immediately when perceive it but not really say what it is with a clear-cut definition. There is an open list characteristic: Open, because (a) some I will mention within this book but it's not like a comprehensive list that I am drawing up but certainly the just mentioned ideas about types of work, class and status, makers of professions, and notions of privilege are on this list; but (b) not all, nor even a majority of these items need to be present in the case of *White Collar academia*. To a large degree, *White Collar academia* is more like a character-type or "unpleasant ethos" that we can see people display in their behavior toward others and in the decisions that they make. It is often characterized by what people shun to do or what they ignore rather than what they actually do. And it is here, with decisions but also with the tension between the secular and the sacred in the way that the tension is political, and exemplified in the deflation of morality to mean religion in sociology, that Carl Schmitt raises his head: *White Collar academia/sociology* is inherently a Schmittian enterprise.

In the last 20 or years, there is a renewed (or continued) interest in Carl Schmitt's work on both Left and Right. While Schmitt plays a role, for example, in Giorgio Agamben's project – Schmitt is both the adversary *and* the lens through which Agamben (mis)reads Arendt and Foucault – I am more interested in the ambiguous position Schmitt has in the work of some post-foundational approaches, where reference to Schmitt serves as gestures of *going with Schmitt beyond Schmitt* on the one hand, but on the other hand in an inherent logic of Schmitt's political theology that is a characteristic of *White Collared Sociology* as encountered in terms of an *implification* of the discussion of morality – *implification* is a concept I introduced in the previous book (Stingl 2015) which describes the process that simultaneously and interdependently implicates and intensifies. This can be identified precisely in the question of secularization – also the topic of Hans Blumenberg's famous study on the *Legitimation of the Modern Age*, which played a central role for my previous book and for my present research context at large. Of course, secularization was the topic that Schmitt and Blumenberg famously struggled over. A main question in this context is whether or not secularization of (political) concepts does delete their theological context; a famous argument is that in order to count as secularized they must retain the core of what they have been secularized from, therein retaining a theological aspect (or, one is tempted to say with an eye toward Bernard Stiegler's discussion of *retention*, a theological programming). Ruth Groh (1998), in a seminal analysis of the anthropomythological gown[14] worn by Schmitt's political theology, presents the exegesis (and apologetics) of Schmitt after World War II – Schmitt's unfortunate and incriminating flirtation with the Nazi's, rivaled in intellectual circles in its foolishness perhaps only by Heidegger's – as a discussion of the role of theology in Schmitt's thought: whether theology does or does not have a central role, as well as among proponents of a strong version, what the actual capacity and consequences of are on a conceptual level. Channeling Ernst-Wolfgang Böckenförde, a former German Supreme Court Judge (*Bundesverfassungsrichter*) and prominent member of the so-called Ritter School (the students of conservative[15] political philosopher [and himself student of Ernst Cassirer] Joachim Ritter, among them notables such as Odo Marquard, Willi Oelmüller, Hans Jörg Sandkühler, Wilhelm Schmidt-Biggemann, Hermann Lübbe, Max Imdahl, and Robert Spaemann), Groh can identify three attempts of an objectivation of political theology (see Groh 1998: 15), which are mentioned as follows:

- *juridical form*, under which she considers the translation of theological terms into governmental-juridical ones.
- the *institutional form*, which is described by her as what serves as the very essence of the terms of faith as it derives in the status, legitimation, and functions of political order.
- and, finally, *an appellative form*, by which is meant the socio-political obligations found within Christian eschatology itself.

Böckenförde, in Groh's account, favors not only the first of the three forms, which is set against the other two which are inherently and explicitly oriented on theological content.[16] Groh argues that therein Böckenförde assumes that we find in Schmitt an attempt to create within the juridical form[17] a(n iterative) *re-casting of theological positions in terms of concept-sociological insights* (ibid.). For Groh, however, this does – while not necessarily being a mistaken observation – not lend itself to reduction: While Schmitt does see political concepts of state that are considered "modern" as being inherently secularized theological terms, he also maintains a normative-appellative character of secularization that retains the potential for re-theologicalization. And it is in this tension that *White Collar sociology* finds one of its points of origin, which we can find in his moment of reaffirming religiousness against any non-theological political, which Ruth Groh (1998: 16) demarcates as if Schmitt saw *secularized modernity* as a kind of a Promethean Age – a trope that, of course, reappears in Blumenberg's extensive work on myth, which is one of his more or less implicit "difficult conversations" with Schmitt. As such, Schmitt's account (in his 1922 programmatic) of modernity as secular means to rebel against God but in the same stroke to de-politicize and to render neutral politics and value. Why? Because *de-theologicalization* of any political order founded and found in the theological means to de-politicize, because once the theological goes out of the window, morality goes out of the window, and with morality we lose, states Groh for Schmitt, the *political idea* itself.

The discourse of sociology can be described in this tension, because few sociologists seem capable to create their sociological position, which is not either stylizing itself as simultaneously Promethean in its wordings yet fashionably neutral in practice, or it is harking back to be moralistically Schmittian in serving as a political theology. On both fronts, we find a healthy obsession with deflating morality to mean religion. Sociology following along this line for which Schmitt is, we can say, the *spiritus rector* of *White Collar sociology*, ends up in endlessly bifurcating – not only nature, but – what they deem as "the social", beginning with the assumption of the existence of such "a social" which confuses *explanans* and *explanandum* to begin with: What Schmitt and *White Collar sociology* share is a foundational approach. In order to go with Schmitt beyond Schmitt, such as Chantal Mouffe brilliantly manages to do, one must understand and shift the discourse toward a post-foundational position (Marchart 2008). Meanwhile, the bifurcation of the social means that, in order to do sociology of the permitted kind, one largely can only explicitly do sociology de-theologized or re-theologized. Attempting an alternative sociology, a sociology which is open for radical Otherness, is nearly impossible. This has to do, for example, with sociologists' obsession with or rather its fetish of an origin story that claims that sociology derives its form, with and after Comte, in a discourse of moral statistics and the emergence of economics; subsequently, it does not account for its more complex genealogy, where we can see how seriously sociology is conceptually and methodologically indebted to the emergence of biology from philosophy (see Stingl 2008/2010; and *Anthropos's Scaffoldings*).

What I find particularly concerning about *White Collar sociology* (and related academicians) is a disconnect or dissociation in their idea of what the academy/university is, what research and higher education should do and look like – visions that often conflict with the actual needs – and often, what, also, the freedom that is preached by certain academicians and their actual decisions and practices reveal but utter conflict. While the majority of sociologists seem to criticize the rise of "contingent faculty" jobs, i.e. the imbalance between adjunct and part-time positions versus full-time jobs (let alone the question of tenure, which is a more difficult discussion[18]), few really *do* anything about it, especially not when they are put in positions of influence. Besides the issue of what kinds of positions are advertised, the issues also include the question of who is hired? "Diversity" and other "first fashionable before getting the hell criticized out them"-terms are rarely realized for example. But: "It's about decolonizing the university!" (Bhambra et al. 2018).

We have been sold on the idea that "more education" is good for you and good for the economy "because a growth; in our obsession with "growth" we mean only "economic growth" – instead of, say, personal, intellectual, moral, justice, equality, cultural, or whatever kind of growth that is not economic. In this context as in many others, we are acting as though economy were somehow a *first philosophy* (while of course hell-bent on denying that economy could be itself a philosophical proposal of any kind) – while simultaneously having effectively changed education into training for the *current*(!) job-market, while not creating enough jobs that either go with the general level of education (level on education on paper at least, for we have watered education down significantly) and/or do not pay remotely enough to live (let alone live well), and the realm of higher education itself being testament to that, since all of that is what anyone familiar with the current discourse will be able to confirm can be treated as factually correct, it is surprising why the very people whose job it is to notice, make explicit, and propose alternative routes, namely social scientists, largely act as the notaries of Late Liberalism, i.e. they take official note that things are (and that they are bad, unsustainable, etc.) and then do not act? I do not see in academia a call supported and followed through by a significant majority, that all university and college teachers across the United States (or Germany, or Britain) should just go on strike until a binding(!) discussion mechanism and actually biding goals for this discussion are set that are intended to make a change toward increasing hiring in higher education, and hiring meaning actual jobs, and in "making education into education again" (in creating new spaces for learning *with* one another, as well as for *thinking with* and changing matters *with* one another) that means that people learn things that make them grow as whole persons and, subsequently, equip them with skills that will enable and empower them to contribute not only to the present job-market but also to future society and its future economy – since the prophets of common sense love their platitudes, here is one: Well, give a person a fish and he'll eat today, train a person to fish, and they will soon empty the sea of fish and with no other skills will go hungry again, educate a diverse community of people with diverse skills and an understanding of the consequences of their actions, and they will build a sustainable society.

But again, why does it seem like the majority of faculty in higher education is aware of this but not acting on it? The reasons for this kind of *akrasia* or (in mutual interdependence, individual *and* collective) "weakness of will" can be described variously and assigned a variety of motives. *Bureaucrasia* is a form of *akrasia* I coined to describe how the bureaucratization and overabundance of regimes of accountancy/accounting lead to the loss of a sense of decision-making. Complacency and opportunism also play all kinds of roles. Finally, I think, the readiness of handy little "excuses" is always an important factor: Excuses are, as Fritz Breithaupt argues, at the start and heart of how we narrate ourselves and our society. In lieu of his earlier work on the centrality of narrative empathy (Breithaupt 2009; Breithaupt and Hamilton 2019) for our understanding of sociality and decision-making, which I have adopted creatively (see for example in: Weiss et al. 2014), I think that the existence of particular narratives for excusing the lack of actively participating in or acting out practices of resistance, delinking, and liberation from the neoliberal, *bureaucrasiastic* governmentality of Western societies is why many academicians don't engage in these practices, while simultaneously and paradoxically engaging in a lot of lip-service talk. This, too, is a characteristic of what I call *White Collar academia*, which in the case of social scientists weighs much heavier, for the issues at stake are the very subject matter of their profession (and here, with a nod to Max Weber, I mean precisely both their job [*Beruf*] and their calling [*Berufung*]). This demarcates also what I have called in my previous book a case of *implification* (implication + intensification), while sociologists and other "higher (aka self-elevated) educators" take notice, i.e. serve as notaries for events that implicate their actions in general lead – seemingly paradoxically – to an intensification of the very implicated process.

And, as mentioned earlier, Schmitt appears to be a kind of *spiritus rector* for these *White Collar sociologists:*

The magical formula is, to utilize Ruth Groh's term (1998: 135), which has however been the apologism of many acts committed in the banality of evil, that of *action under coercion and internal resistance.*

Very much like Schmitt has tried to justify his scholarship during 1933 and 1945 by portraying himself as having had no choice other than to act in the way he did, but that his resistance happened inside of his mind, *White Collar academicians*, when called out on their failings, propose stories of their internal courage, of not having alternatives and being forced to play along (or lose, well what? Especially once you do have tenure?)

It is, for example, much easier to talk about and fight for or against equality for one group, for example the hiring of more women as managers, professors, or deans. The perfidy of these kinds of battles is that under the current hegemony, they are always being fought antagonistically – without addressing the fact that there are too few and too few adequately remunerated jobs in higher education,[19] too few jobs for women, too for jobs for people from lower classes, etc. The fact is, namely, that it is true: There are implicit biases that prevent these individual groups from having equal access to these jobs, and these battles cannot be made into one battle for just more jobs, but must be fought individually; however at the same time, the general (not universal[20]) point that we need more and better jobs in higher education is also true, while it is also true that there are important relations and interdependencies between the general issue and the individual groups' issues. What is for many people difficult to understand or explain is that these aspects are equally true but cannot be reduced to one of them. The problem that exists between the protesters of #blacklivesmatter and those who claim the notion should be that #alllivesmatter is of a similar nature: While it seems like a reasonable proposition that "all lives matter", each of the terms "all", "lives", and "matter" are not unconditional concepts. The conditions different people assume for each concept in relation to the others will lead to different forms of practice: In this way, someone could argue that since all lives matter and they consider any unborn child at any stage of pregnancy a life, termination of pregnancy should be forbidden, whereas under the current conditions of US-American society, the effects of a pregnancy – be it an unwanted pregnancy or a pregnancy with health complications for the mother – might be more likely to affect the life of a black woman in a negative way than a white woman. If that is the case, we can already see how in this example it may be that even though black US-Americans seem to be included in the "all" of alllivesmatter in theory, that black lives do not matter the same in practice. This pattern emerges in various relations and real life scenarios. It is also a misunderstanding (intentionally or through implicit bias) by some people when they read "blacklivesmatter" as "only black lives matter".

Basically, it is important to understand, I think, how theory *and* practice *and* their interrelation of either *blacklivesmatter* or *alllivesmatter* are equally important, and how you cannot really have one without the other. But *blacklivesmatter* activists (hash-tag or no hash-tag) in practice do not deny that all lives do matter, whereas many of those who promote #alllivesmatter in denying that *blacklivesmatter* has relevance already deny *that* "black lives matter" by ignoring factual practices and the differences and the interrelations between different groups in society, not to mention their being obvious to the conditional concepts such as "all". The same happens, of course, with the discussion of gender equality in higher education. Those who call for "more jobs for women" but do not at least recognize that the problem of neoliberalization of higher education has an effect on the number of jobs for women are just as wrong as those who claim that merely increasing the number of jobs and fighting against neoliberalism in higher education is the universal problem and that there is no special problem of gender, race, queer, or any other kind of equality in academia and job-market. Instead of realizing that there is *not* a universal problem but that, instead, there are various individual issues with gender, race, and so on, *as well as* a general problem, to which none of the others can be reduced to, different groups argue antagonistically with one another about who is right and who is wrong. What would really be required is an understanding that the individual problems and the general issue are related, and that in fighting together but also separately more is achieved than in fighting with one another. However, within academia, fighting with one another antagonistically (or creating false consensus that only reifies a status quo) is more common, but it is also the luxury that especially *white-collar academia* as an attitude breeds. What it breeds and thrives on is also a very feudal structure, and it transforms politics into friend/foe-structures in terms of Schmitt's political theology. *White Collar academics*, thus, make one of two choices: not to openly appear to be political at all and use the excuses of "being coerced" or having at least felt they engaged in "internal resistance", or "to be political" in terms of simplified friend/foe-structures (hard versus soft sciences, this sociology school versus that, white guys versus feminists, and so on), which they play out antagonistically and destructively. Neither tactic solves anything, except exacerbating the underlying problem.

The underlying logic here is, of course, the *colonial matrix of power*. Consequently, a decolonial *and* post-foundational attitude will be helpful toward beginning the discussion on the possibilities and opportunities to resolve these issues, but it will mean getting our hands dirty. I am talking about BluesCollarship.

Notes

1 "Nicht genau genug sagen zu können, was ich sehe" (cited in: Bettina Blumenberg 2015).

2 This suffering is, to some degree, about the self and not the Other, it's what oneself does to one's self in that case. I suffer the Other, i.e. their difference, is to induce suffering not by or in that difference nor by or in themselves/"their selves" but only by, because of, and in myself.

3 Fictions, we encounter in the strictly social phenomenological tradition of sociology, i.e. in the works of Austrian banker Schütz and his disciples.

4 I'll let you in on a secret: one of my dream projects, that I still hope to find the time to research and write one day, is a history of the occult roots and imbrications of Western scientific and philosophical ideas – a radicalization, if you like, of Hans Blumenberg's work in conceptual history.

5 There is a very complicated history, regarding the English translations of this book of which there are several and which I won't get into here (see Scaff 2005; Gerhardt 2007; Stingl 2008/2010).

6 While in general a very good overview of the development of sociology and, thus, still an excellent introduction to (the history of) sociological ideas, I disagree with a number of aspects of Randall Collins's *Four Sociological Traditions* (1994; published as *Three Sociological Traditions* in the 1980s). In terms of the history of scientific ideas and the institutional history of the European University, Collins's glossing over important details leads to a number of problems. To name one example, the idea that intellectuals had to "entertain nobles" misunderstands the courtier relationship (see for example Biagioli 2006); the same is true for the role and differentiation of faculties of the university. I also disagree with his assessment that Talcott Parsons's reading of Weber can be reduced to the centrality of the *Protestant Ethic*. On the contrary, the role of the *Wissenschaftslehre* is crucial, I think, to understanding Parsons in light of the history of ideas and of science.

7 Which is very different from (but in a wider framework complements) Sabrina M. Weiss's concept *mindscape* (see also technology by Stingl/Weiss

2014a). [I had originally quoted a key section, but since the publisher insisted on needing permissions for this, readers just have to look it up.]

8 There are a number of conceptual issues attached, I think. One is that contrast with other languages would make the dimensions of ethics, moral, morals, and morality clearer. The other being the distinction between faith and belief is less precise than it should be. German has a similar problem with the distinction between *Glauben* (faith) und *Wissen* (knowing as in "believing that x is true"), which Kant has attempted, among others, to conduct very precisely in his *Critiques*.

9 Not to be misunderstood here. Abigail Saguy is an exceptional and brilliant social researcher, and her study is a very important intervention and seminal for the fields of sociology of health and body and fat studies. I chose her work as an example precisely because this kind of deflation happens among respectable and fun scholars like her.

10 Although there was one exception among the audience.

11 Moreover, this was reiterating the US-American imperialist rhetoric of the Cold War "economic political" expert elites in both the US military and White Collared academia, who were connecting poverty and reproduction through measures such as GDP in the creation of the ill-fated Western modernization theory, creating a model the premises of which *enabled* a narrative that stated that childhood abundance and underdevelopment went hand in hand ... and fed by (and, one could almost say: qua dialectics, feeding) the communism (see Murphy 2017: 35–46).

12 De la Rosa (2014):

> *Die Moderne hat auf der einen Seite einigen Menschen größere Autonomiespielräume gebracht. Auf der anderen Seite hat sie, so wird immer wieder betont, traditionelle und religiöse Gewissheiten zerstört. Auf den Mangel an Gewissheiten reagiert die politische Theorie, so Markus Holzinger, entweder mit einer »wie auch immer begründeten moralischen Codierung« oder »streng kontingenzorientiert und dezisionistisch«.*

13 De la Rosa's paper somehow fails to see this, I believe, because she does not fully engage any of the classic post-war (such as Löwith-Blumenberg [Talay 2011]) or contemporary discussions (such as Dupuy 2013, etc.) on the sacred/secularization-problem, beyond the references to Habermas's suspicions concerning the postmodern thinkers (despite her mention of "contingency", not even Luhmann is brought up), which she reifies through a discussion of Hans Joas, and the reference to Schmitt, who comes up whenever decisionism is mentioned. I had intended to go into more detail on this, but the lengthy [ca. 1,000 word-long] original footnote had to be cut for space.

14 I am, of course, also thinking here in terms of Christoph Jamme's work on myth (1999), which in its whiggish krypto-Hegelianism even promotes loudly that it closes its eyes to *the political* (and Blumenberg) in philosophical anthropology; thus, it's coming second only to Alfred Schütz for whom to be truly a person, one must be able to play tennis and participate in a classic quartet – no wonder he was, like the Koch Brothers, a member of the neoliberal Mont Pelerin Society. The political is, at times, quite personal.

15 It is, however, debatable what "conservative" means in this case. Members of the Frankfurt School, including Jürgen Habermas, qualified a number of dissenting views from their "Leftist" ideology as "reactionary", i.e. conservative, and often fought them viciously. Ritter and his students were a primary target, because they insisted, for example, on the idea that – as Odo Marquard summarized Ritter's view – "future needs provenance" (*Zukunft braucht Herkunft*).

16 One is tempted to say that the appellative and the institutional form attempt to conduct their political-theological business in an ontic fashion, whereas the juridical form of political theology obtains on an ontological level, whereas, from a juridical-pragmatic point of view, its possibility can be doubted as Groh does.

17 On a notably different approach to the idea of the "juridical form", see Foucault (1996 [orig. 1973])

18 While one can clearly see that tenure does protect research and free speech from politics – a lesson the Germans learned the hard way in 1933 and after – tenure should not protect, for example, serial harassers. The question of what tenure can and should be, what an – as I believe necessary – alternative construction of tenure should and could look like, is something we would need to discuss openly over a longer process. I do believe, without a doubt and based on the German historical experience, that a structure like tenure I (or *Beamter*) is an important part of the infrastructure of any republican (–democratic) society; but I also believe that in its current form, it is way too easy to abuse the structure from various angles of power and authority both epistemic and in the organization ontology of higher education. Nonetheless, we need to uphold tenure in its current form, until we have an alternative. However, the current "neoliberalization" of higher education and research with its destruction of the idea of "faculty" itself is not the response but, instead, another expression of underlying the problem, which is responsible for the abuse of tenure.

19 That in many areas, which are precisely those that teach critical skills necessary for the future such as liberal arts and humanities areas of philosophy, culture, and literature studies, we see a concerted decrease of funding in the current *economy of relevance* – precisely because "relevance" is deeply misunderstood.

20 There is a major difference after all between "general" and "universal".

1 Mythical Engines

On Theory, Knowledge, and Technologies

Schmitt Delivered: Re-feudalization and the Economies of Relevance

I claim, and I am not alone in doing so,[1] that we are witnessing an economic if not even a more genreal re-feudalization (Neckel 2010; Supiot 2013; Lotz 2014). My deployment of "re-feudalization" is, first of all, a little more expansive than the classical definition and, particularly with regard to academia and the *economies of relevance*, should be viewed in relation to the co-occurring and co-produced phenomenon of (re-)prebendalization (see: Bakker 2015a,b, and Bakker in: Langman/Smith 2018). The concept of "feudal" that I deploy is more expansive, largely because I include it in the genealogical heritage of the emergence of *assemblages* and *arrangements* of "property" and "political agency" that can be traced back to the Ancient Greek concept of *oikos*.

Sighard Neckel, who is perhaps considered the leading writer to champion the conversation on "re-feudalization" in socio-economic terms, also emphasizes the fact that the question of a "**re**-feudalization" was created in the specific line of inquiry into domination by knowledge discourse by Jürgen Habermas in his seminal – though, from today's perspective, slightly flawed[2] – study on the *Structural Transformation of the Public Sphere* (orig. dt. 1962/2001; Engl. 1999).

Neckel places his operationalization of Habermas's concept within arm's length of the discourse on post-democracy by Colin Crouch (2010, 2011). While I agree with this placement, since I too situate some of my arguments in proximity to Crouch's, and find some of my own points of view and conclusions implicitly and explicitly indebted to him (as well as to Mark Bevir's (/Phillips 2017] and R.A.W. Rhodes's [1994] discourse on hollowing out the state), I do find that Neckel, though I admire his work, misses a few points about Colin Crouch's perspective that seem to me to be necessary in understanding the potential in Crouch's contribution and, more importantly, an effective placement of the concept of "re-feudalization" today. The problem is that Neckel's reading of Crouch runs the risk of placing the pragmatic conclusions of his theoretical apparatus rather more against him than *with* him – wherein "with" means merely a similarity or a shared orientation, or a complementary perspective, not any similarity in identity between their respective approaches.

First, Crouch's discourse – which we must not forget is *arranged* with his publication with the *Fabian Society* (founded in 1884) is part and parcel of a discourse on "from government to governance" (Marinetto 2003) and its institutional move. This was most visibly diagnosed by R.A.W. Rhodes in his seminal paper on the tactic of "Hollowing out of the state" (1994) and in his book *Understanding Governance* (Rhodes 1997). The gist of this "hollowing" out of the state and shift towards governance is, largely, focused on the change of actors (to markets, [accountative/algorithmic] bureaucracies, and [technocratic] networks) that wield power and influence, whereas the power of the state as actor is waning (for a different view on the state, see the Actor-Network perspective proposed by Passoth and Rowland [2010]). The meaning of post-democracy in this context, and in a parsimonious reading of Crouch and co., is basically that this "hollowing out" affects not just "the state" or "the government" but the very substance of democratic institutions without removing them.

In short, institutions such as "elections", "free speech", and a "democratically elected government" remain, but they either have become effectively powerless, or their actual meaning within the system has changed from the meaning shared among, understood by, and tentatively(!) assumed to still be in effect by the majority of voters; meanwhile, in reality, actual decision-making happens in different ways, and power lies with other actors or within other structures. Thus, democratic elections cannot change anything but are still sought out as quasi-legitimation of the structures of distribution in place. These structures of distribution are regulated and used by the technocratic and/or wealthy elites. Ironically, the *constitutive* idea and discussion of (unregulated) freedom substantially benefits a post-democratic status. And equally ironically, many voters suspect and explicitly give voice to their suspicion that "voters cannot change anything": They understand that they live in a post-democratic society, and this notion is deployed, paradoxically, to support the post-democratic state of affairs. Sometimes, voters support this state by resigning and joining the camp of disillusioned non-voters; sometimes by naively assuming that "digital culture" will resolve everything (for details, see: Stingl 2015); and sometimes, by assuming that they will one day become one of the few rich and powerful (through hard work, sheer luck, or divine intervention) and thus profit from the continuation of this state of affairs– a popular media myth in the United States (Shenker-Osorio 2013).

In my opinion (see also my analysis in Stingl 2015), part of the problem is that, all too often, democracy is considered to be a constitutive category rather than a regulative ideal – the Frankfurt School and Habermas, in my view, fall into this trap. Regardless of how democracy is defined (in terms of autonomy, free speech, participation, etc.), it is never fully realized (on this, Crouch [2000] seems to agree, considering it an ideal-type of sorts). Actually, I think that considering democracy constitutively, i.e. taking it for granted – as is pretty much the case with US-American society in a broad sense – actually causes post-democracy to proliferate. Second, I think, as I argued in more detail in *Digital Coloniality of Power* (2015), that considering democracy *a priori* to republicanism rather than the other way round is equally destructive; republicanism (in the first instance, understood as lawfulness and the equal rule of law for everyone) is a necessary condition for any democracy to function. Any democratic society that does not heed to this categorical imperative that democracy (a) is an ideal and (b) requires a foundation in republicanism[3] will eventually become dysfunctional and post-democratic.

Neckel (2013: 41f.) characterizes Crouch's project as illustrating how democracies become de-substantialized and slip back into "pre-democratic times". What he emphasizes is Crouch's narration of the affairs of politics transforming to adhere to the rules and terms of marketing and of markets. While Crouch certainly touches upon that aspect, Neckel does not seem to mention some of the factors and dimensions that are crucial for Crouch's project and in particular since Crouch's publication of *The Strange Non-Death of Neo-Liberalism*: namely, the role that corporate actors play. In the traditional "liberal" account, which is still one that wields a lot of clout – especially semantically – as a kind of folk sociology (particularly in the United Kingdom and the United States), there are two *primary institutional actor-types* in play: state government(s) and the market(s). Crouch elucidates in *Strange Non-Death* that this account must be corrected: (Multinational) corporations are a third actor-type, the interests of which do not overlap with those of markets, as the folk social theory presumes. On the contrary, large corporate entities are necessarily counter-market actors. Their tactics are decisively directed against both governments (whether national or supra-national) and markets. In a sense, they seek to establish monopolies and monocultures, and thus not only defy the logical of the relations between governments and markets but also denounce this logic publicly in multifarious ways, effectively undermining the very conceptualizations of the relations between power, information, and care that are necessary for any functional *democracy* (which I claim must be a *republican democracy*, to begin with). A perverse irony is that, as Crouch notes, even where the privatization of state structures is proliferated and/or private-public partnerships are established, whenever this privatization ends in failure or with excessive costs (for the tax payers), even though the fault most often lies with the private (i.e. corporate) party, the blame is almost always assigned to the public (i.e. governmental) party.

A claim of mine that is, however, generally missed is that these corporations are already quasi-states, and some are on the verge of becoming actual states. This is largely missed because we are very focused on a definition of statehood based primarily on spatial criteria, specifically, of territoriality, and, consequently, on the (Western) idea of (modern) nationhood as well as a spatial idea of sovereignty. My point is that statehood should not and cannot be (and global-historically[4] is not) restricted to these criteria, that there were and are other types of states, and that a major challenge for social scientists is to figure out what statehood actually means and has meant conceptually and empirically. That does not mean, however, that nation-states do not play a role; on the contrary, they are, will, and perhaps should remain for the foreseeable future a crucial form of statehood and political agency, along with supra-national and, hopefully, regional forms of political participation (such as in a future European Union (EU) of internal regional multipolarities and multiple integrative speeds, rebuilt around a notion of solidarity rather than [colonial] economic growth). However, my point, similar to Crouch's concern, is that we must understand the (political) agency of corporations appropriately. In understanding that they are state-like (not simply juridical, let alone natural, persons), the relations between national and regional governments, on the one hand, and corporations, on the other hand, can be much more adequately conceptualized and framed in policies. By failing to do this, we are actually contributing to forms of re-feudalization by allowing and even proliferating the establishment of feudal relations between people and/or small business on the one side and corporations on the other.

This is one aspect that is, due to a limited adoption of Crouch and emphasis by Habermas, not as developed in Neckel's approach. Nonetheless, he does capture many important aspects of contemporary feudalization. One of the foundations of this process is the difference in status groups in society, characterized in the following.

The existence of wealth oligarchies is nothing new, nor is the fact that the distributive order and the market create opportunities that these oligarchies can profit from. However, on top of all that, there are added benefits for members of the oligarchy such as monetary entitlements and derivative powers. This throws out traditional market competition as the way in which social hierarchy was constituted within what was classically considered under such terms to be civic order. Thus, re-feudalization occurs. It means that social groups are now no longer competitors in the game of social mobility. Instead, they "wall-off" incommensurable living circumstances. Social structure itself becomes, well, static and defined by what we can, at best, call *corporative estate-persistence* (Neckel 2013: 44).[5]

In his attempt to conceptualize this development in a manner more accurately than he apparently deems Crouch to have done, Neckel urges us to recall Habermas's original introduction to re-feudalization: It occurred after the public sphere had emerged as mediator between the state and the concerns of the people; it emerged in the very act of making participation possible through a differentiation of the public and the state.[6] It must be remarked upon here that, in the German perhaps somewhat more than in the English, the semantic expansion of the word *Staat* (*state*) coincides with the word *Regierung* (*government*). There is in public and media discourses, but also within expert discourses in policy circles, as well as political theory, philosophy, and sociology, a widespread confusion and blending of these two concepts, even by individual authors and publications. So, when Neckel uses the term *Staatsgewalt* (literally: *the force of the state*) to describe what level of the state is shed from the public in the emergence of the public sphere, translation leads us to cover the aspect of *force* either as *authority* or as *government* – this question is, in my view, one that must separate the ontological (*government*, which also holds or presumes to hold the monopoly for physical violence in a state) from the epistemological (*authority*), which complicates the concepts of Habermas and co. endlessly, and in an aspect that is still not fully resolved, as recent struggles with the "transnationalization of the public sphere" in a discussion initiated by Nancy Fraser show (Nash/Fraser 2014), but also quickly asserts why statehood cannot simply be restricted to mean "territorial nation states".

To return to the original point of Habermas's re-feudalization, he concludes that the commercialization of information publication and exchange in the 19th century and its acceleration in the 20th century led the state to intervene in the public sphere *and* private business interests to intervene in the affairs of states. Neckel sees the *public audience*[7] *artificially kept at a distance*. What was once *a lively and conversant public* is now enrolled in a chatty but *staged publicity*. This refers to a state of *being rendered mute* and subject to mass-media that reprogram social groups in the state of "staged publicity" into "focus groups". The motivational vectors are provided – Neckel (2013: 45f.) seems to agree with Bernard Stiegler (2010) – *as entertainment and suspense*. Instead of engaging in civics, these groups are motivated to give a kind of plebiscitarian affirmation and consent to the status quo. What was once a public display of monarchical power, which, from Kantorowicz to Foucault or Elias, countless sociologists have analyzed, makes *an odd kind of reentry* – however, it takes on a mask or persona: the form of public relations which is found among private persons and corporations. It's a trickster move for "public relations" means, above all, to manage the presentation of individual interests as general ones, and this presents as the process of re-feudalization of and in the Habermasian public sphere. Neckel emphasizes that this is a process, not a status description, and it is nowadays structurally similar to socio-economic re-feudalization. The argument he makes – that the ideology of finance capitalism creates a promise for material prosperity that is also the engine driving the expanding exclusion from wealth – mirrors arguments brought forth in terms of *Late Liberalism* (Povinelli [2016]), *Near-Liberalism* (Ecks 2010), or the emergence of *risky subjects* (Patel [2006]). *Near-Liberalism*

> emphasizes that freedom, emancipation, and autonomy are always held out as future goals, even if they are not realized in the present. It suggests that neoliberalism is not a uniform practice, but that it tends toward nearness and local differentiation. It also accentuates

> that neoliberalism can switch to restrictive modes of governing at any moment. Near-liberalism is not a program to get rid of the social and political from the sphere of economics, but one that seeks to produce new alliances between markets, states, and citizens. These alliances need to remain as flexible as possible, to allow moving in new directions at once, and to allow severing ties form stakeholders that have become liabilities.
>
> (Ecks 2010)

Here, re-feudalization describes (1) the decline of meritocratic realities: In a post-meritocratic society, the meritocratic ideals remain, but the realities are hollowed out, such as when people are told to work hard, get an education, become their own entrepreneur, etc., and lack of success is interpreted as their own fault due to lack of effort, hard work, or similar, while the declining number of actual "opportunities" and inherent network ties among a class of (monetary, social, cultural, symbolic) capital-holders occupying these increasingly rare positions are ignored; it follows that (2) this society is based on the inheritance of (various) forms of capital, which is, however, largely based on or in a transdispositional relation with economic capital, as predic(a)ted by Pierre Bourdieu.

Furthermore, I claim that (3) we witness within re-feudalization (and social closure) a return to prebendalization in specific sectors that are subject to bureaucratic and technocratic practices, specifically those that are most prone to the phenomenon that I coined as *bureaucrasia*, meaning a bureaucratic or technocratic system that promotes among individual actors a form of *akrasia* ("weakness of will") with regard to system transformation; most notably, prebendalization can be found within academia and is directly constitutive of the emergence and maintenance, for example, of *White Collar sociology.* A prebendalist system describes a system wherein those members bestowed with powers (let us tentatively call them *officers* and/or *commissioners*) deploy the resources (in various forms of capital) to benefit "their own" – i.e. benefit their ethnic group, ideologically affirmative subordinates, etc. – therein also securing their dependence on and loyalty to the system, making only officially permitted forms of critique possible and punishing any other (even merely suspected) form of difference and disobedience (including, for example, an explicitly skeptical outlook towards career-paths). According to this system, one must pay their dues and follow conventionalizations.

A major driving force within this neo-feudal/neo-prebendal political economy is the dimension of relevance, i.e. we are faced with is a general economy relevance. In my last book (Stingl 2015), I mentioned *economies of relevance*: for example, in regard to the production of medical relevance in digitalized health care and in terms of the tactics people deploy to stay relevant in online forums, such as being a particular kind of asshole (an example being the brand of misogynist speech that was visible during *Gamergate*).

Let's unpack this concept at least a little bit more and in new directions – although I am sure you already have a few good ideas what this is all about. Let's think of "a general economy" (Bataille org. 1949/engl. 1995; Derrida orig. 1967/engl. 2009), both a political *and* an affective economy, in terms that must be set within a *general ecology*: Erich Hörl understands the concept *general ecology* in connection with "our technological condition" (2015), when

> using the concept of *general ecology* in connection with Georges Bataille's concept of the "general economy" (*économie générale*). Under the technological condition, the question of general economy becomes a question of general ecology, and as such it is to be reformulated as the core of a new political ecology.
>
> (Hörl 2015: fn. 76)

There are several reasons for this connection from a theoretical point of view, which become apparent, I think, when we look at Bernard Stiegler's recent proposal for a *new critique political economy* (2010), or rather its pharmacological critique: Economic systems in general, and the capitalist system in particular, always constitute systems of production of *protentions* (Stiegler 2010: 66).

Protention, as derived by Edmund Husserl in his phenomenological philosophy, is to be understood in the interplay between retention and protention as the interplay between particular phases of perception being retained (Stiegler distinguishes with Husserl the three stages of *primary*, *secondary*, and *tertiary retention*[8]) and the anticipation of the moment to come (*protention*). However, as a student of Derrida, Stiegler's *protention* involves the moments of *Aufschub* (German), of deferral, and of *différance*.

In a related notion on *economies of relevance*, Armando Iannucci recently mused in *The Guardian* that

> Culpability is something big business doesn't recognise. Nor is it something political leaders want to pursue. It has no market value. Banks fail, financial centres crash, Swiss branches help evasion, but no one is arrested or put in any kind of dock. Monoliths such as Vodafone come to gentlemen's agreements with HMRC over reduced payments in back tax. Meanwhile, vans circulate telling foreigners to go home, posters go up telling us to snitch on benefit scroungers and letters are put through people's doors warning them to get worried about their spare bedroom. But dare to express a single doubt over the supreme rationale of having the business community running the whole show, and you're derided as an economic nincompoop, unfit for office. We can launch inquiries into the police, the war and the press, but it's the stuff of fantasy to imagine we'd ever launch a full-blown investigation into why our business community lives under permanent impunity.

> That's because this belief that, fundamentally, we should all be like businesses, has expanded exponentially. It is political life itself. There's nothing left. It's taken on the status of an unshatterable truth: if we are to have any credibility, business is what we must do.
>
> [...]
>
> I'm not convinced. My pet theory is that politicians with no real experience of work outside Westminster are paranoid that they don't look grown up enough unless they act as though they run something. So they run things like business, since business is most of the outside world they encounter. The tougher they talk business, the less they'll be laughed at by the grown-ups in boardrooms they meet daily. Outwardly, politicians act like company directors because inwardly they actually feel like eunuchs. I could be wrong. But if you think my theory pretty mad, consider how sane the reality is we have now.
>
> (2015)

I think Iannucci has described a layer of retentional filters that emerges as this "wonderful" mechanics of one *economy of relevance* today – and by wonderful, I mean with reference to Carl Schmitt,[9] particularly the dimension of wonder so characteristic of the quasi-religious experience that today's business-administrative dogma promises and projects. To be sure, I think that the most revered title that one can obtain today, the MBA from an American Ivy League school, is quite aptly named: Master of Business Administration. The point, often missed when condemning profiteering culture, neoliberalism, economics/economists, and business is that the "business"-dimension is only one half of the problem, and it would not be complete without its other half: the "administrative"-dimension. The point I am trying to make here is that in today's neo-, near-, Late Liberal culture, it is not enough to be profitable in business terms; one must be seen to matter in the administrative, i.e. managerially projectable terms that go along with the culture. Take the example of movie-making and today's blockbusters (whatever that title is worth today). We have come to a point where a movie that cost $200 million to make and earned $400 million, thereby doubling its production cost, is considered a failure, and the making of a sequel to it is unlikely.[10] The logic here is clearly not to be found in the question of profit alone. It involves a question of how much profit a movie needs in order to be considered "profitable", i.e. when is a profit relevant?

This relevance is a complicated matter. While people have discussed the rule of thumb that a movie has to at least triple its production costs, for finance-intensive movies (costing high two-digit or even three-digit million sums in US$), new benchmarks have recently come into existence – for example, due the impact of the Marvel superhero franchise and films such as *Iron Man* and *The Avengers*[11] – increasing expectations to around the one billion US$ mark. Obviously, if the multiplier of production costs to box office success has reached such a dimension, the idea of what makes a successful movie has changed. For example, to even come close to such numbers, a movie will have to play on a minimum number of screens – even within individual multiplex cinemas. So, again, in creating enough attention for the movie in question and "enforce" that cinemas are willing to make a number of screens available for a certain amount of time (thereby also limiting competition), mere money is unlikely to be the sole and probably not even the deciding factor. The consideration of a notion of relevance seems to make analytical sense here[12] to describe the motifs and motivations behind the managerial and administrative tactics, and underlying "algorithms" and decision-making heuristics.

In a similar vein, think about the notion of a "profit warning" for shareholders. While a "profit warning" is often provided when a company will make less profit than expected (again, this usually means that a substantial profit has been made, but it is less than expected), such a warning is also issued when a company expects to make more profit. The point here is that "profit" being made may not be the deciding aspect, even though one would and should, on a common sense basis, assume that to be the case; the point is that the idea of being able to correctly project the profit margin may be valued higher than the profit itself. This refers back to Stiegler's use of *protention-retention* and its importance for our technological condition, wherein we have built for ourselves and become imbricated with a *technological unconscious* (Thrift 2004, as channeled by Hörl 2015[13]). The accounting-and-*audit-culture* (Strathern 2000) that we see at work here has some of it strongest and deepest roots in German and French moral statistics, and the German insurance culture of the 19th century (Simmel 1892; Sarasin/Tanner 1998a; Rodgers 2000;; Rabinbach 1992).

Economies of relevance have another interesting effect in terms of the diagnosis that Maurizio Lazzarato has coined as the *Indebted Man* (2012). I take a few cues from Lazzarato and his interlocutors, Tiziana Terranova and Alberto Toscano, but I veer in a bit of a different direction:

- People are relevant only if they have capital.
- Capital produces relevance because the (political) ontology of production produces expressivities that are based on the notion of relevance. In other words, the order of practices expressed and the kinds of practices expressible in a system such as the (academic) job-market, "*the* Economy", the news-cycle, the heavily financed blockbuster movies, digital health care, etc. are structured around relevance.
- Capital accounts for strong and weak network ties in interesting and diverse ways.
- They are generally relevant if they have a broad mix of capital.
- They can be regionally relevant in terms of a specific sort of capital: monetary, social, symbolic, cultural, health care capital, digital capital, digital cultural capital, etc.

- Thus, we can speak of a "general idea/theory of *an ontology of production*" and of "*regional ontologies of production*".
- Capital can be positive (wealth) and negative (debt).

It is the latter that presents us with an interesting problem. While there is indeed a question not "*that* an indebted person is relevant" but of "*how* an indebted person is relevant" and that the *ontologies of production* have adapted to these types of capital-based relevance by inter-relating the spheres of wealth and debt to one another in increasingly feudal/prebendal ways, the group of people who has neither debt nor wealth become effectively silenced – and, indeed, monetary debt has special relevance here (it should be remembered that Bourdieu saw in economic/monetary capital a more central or model role, considering that other forms of capital translate with one another). An interesting and crucial example from US-American society is the question of student debt. While there is a tension between students from affluent families who do not have to struggle with debt but, instead, have a positive capital and students who accrue large amounts of debt, which seems to be designed to remain with them for extensive periods of their lives, one might be wondering what happens to students who neither come from affluent, privileged families nor accrue any (or not much) debt. While one might think that this "debt-neutral" group of students would be in a good position to move socially upward after college/university because they do not have any debt to pay off and can acquire positive capital, one might also find that their situation is actually more difficult, precisely because they are "capital-neutral" and thus irrelevant. One of the interesting elements of the rise of "*modern* credit" (though it is by no means always only monetary) is that precisely in regard of its monetary dimension, merchants and craftswomen, as Clare Haru Crowston (2013) shows in regard to the 17th- and 18th-century French fashion industry, were in non-monetary debt and owed money simultaneously. One crucial aspect of the interlaced credit systems that Crowston describes was that they were also one large information system. So, the basic analytical observation for the effect of student loans is that it can be expected that students who are debt-ridden after college are small machines (in a cybernetic sense) that constantly circulate money owed to, money to be obtained ("futures"), and money received, therein constantly co-producing information about themselves. Most prominently in the United States, but also elsewhere, this information (and relevance) economy has become institutionalized in the credit information industry, which the *infomedian* John Oliver deconstructed in a segment of his popular television show *Last Week Tonight* in 2016 (see: Lee 2016). To propose a connection between credit score and Deleuze's concept of *control society* almost suggests itself. To be "capital-neutral" in a society that thrives on an information economy in which positive and negative capital are one of the primary modes of producing information marks one irrelevant, perhaps even suspicious and untrustworthy.[14] As a consequence, the figure of the "capital-neutral" person is, as of yet, an underappreciated object of research and scholarship. While much research on social (in) equality and (in)justice is conducted in terms of the role and effects of debt; while much of this research also teaches us that profit and debt exist in various forms of capital, not only in the form of money (not to mention that money itself exists in various forms and temporalities); while seminal works such as Maurizio Lazzarato's *Indebted Man* (2012) give us crucial insights into the workings of present society, there is a bit of a blind spot regarding people who are not within the obvious, aka measurable, rifts between affluence and poverty. It is here, the zone where less measurable information is produced, to which I would like to draw attention. For me, this is, at this point, a methodological intervention: In theories, conceptual apparatuses, choices of empirical arenas, and available methods, this methodological aspect of the great variety of "silent Others" requires far more attention. Any research approach that understands itself as coming "from below" is, in my view, not truly "from below" if it doesn't include the various and heterogeneous groups of people who do not or only barely appear. I have illustrated this point in the example of what I call *digital cultural health care capital*. Having in a certain sense "nothing"[15] and having debt are, thus, two completely different modes of being, which present different kinds of vulnerabilities. And while "having zero" may seem to be an ideal state from the point of view of someone who is in debt, it means a lack of capital, and it means a lack of agency. Already being in debt, i.e. having negative capital, may bestow one with more problematic forms of agency, but it is agency, nonetheless. Anyone who has not already accrued positive or negative capital suffers from a lack of agency. This is why it is much easier to get credit if one already has debt (or significant positive capital, or simultaneously both) than it is for someone who has no capital. Without either positive or negative capital, one is *not relevant*.

Who Is Allowed to Speak (in General), Who Is Allowed to Be Heard (in Particular)

In public and in academia, this issue of relevance leads to the question of who is allowed to speak and who is allowed to be heard. For example, there are swaths of traditional and alternate knowledges that have been swept under the rug or even utterly eradicated in *epistemicides* (Santos, 2018; Grosfoguel 2013). The lead motif that is often deployed, and that is equally often launched against the mere ideas of relativism or constructivism, is that to allow knowledge to be taken seriously, that is, to be either not produced in terms of the institution of science (academia) or not in line with the accepted common sense notion of "realism", would invite chaos: For the epistemological-geopolitical Global North, this "myth of lurking chaos" is an integral narrative, as psychologist Ernst Boesch has illustrated (2002). There is, in the Global North, an ideal at work which harks back to the times of Plato: To be allowed to speak means one has to have good character, and vice versa (as propriety or property?). Today, we can say that "character" translates into (some form of) having capital/credit. The other side to this problem is that, especially within academia, the famous "Matthew-effect" has hit overdrive and created a global structuring effect: With the emergence of the European Research Council (ERC), for example, an independent institution for science based on innovation and excellence was trumpeted, but it has been noted that there are obvious biases in the distribution of the funding (Hoenig 2017). Moreover, the issue is that it is less about good science or the production of knowledge than it is about status (and relevance). But despite these paradoxes, the way in which the ERC evaluates research disseminates among many national research agencies. Similarly, it has been noted that there is a social reproduction in many US-American university departments: namely, that recruitment of faculty is largely reduced to those who come from a small batch of universities that are considered the most relevant, such as Harvard, Yale, Stanford, etc. Over the years, informants have told me that the same is true for "foreigners" who want to be hired at universities across Asia: that they must have visited one of these universities in the United States or one of the few European equivalents, like Cambridge or Oxford. Do we not all know of cases where single semesters, guest visits, etc. at those Ivy League institutions were blown out of proportion on applicants' CVs? The fact that one is advised to portray something in a manner that may be misleading only to enhance the status one has – and that this is, to some degree, accepted practice – is quite telling. It speaks to something that I call – evoking the epistemological issue of "types of warrant" – the difference between entitlement versus justification (see more in the final section of the book). What we are witnessing is the creation of economies of knowledge that are characterized by epistemic injustice. Here, it is important to note that one way of expressing status and relevance is in the form of certain cultural techniques, of which writing style is one. For example, there are certain styles in which grant proposals are written; the same is true of reference letters as well as entire research articles – see here, too, Barthes's question of "readerly vs. writerly texts". These cultural techniques are specifically taught at certain institutions, like the so-called Ivy League universities, often informally by involving students from the start in these kinds of work; "financially and reputation-poor" institutions, however, often do not permit the time to include students in such projects, as several of my informants have confirmed. But the problem is, also, that members of the ever more exclusive clubs become the ones that set the standards. As a consequence, even when procedures are supposedly open, they are often informally biased. As more than one informant gave me specific information, I paraphrase in the following to create a vignette (in this way preserving anonymity):

> *program officer A of funding institution ZYZ has always liked how X from "Ivy League University" writes their proposals and has favored X's proposals:*

> *even though A does not write a peer-review, A communicates to the reviewers what criteria are important for a successful proposal, which matches with how X's proposals look like; subsequently, when the awards are given out, A will allocate resources to X nearly unmodified as A has that discretion. X and their students have gotten all their proposals through.*

Similarly, there is a sort of a bias against research that is "too *inventive*". I deploy the word "inventive" here, rather than "innovative". In our conclusion to *Worlds of ScienceCraft* (Weiss et al. 2014), Sabrina M. Weiss and I identified four main problem avenues for the future of sociology and philosophy of science, and Science, Technology and Society Studies (STS), which we called "Harder, better, faster, stronger". A key issue is and remains that a lot of what is called with mantra-like fervor "innovation" is not very "innovating", i.e. not very novel and/or groundbreaking at all. Innovation is often, when seen with a bit of distance, an incremental increase in the efficiency of an already existing technology or system. An invention, we argued, is something truly new. The simplified example we liked back then was that the problem of traffic-jams and fumes is not overcome by the "innovation" of more fuel efficient cars but by inventing new ways of moving about in masses. Since then, hailed "disruptive innovators" like Elon Musk have made proposals that simply "re-created" already existing technologies and thus are not really inventive, like the subway.[16] In fact, we could say that the world hasn't really made any significant progress at all in the last 50 years. The majority of technologies we are using were dreamt up more than 50 years ago; we are still highly dependent on the same (often fossil) raw materials and extractive industries; and, even culturally, Western pop culture is constantly rebooting the Baby Boomer's childhood programs, comics, etc. The only thing that has seemingly made great leaps is environmental destruction. Of course, this sounds like hyperbole for the sake of provocation. And, perhaps, ten years ago, that, i.e. provocation, would have been precisely the case of why I articulated it that way; however, I cannot help but bitterly register that it now seems to have become an accurate description of the world.[17]

It is difficult to be truly innovative, or rather, inventive, in research. Certainly, research must be articulated, especially in grant proposals (but also in academic job applications), in a way that is communicable and, at best, also intelligible to the people who review it, who may have some expertise in the research field or close to none. But above all, research must seem relevant. Still, relevance and invention can come to a point where they appear to be as good as they are mutually exclusive. To be truly inventive, one must venture "out of the proverbial box", but how can something "outside of the box" be considered relevant? Relevance is, precisely, the box, about which I will I say a little more towards the end of this book with regard to infrastructure. What we also need to consider is the reviewers' position, the logic they are told to adhere to and the demands made on them. Think about it: Most review processes are made without remuneration, for free, and yet they are considered a great responsibility, and it is actually true that reviewers often do take this responsibility quite seriously – that is, unless they put personal interests first, like stealing applicants' ideas or rejecting them for egoistic reasons, such as because they or their friends weren't cited, or the proposed work contradicts their own (and while perhaps rarer than some stories about academic mischief suggest, these cases still occur often enough to be a problem, especially for junior scholars). But the sheer amount of applications (for grants and jobs) is staggering: To have hundreds of applications for one job is decidedly not a rare but an increasingly common event, and dwindling acceptance rates for grants suggest a similar development. But simultaneously, people are encouraged to "apply for everything" when it comes to jobs and to conduct a "shotgun approach" to research grant writing: But if one applies for 50 (or sometimes far more) jobs and writes and sends off 10 grant proposals, and if an increasing number of academic laborers feel forced to do so, candidate/proposal reviewing becomes a task that requires short-cuts to get done at all. The possibilities for reducing the piles of applications are manifold, and, of course, must often remain somewhat clandestine, because the processes in question are supposed to be fair and impartial. But from excluding foreign applicants (even in internationally open job searches) to avoid visa processes to just "halving" the pile[18] to excluding candidates from "lesser known institutions" to even more arbitrary distinctions, such as "color of envelope", I have heard enough stories from informants over the years. Of course, many of the informant accounts and stories that also pop in media are of anecdotal status and their content (aka the "avoidance strategies" they talk of) is often quite heterogeneous. But their existence, their apparent increase, and their persistence are data points we should not ignore, especially as discourse analysts: They may even be, in the worst case, "myths", but the existence of myths, too, tells us something about the "system" in place. In one university I was dealing with, reference to "the system" was made quite often: namely, in regard to the writing of confidential recommendation letters. As one tenured, full professor and at the time department chair in my presence told a junior scholar – a PhD student – asking for such a letter, "You ought to know the system". It turns out that the system he was referring to meant that the student was supposed to write the letter, and after a quick skim-read to see whether it was the "expected standard language", it would be signed and handed off to the department secretary for mailing. This was at a German university, and in Germany, the use of references is far less common than it is in the United States, where academic jobs almost always require three such letters. Now, an increasing issue has been that with the inflation of applications, individual candidates require a swath of these letters, leading to the establishment of the same such "systems", on the one hand, but also to social inequality, on the other hand. Letters should, ideally, be uniquely fitted to the application, but not all professors have time to do this, if they have time at all. Thus, it became possible to use commercial dossier services to get a "standard" letter from a recommender for multiple uses – of course, these letters are less effective in the cases where they have great importance,[19] but they need to be present (often, their lacking or being late prevents an application from being forwarded), and, of course, a mediocre letter from an Ivy Leaguer is often worth more than a glowing recommendation from someone at a lesser known college – relevance is, of course, not identical yet related to "symbolic/cultural/social capital". And even though some universities have begun to ask for these letters only for short-listed candidates, the majority of US-American institutions of higher education and/or research still require them. I believe that the reason for this is, mainly, that they work nicely as a mechanism to reduce the number of applications sent in overall and to reduce the pile of applications that need to be taken seriously in the review process.

But I digress and should finish my point about "inventiveness". In the 2015 Eastern Sociological Society's (ESS) Annual Meetings presidential session with Mario Small, Diane Vaughn, and Christopher Winship, two interesting issues came up in the panelists' discussions: Number one was a question of relevance, namely that because Chicago played such an important role in the history of sociology as the example of "*the* US-American city", any study of urban America would need to make some reference to it, and often, it was still used as an exemplary case, although in fact, it and its neighborhoods are in many cases not representative of the (urban) United States; number two was that reviewers, for example, reviewing grants for the National Science Foundation (NSF), reject method(ologically) "innovative" (aka inventive) proposals because they felt that they were not "competent" enough to evaluate the proposed innovations. With full professional lives between their own research, teaching, conference travels, university administrative tasks, and attempts to have "personal lives", keeping informed about all kinds of developments in theory, methodology, and method is, of course, impossible. Therefore, the approach "when in doubt, reject" is considered safer. After all, even if the review process is anonymous on some levels, on others, it is not, and no one wants to be the person who waved through a proposal that failed in practice – how true progress is possible without the danger of failure eludes me, yet I understand that people like to believe in a myth of the "controllability" of success. Plus, of course, if one identifies a "risky" method(ology) that one cannot evaluate well, it is easy to reject this quickly and thereby reduce one's amount of work. And while this sounds quite sinister, it is increasingly becoming a pragmatic necessity because workloads are increasing. To the speakers at said panel at ESS, I proposed a possible solution, which I am fully aware is not likely to be realized: I argued that, just as for research, paid sabbaticals for working as a grant reviewer would give academics time to evaluate proposals thoroughly and inform themselves about the methods and ideas proposed in truly inventive proposals. But to accomplish that, more money would be required, and more academic jobs, aka "real" jobs,[20] not adjunct positions, would need to be created. Instead, research funding is too often invested in trodden or "relevant" paths rather than new methods and ideas. Even the ERC, which supposedly promotes excellence and innovation, is plagued by this because "excellence", its main criterion, is vague and obscure enough in practice to be conflated with economies of relevance and reviewers' "precautionary strategies", as described. The response that proponents of the ERC often give is that, despite a few setbacks, they still manage to have some independence (from politics) and manage some innovation, and it is better to have at least that much rather than none at all. However, because of the effect of the ERC on the research landscape overall, and the desire of many national and smaller funders to emulate ERC's moral and political economy of "excellence", its global effect is not a positive one (at worst, it is even detrimental), intensifying economies of reputation, Matthew and other prestige effects; creating double binds; etc. To lessen the burden on reviewers – as more and more "hopefuls" think of submitting an ERC proposal, knowing that if they "win in this lottery", they gain enough prestige to have a shot at a "real job" – the main steering mechanism is to make the production of the proposal as difficult as possible and the style of the evaluation such that it does not look for a proposal's merits but for immediate reasons for its dismissal. Applicants spend months on (re-)writing applications so they "fit" the "ways of reading" such proposals rather than on presenting their research. Indirectly and directly, this also contributes to the continuing precarization in academia because it is important to "have worked for" people who already have reputation and prestige, who can also impart the implicit knowledge involved in writing such proposals in the right way: Thus, we are back to the mix of re-feudalization and re-prebedalization. This is, however, incredibly disastrous. Often enough, I have observed that people fall into a state of depression right after having finished submitting a major grant proposal. It may not be too much of a stretch to compare this to postpartum depression. It is in line with the various entries on the exploitation of junior scholars, the resulting mental health problems, and the establishment of a feudal nobility in academia, documented, for example, in *The Guardian* under "Academics Anonymous". But with the majority of proposals rejected and the percentage of real jobs in academia declining, while we witness a dramatic increase in precarious (euphemistically called "contingent") positions, what will become of all these people who cannot help but feel like being in a "*salon de refusés*" (Marianne Weber)?

Giorgio Agamben, in a recent interview (Agamben/Radisch 2015), stated that today, every individual is dependent on the market, and, thus, there is nothing which does not have to be acquired on a market. I think this should make us question what is acquired on the job market, specifically on the academic job market and the "grant" market. It is, to some degree, the job seeker who acquires something – not or not only the employer: The job seeker puts his capital forward to acquire a job (a lease on life or livelihood, a biopolitical perspective might propose); this capital can include symbolic, social, cultural, and other forms but can also be intertwined with monetary capital (one must pay for letters, dossier services, possibly have a car, have a suitable costume [a dress/a suit; here, *discriminatory cultural capital* often plays a role: one buys into fashion as well as into gender stereotypes]), but capital can also be negative in the form of debt since (student loans, for example) can be an indicator that one has "bought" into the Western "career-idea" itself. But in the end, how is all this effectively different from structures of a feudal/prebendal society? In the end, are we not forced to acknowledge the coloniality of power at work, establishing this neoliberal regime that further colonizes the infrastructures of knowledge? Is this academia not just following the logic of exploitation, the rationale of extractivism? If this paradigm (if that is how we want to think of *White Collar academia* and *White Collar sociology* on academia's own terms, i.e. in its own (colonial) terminology) is to be resisted, we need to come up with – what in Juris/Kasnabish's (2013) collaborative project is described as – *discrepant paradigms.*

To create a different avenue, we need to take seriously what Gómez-Barris writes in her book *The Exctractive Zone* (2017: 61) as

> the need persists to find paradigms that acknowledge colonial legacies, elaborate upon contemporary notions of power, and insist on other realms of experience beyond the material. If we understand the current economic rationale that reduces relationality to the legacies of colonialism [*i.e. to "coloniality", as Mignolo&Co would say;* A.S.], and that permeates the affective and relation deficit of the current period – what we can call the "melancholies of late capitalism" [*/late liberalism;* A.S.]

we would make inroads towards a broader understanding of the world, necessary to leave behind the ongoing socioecological destruction that Western epistemologies and Global Northern geopolitics, that coloniality and extractivism, wreak. To reiterate, in my effort, I am not so arrogant as to think that "I" can provide such paradigms – precisely because the atomic individualism of what we know as "academic production", e.g. writing and publishing an academic book, is in its very premise counterproductive; however, this is an effort to continue on a road to provincialize the kinds of scholarship I am entangled with, especially the social sciences, and, more particularly, sociology, and thereby provide an inchoate opening towards perceiving and taking seriously the offer made by decolonial options, to begin learning how to open avenues of conversation with decolonial options without resorting to mere appropriation. This effort is, therefore, strongly characterized by what Anna Tsing (2016) would call "gardening patches".

We Need to Do Discourse Differently: Doing the Provincializing Conversation among Patches

In connection with "how to create a provincializing conversation among patches", I want to pursue a bit further the question of *agonistics* (Mouffe) as a post-foundational approach to political theory (Marchart). In *Digital Coloniality of Power* and elsewhere, I often refer to a seminal paper by Peeter Selg, which he contributed to a debate that began about a decade ago in the discourse on sociological theory and how to debate about its uses and ways. But since Selg defers to Mouffe, his short summary of what Mouffe gifted us to think with is excellent and can be brought back to general use (and, after all, we all are theorists, even if we aren't always or don't wish to be aware of it). Summarizing Selg (2013: 12ff., and improving on how I first thought about it, as I proposed in Stingl 2015), we can identify the following three modes for the political conversation:

1. An *Antagonistic Exclusivism* means recognizing the existence of alternative views, but it denies the possibility of consensus as well as prohibits any interaction and engagement with other views, precisely as other views are marked as the antagonist which must be destroyed.
2. *Deliberation* as *consensual mainstreaming*[21] means that difference does exist but requires us to abandon the notion of dissent. Consensus achieved via deliberation requires us to accept the result of deliberation obtaining with finality. Subsequently, deliberation leads to forms of exclusion for those who continue to maintain dissenting views because dissensus must be given up once consensus is reached.
3. *Agonistic Pluralism* does not shun the existence of dissent, nor does it regard the existence of *dissensus* as something that would lead to an inability to act. On the contrary, a post-foundational position requires the possibility for dissensus to continually obtain, even when, in a specific situation, a decision has come to pass (made by those currently holding the hegemonic position in the discourse) and, thus, transforms dissensus from viewing it as a problem to seeing in it the opportunities.

Mouffe and the postfoundationalism identify in *mere antagonism* the "friend/foe", the "Us versus Them" dichotomies as an absolutist mentality. To follow this means to totalitarize and to render absolute one's own views in the form of being "complete": any *dissens*/disagreement becomes impossible, and thus, we can conclude that constructing the Other means constructing an antagonist, an enemy, but the decolonial view proposes that it is a necessary adversary. An enemy must, subsequently, be construed as "destroyable", but therein, it becomes constitutive of one's identity. To be destructible and to even invite destructibility is a quality the Other is assigned that is inherently necessary to the function of coloniality. But, in order to be able to destroy something/someone and to make that destructible Other constitutive of one's own identity, yes, even to create the idea of identity itself, one must demarcate and make explicit this Otherness of the Other. One must create categories by which the Other can be destroyed or excluded (the deliberative mode is, in a way, a possible outcome of antagonism, and they both share their negative attitude towards *dissensus*, except that antagonism codes *dissensus* as anathema and is thus *exclusive*, while deliberation is *excluding* dissenting voices).

Now, I am interested, for a moment, in a critique of Mouffe – and, of course, there are many as one would expect with a gift of the magnitude and import of Mouffe's – which hinges on her adoption of Carl Schmitt: Alexandre Lefebvre offers an interesting account of Schmitt, which he raises to the claim that Schmitt proper cannot be used by Mouffe in the way she claims (and, of course, this has consequences). I find a number of the claims made by Lefebvre (2005) interesting because they lead us to different connections, for example, to connections between Post-foundationalism, Critical Realism, and issues of epistemic responsibility. Lefebvre's main point is to say that Schmitt's Decionist grounding position (*Grundauffassung* [I use this as a kind of *antagonistic* similar to Kelsen's *Grundnorm*]) is not simply that "there *is* an enemy", but he views

> Carl Schmitt's conception of the political from the position of the *aggressed*. I argue that beneath the apparently active, identifying decisionism of the Concept of the Political (1932), Schmitt articulates the insight that "we" enter our political being only under the threat of existential annihilation. In this way, the genesis of the political occurs when "we" are perceived to be the political victim of an enemy aggressor. Schmitt's Concept of the Political motivates a theory of the genesis of the political as received from the other.
>
> (Lefebvre 2005: 83)

Respectively, *the political* emerges in Lefebvre's account when a "we" is genuinely aggressed, i.e. when it is perceived to be attacked and therefore under an existential threat of annihilation, but while, in the beginning, he speaks of "perceived" first, the threat of annihilation is further qualified as "real" (ibid.). This invites a few interesting thoughts: First, the issue of the "we", which is performatively constituted by either an Other threatening/attacking or by a (autopoietic?) performative construction by perceiving(!) a(n Other as) threat/attack. The latter is, of course, contained in the problem of white, privileged people perceiving a war on Christmas or in their seeing an Other person doing something abundantly normal *while being Black*, such as studying at Yale University (see: https://slate.com/news-and-politics/2018/05/on-being-black-in-the-white-space-of-yale.html). Obviously, the moment people, as an antagonistic "we", talk about "perceived" threats, these threats' status of "reality" becomes questionable because perceptions (may not only be guided by percepts and perceptual frameworks but) can possibly be quite imaginary. This necessitates, perhaps, Lefebvre's further need to qualify that the annihilative threats/attacks he is interested in are those that are "real". But what is "real"? There is, of course, the possibility to take recourse on Lacan's symbolic-real-imaginary or Deleuze's actual-virtual (or even Charles Sanders Peirce), but *Critical Realism* offers the distinction between real, actual, and empirical (Roy Bhaskar). Easton (2010: 123) describes the important distinction as follows:

> The difference between critical realists and social constructionists lies in the acceptance of the possibility of knowing reality in the former case and its rejection in the latter who, in general, concentrate instead on uncovering the constructions that social actors make. Bhaskar (1978) has suggested that critical realism has a stratified rather than flat ontology and this has major epistemological implications. The strata are the empirical, the actual and the real. The empirical domain is where observations are made and experienced by observers. However events occur in the actual domain and may be not observed at all or may be understood quite differently by observers. There is a process of interpretation that intervenes between the two domains. Events occur as a result of mechanisms that operate in the real domain. It is not the case that the real or actual cannot be observed but simply that it may not always be capable of being observed. We see just the tip of an iceberg but that doesn't mean that the invisible three-quarters is not there or is unconnected to what we see. Observation is fallible. It is unlikely to reveal completely and lead to a full understanding of any social situation. Since there can be no definitive criteria to judge the " truth " of a particular version, critical realism relies on the researcher to collect further data that helps to distinguish among alternative explanations and on the community of researchers to debate them thoroughly

Similarly, Carolan (2005: 397) writes,

> Toward this end, Bhaskar arrived at a stratified account of reality, which he detailed as follows: (a) the "empirical," consisting of experiences/observed events; (b) the "actual," involving the flow of events produced either under controlled conditions of experimentation or as uncontrolled "conjunctures;" and (c) the "real" world of causal powers/tendencies and deep structures.

Now, my problem with "the perceived" is, and this extends to the (social) ontological proposition of Critical Realism just discussed, that there is also a (social) epistemological question attached, namely, with regard to Tyler Burge's distinction of types of warrant (see here: Stingl 2017); paired with Mouffe, I'd say that if there is emphasis on what is "perceived", the referred to "we" tell themselves stories that are given warrant by entitlement (which establishes antagonism) and nothing else, whereas Mouffe, with *agonistics*, demands that we distinguish between (imperial) *entitlement* (antagonism) and (democratic) *justification* (agonistics). Lefebvre has no reason to be working through the differences between "perceived" and "real", as I think we must. Of course, his point is not to undermine Mouffe's work but to critique the suitability of *Schmitt proper* for her endeavor[22]: It is a critique that centers on the question of the shared position between them; against the enforced individualism of Liberalism, they posit the importance of group formation, i.e. collective identity, for which exclusion is both a necessary and a generative component. In addition, they center this factuality around the question of *citizenship* – and on this latter aspect Lefebvre's concerns revolve:

> Most obviously, Mouffe is under the constant strain of pacifying Schmitt's terms: she wants to move from "antagonism" to "agonism," from "homogeneity" to "commonality." Underwriting these semantics is an effort to envisage a demos compatible with pluralism. In casting a hegemonic (us/them) political articulation as *immanent to the demos*, she renders "citizenship" a non-exclusionary ground; citizenship provides the field in which constitutive exclusions and pluralistic formations are played out. Unlike in Schmitt where the citizen partakes of a *homogeneous* political association that establishes rights and equality, Mouffe's citizenship is field that structurally inheres and promotes immanent *political difference while preserving universal equality.* For Mouffe, citizenship has no other, no outside; it is the *condition* for — rather than Schmitt's *predicate* of — group formations. This enterprise is precisely what Schmitt criticizes as obscure humanism, liberalism, and parliamentary democracy. Mouffe presses Schmitt into an unlikely service toward "*conflictual pluralism*." She argues that citizenship is universal in that it creates the conditions for a confrontation between *adversaries* rather than a conflict between *enemies*. In other words, there is to be no exterior, no constitutive moment for the emergence of this architectonic of rights wherein we perform "our" interminable variance.
>
> (Lefebvre 2005: 86/87)

Lefebvre's account argues that Schmitt could accuse Mouffe of precisely the mixing up of the two sides of the ontological difference between the political and politics that he accused his contemporary critics of. But, if we follow Lefebvre's account, which suggests that there is a "doubled decisionism" in play – namely, an act of deciding who the enemy is; whereby the demos itself is installed (a Derridean, onto-epistemology of the performative aporia), and installed homogeneously; and how the enemy presents themselves – we end up with a public/state/society that is, because it is, as I said above, dealing only with what it perceives, inherently paranoid. And here, again, the lesson that we learn from Lacan for the individual patient – Zizek, surprisingly correctly, never tires to explain that it still holds true on the collective level – is that, even if the thing one is paranoid about is eventually proven true, this doesn't mean that one isn't paranoid. The fact is that whether there is an *actual* existential threat or not is not important for the Schmittian state/society/public[23]; the perception that there was a threat is what effectively matters for this process, and that is why society-/state-/public-formation in that scheme eventuates a formation that is collectively paranoid at its most constitutive level. This paranoia is, I claim, another variation on the reasons for *the colonial wound* created geopolitically by the Global North, another version of the *coloniality of Being and power.* But here, the second question of Lefebvre, namely how the enemy presents themselves, is an interesting one because of the introduction of looping effects (Ian Hacking): If a society pushes migrants that it constructs as "criminals", as "primitive", etc. from the start to the peripheries of society – often even spatially, geographically, in terms of abandonable, "red lined" urban areas – and does not provide them with opportunities, how can these people not end up occupying precisely the "savage spot" (Michel-Rolph Trouillot) assigned to them, and even adopt the behaviors prescribed to them by society? The "savagery" of the "savages" comes with the "savage spot" as the place they are allowed to inhabit (though never to comfortably exist and dwell in) by the Western hegemons: If you are "the disposable Other", the enemy Other who is destroyable, if no dissensus is allowed to be retained as possibility, what actions are you left with, except the patterns of acting in the very enmity you are originally accused of (for anything you do will be "perceived" as enmity)? The sole alternative would be – and it is the alternative that many accept, rather than committing violence – to be the silent Other: The alternative for the "enemies *made*" is to live as silenced or to be looped back into the discourse, re-*made*, as the very enemy *kind* they were Othered as, an Othering that produces (onto-epistemologically) what it Others in the act of Othering.

"You piss, you live!": *Care, Power, and Information*

I will make a strong, normative claim – there is nothing (the secretly normative) Western social scientists shun more than explicit normative claims. But here's the point: I make it, precisely, *because* I care and because we must care:

> *The Western idea of academia, science, and sociology has corrupted itself (with neoliberalism) because it was corrupted from start (qua coloniality): Its neoliberal*

> *culture of Grant writing, science and technology, an extension of the coloniality of power, is confused and reduced to a* colonial technology.

Here is a short summary of Hans Blumenberg on *Technisierung* and "technological object", which is, I think, important in this context because (1) writings such as Walter Mignolo's recent geopolitical analysis of decoloniality and phenomenology create an interesting context for provincializing Western thinking, and (2) the Western *technocratic-colonial* idea of science is that "science is technology":

> [I]n Blumenberg's 1963 essay on Husserl (...), and in the remarkable "Paradigmen zu einer Metaphorologie" of 1960. An appreciation of this can throw some initial light, too, on the difficulties of placing Blumenberg politically. For phenomenology the essential task of philosophy is to oppose the life-world's universe of pregiven self-evidence to the self-understanding which consists in an exposure of the life-world's 'could-be-otherwise' character. Modem science dismantles life-world certainties too, and as such is not illegitimate. But what is illegitimate is the manner in which this occurs. In the Husserl essay Blumenberg wants to argue that the life-world itself is an indispensable resource for an understanding of what he calls *Technisierung*, the turning of meaningful objects into technical ones. This process is not brought to the life-world from outside, as something imposed upon it.
>
> [...H]e suggests that even phenomenology's 'openness to possibility' can, in a failure of nerve, be elevated into the attempt to realize the possible as what he calls a 'cultural world'.
>
> [...]
>
> In other words, any formula for any 'emancipation' entails an inability to live with contingency, and renders one politically ineffective, supportive of a given state of affairs.
>
> [...]
>
> Here, thinking one doesn't have to live with contingency makes one a dogmatist or an authoritarian. Contemplative passivity and activist terrorism are a product of the same failure to accept a loss of meaning, and are opposed by a refusal to restore it. The idea of infinite progress serves this purpose, the critical self-limitation of reason.
>
> (Turner 1993: 61/62)

Technology and science become absolute metaphor, and, as Turner (1993) cites Blumenberg quite correctly, "absolute metaphor is not 'on the way' to being anything".

What I am interested in throughout this book is an account of technologies of power that is partially Foucauldian and yet also part of another approach, an account of power that I share, surprisingly, with Francis Fukuyama (2013: 350) for we are both "interested in what Michael Mann labels 'infrastructural' rather than 'despotic' power".

Of course, my reading of Mann takes his concepts in different directions. For example, he denied the applicability of his ideas on infrastructural power to the Global South. What I am doing in this book is thinking about care, power, and information through the idea of *infrastructural power*, as proposed by Michael Mann, but *reading* it *que(e)r* through the lens of Foucault in an attempt to provincialize Western/Global Northern infrastructural power.

The different accounts of a "legitimation crisis" that we find discussed by Jürgen Habermas on *late capitalism* (1973 [engl. 1975]) or Hans Blumenberg on Modernity have many roots. In his *Work on Myth* (1996 [orig. 1979; engl. 1988]), Blumenberg offers what can count as a particular impetus behind these different accounts: It is part and parcel of the Modern Age's self-consciousness, to have gotten ever newly serious and down to business with ever new Realisms – be they theoretical, practical, or aesthetical [*I would add: "political", A.S.*][24] (1996: 197; my translation)

This impetus, which is part of the legitimation structure (and hence, the inherent and inherently necessary continuous legitimation crisis) demanded by the *coloniality of power and Being*, is still at work in Late Liberalism and Digital Culture. It is, however, obvious that this means that there are usually several Realisms at work at once, as the recent Greek/EU crises shows, where the Realisms of the EU and International Monetary Fund (IMF) technocrats (certain neo-nationalist economics professors included), the Greek Syriza government, and the European tabloid press and their commentariats diverged, while each of these parties presented their position as *realistic*, denying the others any legitimately realistic reason or justification.

The return of the power of the antagonistic rhetoric has returned, in the United States as well as in what is so mistakenly but therefore also fabulously termed Digital Culture, as I have attempted to outline in *Digital Coloniality of Power*, but it is also back in the EU despite itself, as Denkler (2015) says in an analysis of the rhetoric of (especially second-row national) politicians, who "prefer the easiness of anti-European populism over differentiated truths", and, thus, Denkler can conclude that "The friend-foe schema is back in Europe".

The friend-foe type of thinking and the technocolonization of the social sciences, for example, qua quantification, are, of course, noted and critiqued from within to some degree: Iver Neumann (2014), for example, argued in his inaugural lecture for the Montague Burton chair in IR at the LSE and the subsequent publication of that lecture that the field of *International Relations (IR)* must recall that it is a social science and has delineated what that means. For example, he began with a critique that

> [i]n IR, the discipline's quantitative practitioners are passionate about the problems surrounding data programming, in such a degree that one sometimes wonders if counting is not being substituted for thinking,

> but at least the quants do engage in a debate about methods. Those of us who mostly do qualitative stuff, however, must be severely faulted for having largely neglected methods.
>
> (2014: 337/2338)

I am not an IR scholar, even though I learn from writers and researchers from that field as well as my work and interests bearing on neighboring contexts. But I find that the above statement, as well as much of the rest of his lecture, is not only important for the field of *International Relations* (*IR*) in terms of what is criticized and what is proposed but also seems to resemble the situation in the social sciences in a general sense. One consequence of Neumann's critique is, I think, that the social sciences must always reflect on their concepts; where they come from; how they came to be; and what effects they had and still have, which is often a (neo)colonial effect. A concept, such as the concept of "the state", which stands in for a particular type of polity, is a good example, as Neumann argues:

> IR has so far assumed that polities come first, and that relations between them arise later. If we follow Renfrew, however, this is not the case. Polities arise as a result of relations. Relations are constitutive of polities from the very beginning.
>
> (2014: 335)

Or:

> [t]o say it with Abbott, from the very beginning, polities are things of boundaries. That is good news for those of us who want relational sociology to play a larger role within the discipline. The same point may be made about non-sedentary polities. More than two centuries ago, Hegel pronounced any state worthy of the name had to be sedentary. As a result, social sciences have neglected the study of non-sedentary polities.
>
> (2014: 336)

As I myself have done for the past few years, Neumann – following the example of Marcel Mauss – suggests bringing the body back in and taking it seriously this time:

> In IR theory, the physical body remains under erasure. I would argue that, as a discipline, it is just about time to excavate it. Let me use an example close to hand, namely myself. What you see before you is a performative body, me (not mine; me). It is a body that would not have existed if it had not been for the physiological fact that I had a necrokidney implanted five and a half years ago. [...] If discourse produces the social fact that is the body, it does not produce

> the physical body from scratch. The physical body remains an ontic precondition for the performing body to exist. The necrokidney was an actant that made things happen, independently of performativity. You piss, you live.
>
> (345/346)

Besides allowing ourselves to talk more openly in the social sciences about pissing, shitting, eating, fucking, and dying, what we need to do is lead a patient and epistemically *response-able guerilla warfare* against the technocratic occupation of the sciences, humanities, and social sciences, in particular, against White Collar sociology.

Against colonial technology and qua provincialization and decolonial options, we need to help people understand what is going on with neocolonialism and capitalism: help them, for example,

> to understand an accumulation regime unable to construct longlasting modes of regulation, there is only one alternative: assume a new point of view that immediately focuses on the problems of command and power. This new capitalism needs a social control compatible with democratic societies where order is based on the formalized participation of great masses. One of the new characteristics of the financialization process that involves us is its mass participation, its formal democracy. [...T]he American economic model—intended as ideal-type—and I will propose a personal reading of the financial crisis (§5): the roots of the crisis are to be sought in the instabilities in the new accumulation regime that is characterized by a dominant technological paradigm. The new technological paradigm started with the crisis of Fordism and the so-called Smithian division of labor. In the new division of labor, along the whole productive assembly line, knowledge plays a key role in the redefinition of the capital-labor relation.
>
> [...]
>
> In other words, the wealth-effect that supports the financial world's thought process depends on the degree of subsumption not only of labor, but of life itself and finance (this is where biopower's feature lies). In macroeconomic terms, this dependent relation is translated in the growing liquidity that financial markets attract starting with private savings that were previously invested in State bonds. Even so, this growing liquidity is not enough by itself. There is a second explanatory variable necessary so that the wealth-effect continues: it is the common sense that forms between individuals about the reasons that should explain the production of money by means of money, and about the hierarchy of needs to satisfy in order to maintain an acceptable social status. The logic of valorization leads to the transformation of social relations. More precisely, this means that

> financialization is a form of socialization that makes liberal governmentality evolve.
>
> (Lucarelli in Fumagalli/Mezzadra 2010: 120, 127)

Provincialization means, in contrast to old or new capitalism, to also regain hidden, buried, and lost critical potential in our own Western/Global Northern culture:

> Calling attention to this creative expansion of our concepts, I want to suggest an alternative not only to Habermas's cognitivist dismissal of the third Critique as relevant to political community but also to Lyotard's celebration of the Kantian sublime as a critical response to Habermas: the affirmation of a differend or *Widerstreit*, that is, a conflict that permits no resolution whatsoever. It is important to stress the impossibility of achieving permanent resolution of any conflict we would call political; but Lyotard's reading tends to foreclose any possibility of a politically mediated agreement about community whatsoever. Lyotard excludes more than the disputieren that Habermas endorses in a politics oriented toward mutual understanding (i.e., the exchange of proofs). He excludes as well the kind of agreement peculiar to democratic politics, namely, the *übereinkommen* that can be reached through *streiten* as persuading speech. Indeed, what for Arendt is the contingent achievement of this interlocution (i.e., the constitution of community through shared judgments) is for Lyotard just one more dangerous illusion of "empirical realism." Lyotard sees something crucially important that Arendt's other critics miss: the imagination, considered in its freedom, opens a question of community that cannot be settled by a practice of politics centered on the exchange of proofs, for such a practice tends to conceal how we misunderstand precisely at those moments when we understand and also occlude from view the source of the misunderstanding. But Lyotard offers little in the way of an alternative conception of community, other than saying that it is always in process, always anticipated but never reached.
>
> (Zerilli 2005: 181f.)

My motto here is to open up Western social science by allowing space for creative misunderstandings and equivocation. In such a way, Hans Blumenberg's daughter Bettina Blumenberg describes her father's writing style, which mirrors the mission of this book, in a wonderful and wonderfully accurate way: namely, that it

> requires prior knowledge(s), nothing can be had with him without prerequisite, that is pleasure of the text.[25] But also: say, what I see.

> This program means: To lift/create[26] knowledge from the imagined images that are not evoked linguistically.[27] He ignites the readers' power of imagination, but he does not satisfy it; rather, he assigns them the mission to work for themselves. Not everything must be made explicit, there must remain room for thought, room for work (my translation).[28]

The conceptual-methodological approach of this book is best captured in something that Sabrina M. Weiss wrote a while back (in: Weiss/Stingl 2015: 97):

> [W]e practice three gestures on the basis of reconceptualizing mass-media/pop-culture as transmedial/translocal discourses: 1. We interrogate Western modes of difference through a wider concept of Otherness, 2. we trouble difference by looking for Otherness in transmedial cultural products, and 3. we consider that difference/otherness as modes that can enable participation.
>
> 1 *Difference/Otherness*: We posited that difference is not the only mode of Other- ness. Identifying within Western imaginaries the two main modes of passable, colonial Otherness as hierarchical and transitive difference, we showed that these colloidal imaginaries are beginning to become liquefied through transmedial discourses and that there are tiny seeds potentials in Western culture (the Weird) to think decolonially. We accepted that Maria Lugones and those who participate in thinking in decolonial options have a point in claiming that in the coloniality of power difference was introduced as the one way of Otherness that was "acceptable": Acceptable meaning, acceptable to be part of the world of practices, even if those practices meant to be excluded. On the other side, there was the kind of Otherness that could not be thought. This kind of Otherness was (and still is) silenced by the coloniality of power.
>
> 2 *Transmedial imaginaries*: We wondered what kinds of imaginaries are possible to enact – enactive meaning that thinking, perception, and practice are not viewed as separate – relations to Otherness in ways that allow for conceptions of "being different-from" that are non-hierarchical and non-transitive.
>
> 3 *Decolonial Biodigital Civics*: We believe that these imaginaries lead to new forms of political discourse, solidarity, and integration, obtaining as enablements to participate by way of being Other.

So, what I am doing in this book is to *mycelially* "think with" *slime molds* and *crochet reefs* as the "forms" of a recursive, wound up, messy journey through different times and spaces, in a decolonial attitude, showing where a provincializing of Western academic discourse must begin by deploying in itself Hans Blumenberg's work on *myth* and the emergence of a philosophy of *Lebenskunst*, which Wilhelm Schmid found in Michel Foucault – but with the twist that *Lebenskunst*, which must constantly be exercised (*askesis*), is not merely *ethos and aesthetics of existence* but should be seen as an ethics as a ways of knowing and an *aiSthesis* (Mignolo), and in academic discourses means to critique disciplinary decadence and White Collar academia (especially in sociology) by postulating that we need to come together and develop further an idea of *BluesCollarship*, to create new spaces of *learning together with (radical) Others*, and to think not only in problems but in *opportunities*.

Notes

1 Although, I think, I am more honest about my own position in this development than many proponents of this claim in stating that present academia itself is an abode for this re-transformation/re-colonization, including and perhaps even perpetrated, in particular, by self-proclaimed "liberal", "Left", or "Critical Theorist" scholars.

2 I refer, among other concerning aspects, to its missing feminist, embodiment, (de)colonial reflections, which, for example, Sabrina M. Weiss and Dipesh Chakrabarty, among others, point out.

3 It's ironic how little the prominent members of the US-American Republican party seem to understand and heed republicanism. Consider the case of the Kentucky clerk Kim Davis, who refused to issue marriage licenses to gay couples in 2015 and was held in contempt and jailed when defying the Supreme Court, where Republican presidential hopefuls came out with clearly anti-republican sentiments, as referenced in Blinder/Lewin (2015), among others. A new level has been reached with the advent of Donald Trump's candidacy for the presidency (the famous "elevator ride") and the subsequent Trump *administration* (or lack thereof).

4 But we must consider the fact that *historiographic* concepts, such as history or global history, depend here on our appreciation of both a provincialized and a decolonial point of view.

5 I had originally quoted the German text by Neckel several times and provided my translations into English. My publisher's opinion was that any such translation of even shorter quotes requires permission from the original's publisher. This practice is new to me and seems counterproductive, yet I concede and have removed the quotes and translations.

6 As McCarthy (2002: 35) notes, this is an aspect that is inherently Hegelian: "It was Hegel who recognized the importance of viewing the state as institutionally separate from civil society, which embodies the material and private conditions of life in the market, contracts, private property, family, and so forth".

7 The word *public* in German can be translated from English as both *Öffentlichkeit* and *Publikum*; *Öffentlichkeit* refers to the public with a capital P and, thus, a public sphere, whereas *Publikum* nowadays means either a universal, general, or specific audience: A specific audience is, for example, the audience attending a concert by the Canadian band Rush; a general audience could be people who have a radio and randomly tune in to a station playing either news or the Rush song "Big Money"; and the universal audience would be everybody capable of listening *or* everybody being affected *or* having a stake

in communicative content, specifically if it concerns a political kind of content. This differentiation is, of course, only a tentative description to better illustrate the minutiae in play when using these terms and concepts.

8 In a short, but certainly not sufficient, summary: What is *retained* primarily from a perception becomes a secondary filter for further perception in what can be retained from them, while these retainers are condensed, for example, externally in the form of media; and that is turning into a third level of filters for what can become retained in perceiving (and how and why).

9 Schmitt holds, in terms of political theology, that, for jurisprudence, the "state of exception" takes on the same role as "wonder" (or miracle) in theology.

10 An exemplary movie one could be referring to is Guillermo del Toro's *Pacific Rim* (http://io9.com/dont-hold-your-breath-for-that-pacific-rim-sequel-1731061100; last accessed on September 16, 2015); even if – with its success in Chinese markets – a sequel without Del Toro was eventually produced, starring a prominent Chinese actress (not for "diversity" but for "relevance" for the Chinese market), it was, ultimately, critically and commercially a failure.

11 According to www.boxofficemojo.com/movies/?id=avengers11.htm (last accessed September 16, 2015), the first *Avengers* movie in 2012 had a production cost of 220 million US$ and grossed 1.51 billion US$ worldwide.

12 I do not mean, however, as I have argued repeatedly, a concept of relevance, as was suggested by the Austrian banker Schuetz and his disciples in social phenomenology, which is (due to a text-centric and optico-centric reduction) – as Talcott Parsons stated – the *fictional* type of philosophy.

13 Hörl (2015, quoting Thrift):

> It is precisely these new, excessive, mostly electronic "object frameworks" that today guarantee the recursivity of the world and are to be understood as the technological unconscious. According to Thrift, they bend "bodies with- environments to a specific set of addresses without the benefit of any cognitive inputs. The technological unconscious is therefore a pre-personal substrate of guaranteed correlations, assured encounters and therefore unconsidered anticipations." Through the implementation of intelligent environments the surfaces and textures of everyday life are intensified, automated, and powered by all kinds of software-controlled devices. This results in the genesis of a "process reality" that anchors "more and more of what was regarded as "human" in the "environment" in the form of small cognitive assists but which are drawn on pre-cognitively," and therefore "this new technological world is working directly into our unconscious."

14 In the United States, this is correlated with the possession of a driver's license and a car. Somebody who is not driving and does not wish to is immediately judged a suspicious individual, i.e. there must be something wrong with that person. Both I and former collaborators have made this observation.

15 This, too, renders the notion of "having nothing" heterogeneous; in other words, there are several types of "nothings". None seem, in current society, desirable.

16 https://mashable.com/2018/05/18/elon-musk-subway-invention-twitter-lols/?europe=true (last accessed November 15, 2018).

17 Take, for example, the song "Blood on the rooftops", published by the band Genesis in 1977 on their album *Wind and Wuthering* (I will not reproduce the lyrics here because copyright is a headache, while everybody can look them

up on the internet), and ask yourself: Which part of these lyrics would have to be changed in order to fit the contemporary situation?

18 This strategy also "works quite well" for grading exams, as I was told some years ago by a TA working in a German University's economics department: Because they had hundreds upon hundreds of exams to grade, making detailed reviews impossible, piles the size of the "expected" rates of pass/fail were created, and reviews were created by skimming, paying more attention only to papers in either pile that on the immediate surface seemed to belong to the other pile. This is, of course, just an anecdote. Nonetheless, the existence of such anecdotes makes them into interesting data points.

19 What counts as the main document(s) for an application process changes by discipline, university, and composition of the selection committee. But, overall, "the myth of 3Ps" – pedigree, publication, and politics – is a widely accepted account.

20 Of which we have too few, anyway. And it is bewildering, given that many universities, especially in the US, pocket overheads on most research grants, at times exceeding 50% of the grant.

21 I have a bit of a dislike to van Bouwel's and Selg's use of the word "mainstreaming" because it is used nowadays in politics as a kind of weapon, like *identity politics*, precisely by antagonistic actors to silence both deliberative and agonistic voices.

22 Lefebvre (2005: 86):

> Mouffe is, of course, aware that her agonistic liberalism departs significantly from Schmitt's political projections. Instead, she is satisfied to use Schmitt to probe the deficiencies of liberalism and import his polemical ontology as a reconstituted liberal endeavor. My aim is not to critique her project but only to ask whether Schmitt suitably grounds this ambition.

23 I gloss here over the fact that there are, if we are required to have an even more in-depth discussion of the inner workings of Schmitt's theory, important distinctions between State, the Political, and the public in the way that, via the constitution of the public, the State and the Political presuppose on different levels that mark each other "reciprocal teleological poles" (Lefebvre).

24 "Jedenfalls gehört es zum Selbstbewußtsein der Neuzeit, mit immer neuen Realism – theoretischen, praktischen, ästethischen – immer neuen Ernst gemacht zu haben".

25 Possibly, a reference to Barthes, who not only wrote an essay titled *Le plaisir du texte* in 1973 (distinguishing pleasure from jouissance) but also insisted on the distinction between readerly (*lisible*) and thus closed and writerly (*scriptible*), and thus open texts, i.e. the reader of a writerly text is an active and not a passive reader, of whom effort is required. The reader, enacting the writer's effort, writes the text therein anew and, of course, differently.

26 The German word *schöpfen* is cleverly applied here for it can mean either "to scoop" or "to create". But that is no mere accident: After all, the creation of the world [*Weltschöpfung*] requires a ladle to scoop it up from the *ur*-soup [*Ursuppe* or "primordial soup"], to use Alexander Oparin's 1924 term.
Additionally, while philosophically, *Erkenntnis* is usually translated as "knowledge", such as in *Erkenntnistheorie*, which is taken to mean "epistemology", theory of knowledge, *Erkenntnis* can also be translated as "cognition" or sometimes "realization".

27 Although the translation for *sprachlich*, which also means "by means of language", would perhaps better be translated as "denotatively" here.

28 Diese Beschreibung setzt Wissen voraus, alle Texte Hans Blumenbergs tun das, voraussetzungslos ist bei ihm gar nichts zu haben, das ist seine Lust am Text. Und auch: sagen, was ich sehe. Dieses Programm bedeutet auch hier: Erkenntnis schöpfen aus den vorgestellten Bildern, die nicht sprachlich evoziert werden. Er entzündet die Phantasie der Leser, befriedigt sie aber nicht, erteilt vielmehr den Auftrag, selbst zu arbeiten. Es muss nicht alles ausgesprochen werden, es muss Raum zum Denken bleiben, Raum für Arbeit (Bettina Blumenberg 2015).

Part B

(As in *Belonging* and *Being*) – Under the Spell of a Myth

Oikos as Operating System

2 Matters of Fact, Matters of Interest, Matters of Care

Caring for the Blues

"The blues", Lewis Gordon (2014b: 18) wrote recently, "is the leitmotif of modernity". My subsequent line of inquiry attempts to understand the importance of the blues in relation to the situation of knowledge and cultural labor in contemporary society and to understand why and in what ways these relations, respectively, the *poetics of these relations* – to channel the worlds of Édouard Glissant – are matters of care, never losing sight of the fact that what we call ethics are a ways of knowing and doing, and that, as a task for the future, we must figure out how to re-express knowledge and cultural labor in terms of a naturalized, situated, integrative care ethics – care ethics that can be obtained in the form of cognitive cultures in the 3E's registers (embodied, enacted, extended; see on this: Weiss, Stingl, and Weiss/Stingl in: Weiss et al. 2014; also Stingl/Weiss 2014, 2015, 2016). To that effect, I am interested in the *assemblages* and *arrangements*, in the *enablements* and *constraints* for that line of inquiry, for those matters of care. This is an attempt to talk about these *enablements* and *constraints* by (*mycelially*) *thinking* them *with* hyperbolic forms and slime-molds: Thus, there is no linearity in the way my "analysis" works; yet it is careful in the attempt not to fall into conceptual imputation or presentism; but instead, it is an attempt to overcome the sociologists' fears of equivocation.[1] Quite on the contrary, I take up Eduardo Viveiros de Castro's engagement with equivocation and deploy it as an act of provincializing the Western *assemblages* and *arrangements* for (and stretching from) Global Northern knowledge and cultural labor, keeping in mind the main insight from what I call *semantic agency theory* (SAT [Stingl 2011]): That the history of knowledge and culture is a history creative misunderstandings. Here, I am interested in the creative misunderstandings and equivocations of the *oikos*.

Knowledge and cultural labor are not often considered in terms of care work; but care, I claim, is an inherently important dimension in them. But care work is widely and with a long-standing tradition – politically and intellectually on the Left as well as the Right – either considered and constructed as "dirty" (the bluest of blue collar work) or it is not spoken of at all and effectively othered and silenced. It is largely thanks to feminist and postcolonial scholarship that care work is slowly taken more seriously as an object of study and that its prevalence and ubiquity are recognized, or rather: its ubiquity is not necessarily the reason for it being overlooked by social scientists. However this does not count for all kinds of care work equally. The kinds of care work that have been investigated first and more extensively, such as various kinds of health care work, have come to the attention of researchers largely because the health care system began to fail, and not only to fail on the scientific or administration front, but also in terms of the organization, resources, and technological, social, and political demands made on actual care labor(ers). But care work happens in many places, in many forms, and cuts across in diverse temporalities. Joan Tronto (2013) has done impressive work in arguing, in what can be characterized as a good example towards doing empirical ethics, that care work is at the heart of a functioning democracy and that is why it should receive more attention, especially in the United States, than it usually does. This requires, which I think is obvious, an approach that is not conducted "from above" but that takes seriously those it considers and who consider themselves as coming "from below" and then goes – in my sense "before and beyond" – on to dispel this hierarchy: The sneaking suspicion, that knowledge labor and cultural labor are but should not be rendered in terms of hierarchies, has even been considered by an Enlightenment figure such as Immanuel Kant who is so quoted from a note he left behind in Simmel (19 in GA 12[2001]: 9, a note which I would have quoted and translated here, but supposedly translations even of Kant or Simmel require permissions, my publisher insists), who would confess to having learned to honor the human being and rights of humanity in terms shared with the "common worker". Of course, this still rather lushly ambiguous Kantian confession suffers from the geopolitical context in which it was uttered, i.e. it suffers from and in its coloniality. Kant in this note, which Simmel delivers us,[2] is, after all, evoking a "humanity" – and it should not be forgotten that Kant hails at the advent of anthropology and biology (and biological racism) as institutionalized discourses and academic disciplines – that bears the colonial mark of the *humanitas/anthropos* distinction, the mark of racism.

This distinction is about how to deal with Other(nes)s. The Western imaginary of *Otherness* is in-ground in the *coloniality of power and Being*, which constructs Others as follows: "'Killability' and 'rapeability' are inscribed into the images of the colonial bodies" (Maldonado-Torress 2007: 255); "[T]he interrelated spheres of the colonial matrix of power (economy, authority, gender and sexuality, and knowledge/subjectivity) operate at the level of the enunciated while patriarchy and racism are grounded in the enunciation" (Mignolo 2009: 5); "[T]he making and remaking of the colonial matrix of power: a complex conceptual structure that guided actions in the domain of economy (exploitation of labor and appropriation of land/ natural resources), authority (government, military forces), gender/sexuality and knowledge/subjectivity" (Mignolo 2009: 19); and

"Quijano (…) identified the key contours of the colonial matrix of power as consisting of four interrelated domains: control of economy; control of authority, control of gender and sexuality; and, control of subjectivity and knowledge" (Ndlovu-Gatsheni 2013: 37).

This Western imaginary permits – exemplified in *science fiction* or in the Western academic discourses such as very prominently in the "philosophy of mind" – only two forms of Otherness: First is the "invert" or, thus, the "alien" who is the opposite of "the human", but therein often re-invertible and humanizable, either by being converted or by discovering their "human or higher qualities"; second, the "zombie" as the mindless and, thus, utterly "killable" kind of "alien" (see the exposition in Stingl/Weiss 2015). The non-permitted forms of Otherness are thereby effectively silenced: By this, I mean that they are not just excluded precisely because exclusion is a process that imparts subject positions and subjectivities on the excluded which *re-forms* them to become at least potentially (motivated to be) included, and, thus, serves as an inclusion-and-difference-paradigm/ machine[3] that produces *particularly* different *kinds of people* (see on "making kinds": Hacking 2000, 2006) meanwhile engaged in the simultaneous Othering and silencing of radical Others. In *un-forming* for example sociological thought, we can obtain a decolonial power by

> turning Descartes's dictum inside out: rather than assuming that thinking comes before being, one assumes instead that it is a racially marked body in a geo-historical marked space that feels the urge or get the call to speak, to articulate, in whatever semiotic system, the urge that makes of living organisms "human" beings. By setting the scenario in terms of geo- and body-politics I am starting and departing from already familiar notions of "situated knowledges". Sure, all knowledges are situated and every knowledge is constructed. But that is just the beginning. The question is: who, when, why is constructing knowledges.
>
> (Mignolo 2009)

In this line of flight, the tension can be uncovered that exist between and because of *anthropos* and *humanitas*. Mignolo proposes a different attitude:

> Because border epistemology is the epistemology of the *anthropoi*, who do not want to submit to *humanitas*, but at the same time cannot avoid it.
>
> […]
>
> "Immigrant consciousness" is located in the routes of dispersion of decolonial and border thinking.
>
> (Mignolo 2009: 131)

This *anthropos* stands for the "Other", especially in academic debates about alterity; making it explicit troubles the ontological status of the *Other*.

The *Other*, made into *anthropos* by way of being enunciated, is "not given" agency in or by these debates, because the *Other* depends on the one who enunciates him/her/it as *anthropos*, carved from the discourse on belonging that *inhabits institutions* (Hallet/Ventresca 2006; Hallett 2010;). The *Other* enunciated and invented as *anthropos* is not part of *humanitas*; it is an "imposition" on those who enunciate (active) as well as on those whose agency is thus denied in this act.

> [T]o impose the *anthropos* as "the other" in the collective imaginary, it is necessary to be in a position of managing the discourse (verbal, visual, audal) by which you name and describe an entity (the *anthropos* or "the other") and succeed in making believe that it exists. Today, the *anthropos* ("the other") impinges on the lives of men and women of color, gays and lesbians, people and languages of the non- European/ US world from China to the Middle East and from Bolivia to Ghana.
>
> (Mignolo 2013: 134)

Through "epistemic disobedience" we can critique what and who is subject to or being subjectified in terms of the *colonial matrix of power*. If *radical Otherness* and *border thinking* are given room in our academic discourse, if a new form of solidarity is established, then we might see that how it was our very own mechanisms that took away the agency of silenced Others; we will also learn how and why these mechanisms of silencing function. The variety of Others, whether suppressed into the mode of inclusive exclusion of *humanitas[anthropoi* or radical Others, faces a room that was already spacialized as territory in the terms of urban normativity, begging the question how to dwell in these spaces:

> We, the *anthropoi*, who dwell and think in the borders with decolonial awareness, are already on the way to delinking, and in order to delink, you need to be epistemically disobedient.
>
> [...].
>
> Our senses have been trained by life to perceive the difference, to sense that we have been made *anthropoi*, that we do not belong or belong partially to the sphere and the eyes that look at us as *anthropoi*, as "others." Border thinking is, in other words, the thinking of us, the *anthropoi*, who do not aspire to become *humanitas*, because it was the enunciation of the *humanitas* that made us *anthropoi*. We delink from the *humanitas*, we become epistemically disobedient, and think and do decolonially, dwelling and thinking in the borders of local histories confronting global designs.
>
> (Mignolo 2013: 137)

Where can *radical Otherness* dwell then? Mignolo sketches options, such as places of non-thought, non-European places, where epistemic disobedience and delinking suspend Western wool drawn over our eyes (2009: 2ff).

How, then, can we dwell and create dwellings "before and beyond" the *oikos*? With care, I guess (see also the interesting discussion of the works of Merleau-Ponty and Tim Ingold by Ramirez Baretto 2010).

Catching Fish in Muddy Waters: Before and Beyond Latour

In 2011, I first talked with a former collaborator, Sabrina Weiss, about the idea that Bruno Latour's distinction between *matters of fact* and *matter of concern*[4] needs to be supplemented by the notion of *matters of care*. We developed a discourse that led to the first formulations in publications on ADHD (2013) and social psychological theories of mindfulness (2014). I was thrilled to learn, following these publications, from friendly commentators that Maria Puig de la Bellacasa (2011) had published on a paper discussing *matters of care*, which I eagerly picked up. Her text and her perspective are important contributions that impressed and inspired me greatly. I want to emphasize how much credit should go to Weiss for making me able to express these ideas at all where it is not possible to really disentangle who contributed what in years of constructive discussions[5]; third several discussions with students whom I taught in class have also helped me (re)shape ideas, by allowing me to expose students to arguments as well; therefore, finally, my account of *matters of care* should be seen as one possible contribution in a much larger discussion which we share with an attitude of striking up and creating a conversation towards making things better. Thus, in talking about matters of care in a way that takes care seriously, we precisely cannot think of care as a competition let alone an unfriendly, non-cooperative one; our ideas should add to one another as much as possible rather than take away. Our different yet converging accounts both contribute different things, which are complementary at the end.

This discussion should, however, begin with a little critique of Bruno Latour whom Bellacasa references – after all, it was Latour who made famous the distinction between *matters of fact* and *matters of concern*. Most prominently, the distinction was con-figured in a paper asking the question "Why has critique run out of steam?" (2004) and the Spinoza Lectures (2008 [orig. 2005]), and my problem begins with the very question that Latour asks in the title of the prior. While not merely an attempt at rehashing the argument made by Habermas in *Erkenntnis und Interesse* (1968), which in English was titled *Knowledge and Human Interest* (1971), the major switch-up in comparison, though, between Habermas and Latour is that Latour seemingly removes the "human" out of interest. For one, I have never been fully convinced that Latour manages to remove "the social" from his theorizing as completely as he wishes to – Graham Harman (see for example 2012;

Gratton/Harman 2010) has repeatedly and convincingly made the point that Latour's later introduction of "the plasma". serves as a kind of stand-in for some of the respective problems in Latour's metaphysical frame . Second, I am not so quick to abolish "critique" as a mode of inquiry as Latour (and some recent philosophers[6]) has done: Of course, Latour is well aware of the differences between critique, critic, criticism, and critical theory, and whenever he glosses over them in his work, we suspect he does so deliberately; and, of course, Latour says that the problem is rather that critique used to mean and be conducted as a reductionist version that focused on *matters of fact* only, when it should have included *matters of concern*: But at the end of the day, Latour seems to want to do away with "critique" rather than provincialize it. My response to his title-question has always been that critique has run out of steam because nobody does it anymore – not critique but STEAM: By evoking STEAM in capital letters, I refer, of course, to the distinction between STEM and STEAM, in education, higher education, and research grant focus on "science, technology, engineering, and maths" without or with the arts (i.e. liberal arts, humanities, social/cultural sciences included). Stems, I argue, are different from steam, because stems, the more "robust" – to use one a formerly quite popular grant-writing bullshit bingo buzzword – and rigid they are, will break, whereas steam can move into places and make things moving: To paraphrase Bruce Lee[7]: "Be like water, my friend", here: "Be like STEAM" (see also on Liberal Arts and the pluriversity in terms of infrastructure towards the end of this book). Thus, Latour builds his alternative to (his account of) Kantianism[8] through Whitehead:

> Whitehead is the only one who, instead of taking the path of critique and directing his attention *away* from facts to what makes them possible as Kant did; or adding something to their bare bones as Husserl did; or avoiding the fate of their domination, their *Gestell*, as much as possible as Heidegger did; tried to get *closer* to them or, more exactly, to see through them the reality that requested a new respectful realist attitude. No one is less a critic than Whitehead, in all the meanings of the word, and it's amusing to notice that the only pique he ever directed against someone else was against the other W., the one considered, wrongly in my view, as the greatest philosopher of the twentieth century, not W. as in Bush but W. as in Wittgenstein.[9] What set Whitehead completely apart and straight on our path is that he considered matters of fact to be a very poor rendering of what is given in experience and something that muddles entirely the question, What is there? with the question, How do we know it? as Isabelle Stengers has shown recently in a major book about Whitehead's philosophy. Those who now mock his philosophy don't understand that they have resigned themselves to what he called the "bifurcation of nature."
>
> (2004: 244)

Bellacasa (2011: 86) draws from Latour's Whiteheadian approach to the "*thing*" the following line of inquiry for her own approach:

> Our modes of thought as well as our research ethos affect the politics we attribute to our objects. This means that, ultimately, every *Dingpolitik* – Bruno Latour's (...) name for the politics of things – denotes a *thinkpolitics*. Ways of knowing, theories and concepts have ethico-political and affective effects on the perception and re-figuration of matters of fact and sociotechnical assemblages, on their material-semiotic existences (...). Ways of studying and representing things can have world-making effects.

Following Latour's attack on our modes of thought that are occupied merely with "*thinginess* of things" (2004: 237), Bellacasa deploys the Deleuzian term *assemblages*[10] to move the attention to objects away from being in the *thinginess* and towards "knots of social and political interests" (2011: 86) given the latter an ontological import. The danger, however, lies in the question if this is all merely a reduction to a linguistic or, perhaps, semiotic enterprise, which would not be entirely unsurprising with regard to Latour's repeated referral to his debt to the work of Algirdas Julien Greimas. Says Bellacasa,

> [t]he ethico-political sensibility of accounts dedicated to these intricate agencies is well represented by the re-baptism of matters of fact into "matters of concern" (...). This naming can help to emphasize caring responsiveness in technoscience in an integrated way, within the very life of things, rather than through normative added values.
>
> (Ibid.: 87)

The rendering of things into *matters of concern*, as Bellacasa brilliantly summarizes Latour, does first of three things to things, which all hark back to returning to "things" their agency: "Things beyond thinginess" have lives of their own, independent of the human meddling; subsequently things have politics of their own, i.e. they have participation in political discourse and practice that "involve us humans" (which then should be phrased as "in political discourse and practice that involves humans, too"); finally, "things beyond thinginess" must have their vulnerabilities exposed and respected, instead being rendered into discourse that "construct" them and thereby leave them disempowered. *Matters of fact* therefore "are" *matters of concern*, because by *mattering* "things" merely as "fact" their agency disappears. Bellacasa reads *matters of concern*

> as representing a vision of STS's knowledge politics. First, an aesthetics: the way STS presents things doesn't split affects of concern and worry from the staging of their lively existence. Second, a thingpolitics: its

> representation of things gives them a voice as embodied concerns in the "we" of the democratic assembly. Third, a respectful ethos of knowledge production: its critique when *explaining* things doesn't reduce technoscience to a struggle for domination. In this sense to account for concern is a material-semiotic gesture of inseparable think-politics and thing-politics.
>
> (Ibid.: 88)

She goes on to set up a line of questioning that sets up the transition of types of questions for which concern is a response to those for which care is the type of response:

> How does respect for concerns in the things we re-present encourage attention to the effects of our accounts on the composition of things? Exhibiting entangled concerns at the heart of things increases the affective perception of the worlds and lives we study beyond cartographies of interests and practical engagements. In this sense, the staging of a matter of fact or a sociotechnical assemblage as a MoC is an intervention in its ethico-political becoming. And in the context of a troubled and strongly stratified world, do we not still need critical approaches to play a role in the assembling of concerns? The notion of "matters of care" attempts to respond to these questions. Caring involves a notion of doing and intervening.
>
> (Ibid.: 89)

But her subsequent subheading "Adding care to our concerns" (ibid.) illustrates where possibilities for different, equally generative accounts emerge. While it is not immediately clear whether or not *matters of concern* emanate from *matters of fact* initially, *matters of care* clearly seem to emanate from *matters of concern* in her construction, while also being of a related but different quality – this could be a hint that Latour's deployment of *states of affairs* is at the very bottom of things, a concept Latour references in relation to Whitehead, which is, however, to some degree part of the *equipment* that Whitehead would have encountered the world of his erstwhile collaborator Russell as an integral part of the *conceptual framework* (see also Kann 2001; Milkov 2002; Smith 2004; Shaviro n.d.; Stengers 2011), but also in William James's work who was an avid reader of Rudolf Hermann Lotze from where the term can be derived as having forged within a *theoretical vernacular*:

> Reality is not defined by matters of fact. Matters of fact are not all that is given in experience. Matters of fact are only very partial and, I would argue, very polemical, very political renderings of matters of concern and only a subset of what could also be called *states of affairs*.
>
> (Latour 2004: 232)

For Latour, the situation is to obtain, thus, on (given) *states of affairs* (which appear to equate with reality), matters of concern from which *matters of fact* are derived politically – herein I agree partially with Latour.[11] For Bellacasa we can say that in her account *matters of fact* (or "sociotechnical assemblage") can be "staged" as a *matter of concern* and that seems to turn Latour's hierarchical relation between the two matters around. Yet there are preceding states of affairs for both. The move to care is one of degrees or to use a more Deleuzean term – intensity.

> We can think on the difference between affirming: "I am concerned" and "I care". The first denotes worry and thoughtfulness about an issue as well as the fact of belonging to those "affected" by it; the second adds a strong sense of attachment and commitment to something. Moreover, the quality of care is more easily turned into a verb: to care. One can make oneself concerned, but "to care" more strongly directs us to a notion of material doing. Understanding caring as something we do extends a vision of care as an ethically and politically charged practice, one that has been at the forefront of feminist concern with devalued labours.
>
> (Bellacasa 2011: 89/90)

So she defines her task as one that is not to "blend together under a general theory of better 'caring knowledge', but give some clues as to what it can mean to represent matters of fact and sociotechnical assemblages as matters of care" (ibid.: 92).

> [A]ttaching the notion of "matters of care" to such a socio-political vision is a thinkpolitics. Staging and representing a sociotechnical assemblage in this way can provide a better account of a thing, but it also gives ethico-political significance to particular socio-material practices by generating care for undervalued and neglected issues.
>
> (Ibid.: 94)

There is a difference in that I do not consider *matters of care* as doing any kind of representation, precisely because for me ethics is a way of knowing and doing, and a care ethics is precisely doing in a way that can never be considered "representative" (politic) or "representational" (semiotic). Care is already political, in my theory (SAT), because it is a negotiation of what could be foundational for our discourse – I construe this precisely from a post-foundational point of view (Marchart 2008; Mouffe 2013) – and this is also inclusive of "non-caring" as an ethical way of knowing and doing framed in terms of care. My concern with *matters of care* is that – if phrased into the terms of a discourse on representations – once again only the included and excluded matters of care are accounted for, but that therein the *silencing* and *Othering* of *silent Others*, i.e. that *radical*

Others are left out. Precisely those who haven't been "*significant*-ly" erased but "merely *othered* and *silenced*", because they were, in a sense, "not even worth (to care for their) erasing"; precisely those *radical Others* might not be included when the notion of *matters of care* is put into actual practices. In evoking notions of the "unloved" or "most marginalized" (94), Bellacasa embarks on an important project; my question is, however, that to "unlove" or "not love" someone is often a way of knowing and doing/making someone, just as is marginalizing someone. While it seems to say that there are those who are so extremely marginalized that they have been pushed, for example, the "furthest away" from care, not only where these marginalized subject marginalization but they are still being constructed by geometric and, thus, geopolitical metaphors (most, furthest, etc.). In doing this, we once again construct an inclusion-and-difference-engine of one kind or the other. While I am fully aware that the problem cannot be completely avoided, to some degree we will always construct such inclusion-and-difference-engines and always *Other* and *silence* someone, my effort here is to establish a critically realist understanding of this problem that leads to a (post-foundational,) naturalized, concrete, integrative, situational care ethic that can be *mindful* of it. What I am proposing is that we understand that *matters of care* are political, i.e. they are first and foremost political: Bellacasa's focus seems to be on "sociotechnical assemblages" and "ethicopolitical significance", where I myself am, instead, following a post-foundational view wherein *the political* comes first,[12] *the social* and, thus, also the sociotechnical comes after (so, in other words, the question between us maybe is not so much "do artifacts have politics?" [Langdon Winner], but "in what way and when" do they have politics?): In other words, decisions have been made, and it should be understood here that when I say "*decisions*" I mean first and foremost that *agencies* in light of *affordances* have been *actualized* by *agents* (human and extra-human) in the form of *narratives* – wherein "narratives"[13] mean the sequentialization of (at least) two events in time.[14] From this motion forward, my approach to bringing care into play is about an *attitude* or *ethos*, if you like, but which is not so much about representation. What I am aiming for is this: *Matters of care are comprehensive*, comprehending the *silenced Others* (as "genuine Thirds"); actually practiced care will always mean that the extension of "care" is for those *included* in the practice of care (those actively cared for) and those *excluded* from the practice of care (which has the intension of those who are actively uncared for and those passively ignored), which leaves the *silent Others* out. But they, too, *matter* even if they are not *mattered by care* (but *silenced* and or *Othered*).

The point is, too, that this *silencing* and *othering* may not always be passive, but it can be active – for example, whole groups of people who are not only excluded from digitalization of health care and, thus, from health care, may be comprised of different kinds of people, those who the system of creating digital health information does not reach and those who actively resist[15] participating in any digitalization of their (informational) lives (see also Chapter 4 in Stingl 2015). Thus, decisions happen on the side of those who "care" as well as those who are *silenced* and *othered*. If you think this a mess, then: yes; and that it is precisely the problem that *matters of care*, as I see them, always point to the mess and messiness which cannot be fully disentangled. My point is, therefore, that we must be aware that in our actions *on* and *with care*, as "inclusive" as we aim to be, not only will we not reach all the excluded ones (human, extra-human, relational, etc.) but also we will always "in caring" *silence* and *other*. What I aim to stress is that we need to be aware of this and that this requires of us a more profound *response-ability*. Of course, it is more comfortable to think that by caring for the "most marginalized" were doing "good" – although that alone harks back to the question of the requirement of making surprisingly volatile "goodness claims" in the first place (see on this question Dussel 2013). But what *matters* is that we accept that even in doing so, even in *caring* we "do" *othering* and *silencing*, which may or may not be (and *that* it is difficult to figure out which) harmful to these *silent Others*. To be *mindful* of this – and this is part of why I argue *mindfulness* "is" a *care ethic* (e.g. Stingl/Weiss 2016) – is important, because it means that we are not ignorant or naive towards our actions, that we can remain open to alternatives, liberation, and resistance by delinking when required (rather than surrendering to inertia or *bureaucrasia*[16]). To do care work in the full awareness that we not only may be but we will have missed something or someone, and, furthermore, to have missed those probably precisely because we are caring, is an uncomfortable thought (it is uncomfortable because we care, of course). But in being *mindful* of that messy aspect, we remain also open to do better if given the chance. The potential danger of any care that confuses its extension with the *comprehension*[17] of *matters of care* is that it keeps reifying the kinds it makes, even if the kinds are relational kinds, and not only maintains but also accelerates *silencing* and *othering*, particularly when a situation has arisen that silent Others would require *care* to *matter*.[18]

Another quite political aspect that I follow, something I have alluded to, is the difference between *sensibilities* and *vulnerabilities*. I think it is important to understand that many *matters of care* that move around *sensibilities* are silent on actual *vulnerabilities*. Here, *temporalities* come into play – which is one of the lesser conceptual-empirical and methodologically appreciated dimensions in political and ethical theories. For example, a previous *vulnerability* cared for may have morphed into a mere *sensibility* (the opposite, e.g. by way of kind-making, is also possible). But in this transition silencing and othering may have happened. However, established politics, practices, concepts, and institutions of care built around or *mattered* by *matters of care* stand in the way of a re-rendering, precisely because any explicit re-rendering requires warrant. Since in *silencing* and *othering* we do not really perceive *silent others* (we do not even perceive them as excluded[19]) the care regimes practiced turn into warrants for certain beliefs (and actions); but not any type of warrant but the kind of warrant that, with Tyler Burge (1993, 2003; also: Graham 2012), we can call *entitlement*. From Burge we learn about *entitlement* and *justification* as two types of warrant[20] for a belief for something to be true. Elsewhere I have tried and explained how this plays into the exercise of epistemic authority and the question of epistemic (in)justice (Stingl 2017 forthcoming), bringing forth an argument that the exercise of epistemic authority is sometimes actually necessary – but should never slip into epistemic paternalism/imperialism, which is a kind of epistemic authority that sacrifices justification for entitlement. Instead, I argue that a care ethic demands that what is upheld, even in the exercise of epistemic authority, is in relation to the right of justification (Rainer Forst), a (mutual) "right to being justified to". This is precisely how and why I can defend my approach against anyone who attempts to occupy it for any political purpose.

To reiterate, the main difference in accounts of *matters of care* is the proposal of an *ethos* wherein *silencing* and *othering* are understood as something other than marginalization, and that even, yes precisely because of care, *silent Othering* can happen and that it is inescapable that it happens; subsequently, this should be "moralized" but seen as a source for permanent provocation and permanent opportunity, which do not fall into any preexisting conceptual frames. But however small the degree of difference, for the theory and the kind of critical purchase it offers, it *matters* and in terms of *mattering*, it can make differences in actuality. It can make a difference, because sometimes the way we talk about matters of care not only silences and others but also affects how *silent Others* already care (or: make care *matter*) themselves – for another or even for us. *Matters of care*, from my point of view, do not add to *matters of fact or concern*; *matters of care* "are" the "*mindfulness* for" the negotiations for the foundations on which facts or concerns are decided. Therein, I, too, understand, care as an ethos; but for me, it is not so much a "transformative ethos". Bellacasa offers words of caution, too (not to moralize care, not to use it in terms of categorization), and herein we can sense her worry. And it may be important here to recognize that therein I have not "arrived" anywhere, because there is nowhere to arrive: That there is no position (of/for care) at the end, taking a position as "representing", is something else; I am interested in an openness towards spaces (and temporalities) that are not positions (or places) that can be taken. I look at those places where there seems to be nothing and no one who can (or wants to) be represented: This approach to care is, however, just making everything a bit messier.

(De)Colonial(ity)

I have been actively seeking out decolonial options for the social sciences, and perceive my own role in this effort largely in an effort of provincializing Western philosophy and sociology. This means, in my view, not to reject Western sociology, but to understand its productive capabilities and its geopolitics, in other words, to understand the kinds of information it selects, the power – with Foucault – it is entrenched in, and the care it *enables* and *constrains* through *assemblages* and *arrangements*. What sociology does productively cannot be expressed simply in moralized terms, but it is important to acknowledge that what sociology has done and continues to do may not all be good (for all) but it can be and has been harmful (for some), but – in geopolitical terms – one thing it is not is baldly "objective" or "neutral"; any science, which makes that claim is horribly naive and ignorant. Nothing is (geopolitically), perhaps, more harmful or an indicator for harmfulness than claims of absolute neutrality, purity, or bald objectivity. A "strong objectivity" is, instead, a kind of objectivity that – unlike bald claims to objectivity – manages to reflect the reflect positionally and conditions of the production of the knowledge and (truth-)beliefs[21] and thus unveils ways in which the knowledge, while useful, is not baldly objective (Harding 2008). Social science has played a colonial role in the past and is not free of neocolonial effects, in part because it has neither adequately reflected its original colonial role nor its internal mechanisms of wielding epistemic authority imperialistically. In its ways of knowing and doing, in the adoption of the academic habitus in vernal and the continued production of a particular sociological sub-habitus it has retained the coloniality of power as (what with Bourdieu we can call) a hysteresis effect. Whenever sociologists go as far as making colonial errors explicit, they have – often enough – ended up with the paradox of implicating themselves while simultaneously and related to the implication intensifying the problem. This paradoxical effect, which we need to avoid if possible, I have called (in looking for a general term) "implification" (Stingl 2015). As Walter Mignolo (2014) has shown, the efforts of social scientists who were trying to build a new and better social science which climbed to be aware of, such as in the effort of the so-called Gulbenkian report in 1995, lead to more Re-Westernization instead of taking into account Southern Theory. This is, of course, not entirely lost to all Western sociologists, as efforts led by Michael Burawoy show (e.g. Burawoy/ Behbehanian 2012). And yet, while I would agree with some critics of Burawoy and his collaborators that sociology does not need "saving" – which indicates yet another quite problematic discourse of its own –, I disagree entirely with the sentiment of these same-some critics like Mclaughlin et al. (2005) who claim that reflexivity and critique

> are too ambiguous and value-laden to allow for empirical investigation of the different major orientations of sociological research and the ways the discipline can address non-academic audiences. Debates about the future of sociology should be undertaken with empirical evidence, and we need a theoretical approach that can allow us to

> compare both disciplines and nations as well as taking into account the institutional context of the universities in which we operate.

Such statements run the danger of conflating an attitude of *defaitisme* with disciplinary decadence. They are precisely not inquiring after the conditions and effects that lurk in terms of "evidence", "empirical", "disciplines", "university", or "nation".

As a general theory from the West/North about and for the West/North, social science can be useful but it cannot be universalized. Social science is a way of making the world we experience intelligible and communicable, but it is also a way that is inherently based on the emergence of the Global North and only accounts for the Global North and Western knowledges. It can be helpful to make intelligible and communicable the world that is not the Global North and non-Western knowledge but one must bear in mind that the world and this knowledge, and the way people that live outside Western knowledges experiences the world, do not correspond to our social scientific knowledge – even if social scientific knowledge may be the only useful way for us Northerners to understand it. Furthermore, we must be careful in our interactions not to colonize through social science's concepts and methods, while acknowledging that in the past we have done so to great effect – making things yet even messier. At the same time, conversations with Southern theories help us interrogate and trouble social science, and make it better. In short, I do not think we need to save sociology; I think we need to adopt a position of humility and continually attempt to make it better.

A first and most important step here is, of course, to understand what is meant by coloniality and decoloniality in relation to postcolonial sociology and to decolonial options (see also from a different angle and in more detail in some aspects, Stingl 2015: 1–102). To do so, an array of decolonial writers could be cited, but I will merely emphasize Walter Mignolo's work here (and point readers to a very concise depiction in a recent interview [Mignolo/Kaganof 2016]). So, these next few paragraphs owe themselves in no small amount to Walter Mignolo, which I need to point out and emphasize because I am neither seeking to appropriate a discourse others have built and led explicitly, nor, however, would I – on the other hand – fall into a misconstrual of these important ideas. So for some, I might chug along too closely to Mignolo's original texts, for others, trying not to simply "plagiarize" might already be a colonizing move. In walking on a thin line of thin ice, one can only fail and breach, but in our own "Archer"-moment, we could resort to live dangerously in expletives, and say something like "what the shit!", and, thus, let's break the ice, shall we?

Decoloniality, as I read Mignolo, is a wonderfully messy attitude, a fuzzy movement, a way of knowing and doing – Western scholars familiar with Foucault (and that decolonial scholars, including Mignolo, are certainly also familiar and to some degree sympathizing with Foucault is also known) might speak of this an *ethos*. I would say that, instead, what we need in social science is an *ethos* which has decolonial options. Now for Mignolo, what matters greatly is that we practice (or for Foucauldians: exercise) epistemic disobedience and border thinking. Through these practices, one can see that in the standard and canonic ways of knowing, feeling, and understanding certain acts of classification have been inserted and are deployed by those in power to do so. Disciplinary academic knowledge, precisely in its dependence on classificatory practices, is in trouble when it assumes (that it can conduct) leadership, whether epistemic, ethical, or political. One must go, instead, to a level of "life, thinking, and doing", wherein

> [s]aying that decoloniality is an option, means two things: it is a notion, not a mission. And second, once you present decoloniality as an option, an option of knowing, understanding, emotioning, living and being, everything else is an option. We live among options, options are systems of beliefs, ideas and disciplinary formations that make you believe that the world is as a given option says it is. That is everything being an option means that we do not see what there is but we see what we see. Period.
>
> (Mignolo/Kaganof 2016)

To a degree, decoloniality needs decolonization to be in the world. And, I think, Mignolo is not wrong to argue that this cannot be "post-coloniality". To refuse for oneself post-coloniality should not be seen as a rejection or de-valuation of postcolonial scholars who have done important work, nor should it mean that the current attempt at postcolonial sociology should be rejected – on the contrary, I think, postcolonial sociology is the best we can currently, practically achieve within the social sciences on the way towards creating decolonial options for the social sciences. But there are clear caveats and discontents in the notion "post-colonial", beginning with various questions that the prefix "post-" necessitates. Far from eliciting productive equivocations, the ambiguities of "post-" run dangers of being geopolitically instrumentalized: The danger in saying that "we live in post-colonial times" wherein "post-" is understood as "after" requires little explanation. But some tendencies among postcolonial theorists are also worrisome: There are some who think that a postcolonial attitude means to return to an earlier and, thus, "purer" precolonial stage. But what does that entail, really? And can we really know what it was like in any given place or community before colonialism, precisely since much of the knowledge and ways of living were not only destroyed in the various epistimicides of colonialism, but later re-capture of the remaining traces of these knowledges and ways of living have been remediated through colonial concepts and classifications (such as those of the social sciences)? Post-colonialism in this register cannot but end in doing one of two things – or even both simultaneously –, which Mignolo has identified in the presently more generally observable gestures of Re-Westernization and De-Westernization. But even De-Westernization in its concrete resistance against the classifications of the West, thus, cannot but orient itself concretely in its "against-ness" on the classifications of the West. In the academy, much of the post-colonial discourse has, unfortunately, also become canonized. Without wanting to devalue the important and admirable work that scholars such as Edward Said, Gayatri Spivak and Homi Bhaba have done, but when "post-colonial theory" is almost exclusively reduced to discussions of their work, such as in the most recent "completely revised and expanded" edition of a German textbook (by Castro Varela/Dhawan 2015) which spends 193 out of 326 pages of its actual content on the detailed depiction of only these three scholars who spend the majority of their adult lives and academic education and career at British and US-American (Ivy League) universities, one cannot but ask critical questions about post-colonialism's blind spots and limitations when it comes to the change it seeks to affect and the opportunities it hopes to generate. One part of post-colonialism's problems is rooted in its engagement with colonialism, whereas decoloniality is a disobedient complement to coloniality; "coloniality is not colonialism".

> Colonialism is historical manifestations of colonialist: Spanish colonialism, French colonialism, British colonialism, Russian colonialism, Soviet colonialism, etc. Coloniality is the common logic of all of them. But we do not talk about coloniality in the Roman Empire, for example. The Roman Empire functioned with a different economy, not the economy that liberal and Marxists call capitalism.
> (Mignolo/Kaganoff 2016)

"Decoloniality is an option that operates in two venues: the analytic of modernity/colonialist (how it was formed, transformed, manipulated, disputed) and, on the other hand, envisioning and engaging in the collective building of decolonial futures" (Mignolo/Kaganoff 2016).

Mignolo further emphasizes that *thinking* and *doing* cannot be separated, which can be accounted for in terms of decoloniality viewed as an attitude or akin to something that in the Western frame of mind could be understood as *ethos/ethics*, because an attitude could be seen to *inform* generally both thinking and doing; more importantly, however, "attitude"/"ethics" are – as I see it – *ways of knowing and doing* in general: Words like "attitude" or "ethos" help us understand and talk about differences in our intertwined ways of knowing and doing. Mignolo, thus, alerts us to the question that Westerners, both scholars and their "anti-intellectual critics" often evoke – the question for the difference of theory and *praxis*, and challenges the usefulness of the distinction, calling for it to be undone.

A crucial point about the tension between theory and praxis is the Western epistemological and geopolitical need for "classifications" (see on classification and the return/persistence of "class" and "identity" Stingl 2015: 103–144). Any theory in order to be considered practical[22] must be *relevant*. *Relevance* is usually gained or attained by either producing or reifying classifications in today's *economies of relevance*.

> ***Excursus***: Consider for example the way that "intellectual merit", "broader impact", "expected outcomes", "innovative aspects", "gender dimensions", "interdisciplinary aspects", "position of professional maturity", "career development/mentoring plans", "exploitation of results", or "communication to target audiences" are described in research grant proposals. In order to write successful proposals, it is not enough to simply describe or propose practices or formulate an interest, an open lines of questioning. Instead, one is immediately geared toward formulating one's research within classifications (underlined) – for example: even if one would assume that "gender dimensions" aims at equality, but it immediately demands to think in gender*ed* dimensions; similarly, in pointing out interdisciplinary aspects – let alone that for most major research funding one must choose a disciplinary focused panel or program expecting reviewers that are only conversant in that area – one is immediately required to think about existing disciplinary boundaries of one's project and construct isolated "aspects" where these boundaries may be affected, but with that outset a project cannot be integratively interdisciplinary or transdisciplinary, because demarcating "interdisciplinary *aspects(!)*" requires one to assume a position from *where* interdisicplinarity can be identified and that position – in order to be identifiable for a *accountative* research system – can only be a discipline. This issue of classification goes even deeper, especially in sociology. Take the following statement that a highly ranking American social theorist explained during a workshop conference a just a few years ago, to explain the difference between the uses and relevancies of social/sociological theory in the US and Europe: *In the US, sociological theory and research must be generally geared towards questions of inequality and social justice.* There are various layers of classifications here, the classification of "theory" being "not praxis" – and thus having to prove its relevance – included. And let me reiterate this: In using such statements for analysis, I am not interested in their generalizability or representatvity – another set of classifications in themselves – but in the ways (the how, when, and why) that they are possible in the first place *and* in what they do once uttered. There are classifications about US and Europe, but – without being explicitly referenced – the lack of alternative non-Western views is implied. And more importantly, "inequality" and "social justice" not only indicate that the existence of inequality and

> injustice is already accepted in this statement, it is further implied that theory – because we can see that inequality and injustice are abstractions, which cannot be globally or generally changed – theory must likely be geared toward some "achievable" impact on inequality/social justice, which will implies that further classifications are inherently part of the process of *doing theory.* But therein, we immediately will produce or reify the people that will then (continue to) exist under these classifications. I am not saying here, that this must be inherently and always "bad", often enough that is indeed the best we can do as social researchers in a present moment; however, we need not only be aware that that is what we are doing – many sociologists are not even aware, while others make their awareness of this problem explicit only in term so of lip-service – but we must be open to all efforts that work on different paths. The latter is the larger problem. Genuinely innovative research ideas, which aim to overcome these issues, often face responses by established social researchers of the kind "That is such an important project, but you will never get funding, because >>INSERT: random classification story<<" or "That's so needed, I wish I could do that, good luck with it". What researchers working on these projects need to hear is, however, responses such as "What can I do to help you make it work" *and* that such a commitment is upheld. But, as I have learned in conversations with countless junior and senior social scientists in the US and in Europe over the past decade, it seems rare that commitments are made, and even rarer that they are upheld. In approaching research infrastructure with a decolonial attitude, we must make explicit and render inert the inherent classifications and economies of relevance that guide the *thinking and doing* of research financing.

In that regard, "delinking from the logic of classification" is an important task for social science, because we cannot hope to ever challenge "the content of classification" (Mignolo) otherwise. If the task of sociology (or STS) is not to just be the notaries of contemporary society, but to genuinely contribute to change, they must understand that the task is one of challenging the content of classifications: If a sociologist of inequality wants to challenge inequality, s/he cannot keep reifying the classification. Many classifications are so "normalized", however, that even many trained sociologists do not even question them overly much. While LGTBQ movement and recent transgender discourses have certainly created an awareness that not everyone fits neatly into bifurcated sex-/gender-categories, at the same time, the male-female dichotomy remains the baseline of many of these discussions: Even sociologists of sex and gender often operate on the basis of studying how the groups of people they study differ from the male-female model, which is a completely different approach from beginning one's line of inquiry with the idea of variability. How many researchers who can be called "inclusive" and "open-minded" will actually take seriously the idea that the possibilities for "sex-expressions" that biology already gives us, including the different intersex variations, Klinefelter, Turner, etc., are expressions of variability and not "syndromes", "chromosomal aberrations", "conditions that affect the development of males/females", etc.? In thinking about sex and gender, we must acknowledge that in other times and places, a greater variety was/is already possible and that the male-female bifurcation of "human nature" is a production of Western coloniality, as for example Oyeronke Oyewumi (1997) and Maria Lugones (2008) among others have shown. We find similar classifications that unleash the forces of coloniality, when talking about racism, classism, etc. And merely "addressing" them as problems, even with the goal of creating more equality or justice, is not enough and sometimes even detrimental to the goal, as we can learn from Walter Mignolo, because the (disciplinary teleo-)logic is not suspended but reified, and, at times, as I have attempted to *think* by employing the novel concept *implification* (Stingl 2015), even accelerated and intensified: This can be seen in the debates on genetic research in biomedicine and the conceptualization of "race". While there is much discussion here that *implicates* the various ways in which "race" as a construct and concept is problematic, faulty, and harmful in its effects in biomedicine and health care, at the same time there is an *intensification* into genetic research using the classification of "race" which proclaims to do create better health outcomes – targeted at, for example, black people. This paradox, that to *implicate* something leads to its *intensification*, which follows along the logic of classification as described by Mignolo, is what I have coined *implification* because numerous instances of such paradoxes can be identified. Using *implification* strategically as a tool to *think with*, we can identify, I hope, such paradoxical and inherently negative developments more clearly and begin to unravel them. The point about identifying classifications and challenging them, in terms of decoloniality, requires an approach, as Mignolo shows on the example of "*the* university", of *decolonizing knowledge*. As Mignolo explicates, with regard to classifications of race and gender, part of the problem of the application of knowledge is that its epistemic power is eventually confused with ontology, especially if an epistemic kind is constructed as inferior, the inferiority will be assumed as being ontological. This is why a decolonization of knowledge is required, but we must understand what it means to decolonize something, such as decolonizing an institution such as the university. This action is one of understanding how the thing, in Mignolo's example the university, came into existence in the first place and what role "coloniality" had in this process in the form of "because of coloniality/modernity". Mignolo makes it clear that this cannot stop with "me", i.e. it is not something that can happen in solitude or monologue, but only in togetherness. We must *learn* about this and *think* this *together with Others*. Decoloniality is about the "*with*". And this is, precisely, what I claim is missing in contemporary academy, even in so-called "outreach activities" and "broader impact" there rarely is the notion of *with* but the direction of information flow is always one that includes hierarchical classifications of "above and below", in other words a "*from*-to"; even when the talk is about doing something "from below", we reify a preexisting classification.

The point is, however, to enable experiencing (including the learning experience) together *with* others, which is not possible as long as we classify along a demarcation of above-allow. And since this is inherently about experiencing, Mignolo inquires after the "aesthetic" in play. His point is that aesthetic is formal and, thus, sterile and static: its *aestatic*, if you allow me this play of letters. But it should be fluid, a process, "the aestheSis that both emerges from the colonial wounds and as the decolonization of aesthetics and the liberation of aestheSis bring about communal colonial healings" (Mignolo/Kaganof 2016).

And this is how we should approach the relation between disciplines in the academy. Mignolo makes, and neither is he the only one to make it and not only does he make it in his conversation with Kaganof, a point that resonates with things that I have talked about in previous publications with regard to problem – rather than discipline-oriented thinking (see, besides various blogs on my WordPress blog since 2010: Weiss/Restivo/Stingl 2014), with making an even further move from problem to opportunity thinking (Stingl 2015), and by opting for an integrative interdisciplinary or transdisciplinarity (instead of "merely" associative interdisciplinarity).

I have described decoloniality close to notions of "attitude" or "ethos", and Mignolo (in: Mignolo/Kaganof) in saying that "to think, and to think well, doesn't require disciplinary demands and presuppositions" *does* something rather extraordinary for me, because he simultaneously catches up to a deeply antagonistic debate that has been raging within the Western academy for at least two centuries and for which the name of Auguste Comte has become an ascribed and fictional point of origin – it is telling, by the way, that the more "disciplined" and "disciplinary" a discipline is, the more Whiggish in accounting for its history and the more focused on telling the story of a (causal) point (or person) zero it seems to be, while denying the fictionality of this kind of self-description, precisely because the fiction requires "factionality" to serve as an affirmation of the legitimacy and the factionality of the discipline. Ambiguities and equivocations are, thus, a huge problem for disciplines that are hell-bent on giving an account of themselves: Sociology is a good example, precisely because it is evoked in the very historical context that Mignolo's discourse on disciplines reminds us of, namely the dispute between "the two cultures", following the famous work by C.P. Snow, the dispute between literature and natural science, which is the context in which sociology emerges as a "third culture" (Lepenies 1985). Lepenies's seminal and still highly productive account of the historical context of the emergence of sociology in three cultures (France, England, and Germany) as a third culture does not necessarily require deviation from what I call the standard account of sociology, which is a very typical academic discipline in that and almost every other regard – the standard account basically relating sociology to the emergence and disciplining of economic thought in *political economy*, *Nationalökonomie*, *Staatswissenschaften*, – and, in the wake of Dilthey and Paul Graf Yorck von Wartenburg, as successor to philosophy of history –, and *moral statistics* (*Moralwissenschaft*) in academia in the 19th century. This standard account has little place for the history of biology, which I argue (Stingl 2008/2010; *Anthropos's Scaffoldings* forthcoming) is, at least, the other half of that "origin story" and accepting that leads for various reasons to accepting the need for equivocation. However, the problem with accepting this more complicated account would require facing not only equivocation but also sociology's inherent coloniality. The problem is that sociologists like to evade the heritage of biology that led to Social Darwinism. The problem only is that the road to Social Darwinism is not all that biology brought into sociology, that it brought many conceptual ideas with it that are actually quite productive and do not need to lead to Social Darwinism at all, but some aspects of biology lead to other elements of the "dark side of modernity" beyond Social Darwinism. In order to understand itself, its potentials (for good or ill, if you like), and its province (after having undergone provincialization) requires the recognition of this more comprehensive account and of letting go of the fear of equivocation.

Mignolo's *political* brilliance, in relation to this history and in evoking the cultural memory "two (three) cultures", lies in the fact that despite evoking the underlying antagonism, he moves beyond it at the same time, creating a playing field that is more agonistic and open for a plurality of hegemonies – to use the language of Chantal Mouffe. The point is precisely not that disciplines or that the "natural science attitude" must be eradicated; instead, the point is that other attitudes are possible and that the attitudes and practice regimes of the natural sciences (of STEM), which many sociologists attempt to mimic (or even in denying manage to inherently intensify – a paradox I coined *implification*), are at the same productive level within their own limits but we also must understand that these regimes' epistemic reach in the past has been an overreach (*coloniality*) and that the colonial effects were often harmful to Others, for which they – the sciences in general and the social sciences in particular – must accept responsibility. Mignolo's move is to say (ibid.) that "the disciplines impose restrictions on your creativity. That is why art and literature [...] offer the possibilities of thinking beyond disciplinary boundaries". Mignolo does not insist that knowledge and cultural laborers *must* do this, but that it is important that they *can* do this and are aware of and free to pursue that possibility, but "[I]t is not enough to question the disciplines if your questioning remains within disciplinary rules" (ibid.).

Lewis Gordon characterizes what it means to "take responsibility for the production of (social scientific) knowledge(s)" through such transdisciplinarity and through the "teleological suspension of disciplinarity" (Gordon 2014a).

A disciplinary knowledge, which Gordon reminds us, is already restricted in accounting for itself in the singular, tell its story – with paternalistic, imperialistic epistemic authority – through tropes of "unification" and "centrism" (2014: 81f.). The Western academy centered in the uni(fied)versity likes to tell its story around the story of philosophy and social science knowledge (and science) with Descartes (1596–1650) (and Bacon [1561–1626]), but in analogy to the post-foundational insight that that there are no perennial philosophical and social fundamentals, that there is *the political* – and from the post-foundationalist angle *the political* is the always looming antagonism –, Gordon can state the following:

> Enrique Dussel is a member of a community of scholars who have questioned the logic of self-reflection offered by the most recent stage of centric productions of knowledge. The philosophical framework of such rationalisation is familiar to most students of Western philosophy: René Descartes reflected on method in the seventeenth century, grew doubtful, and articulated the certainty of his thinking self in opposition to the fleeting world of physical appearance. A result of such intellectual labour is a shift of first questions from meditations on what there is to what can be known. This focus on epistemology as first philosophy charted the course of philosophy in modern terms against and with which contemporary philosophers and social theorists continue to struggle and grapple. For political thinkers, the new beginning is a little earlier, in the late fifteenth century – through early sixteenth-century reflections on politics by Niccolò Machiavelli. Against these intellectualist formulations of modern life, Dussel raises the question of its underside, of the geopolitical, material impositions and the unnamed millions whose centres collapsed not simply from the force of ideas but sword and musket. That modernity was ironically also identified by Machiavelli but is often overlooked through how he is read today.
>
> (Ibid.: 82)

It is important to take notice of the near-eradication of the "ironic" account of Machiavelli, an aspect that for example Erica Brenner (see for a short summary: Brenner/Marshall 2014) has emphasized. Even and especially Latour, who also focuses largely on *The Prince* rather than on a more comprehensive account of Machiavelli, takes Machiavelli too seriously and that, in my view, is a large problem for him and those that follow his version of Actor-Network Theory (see Stingl 2015: 153). So in Lewis Gordon's sentence here, there is this important ambiguity in his use of "ironically", an ambiguity that becomes precisely equivocal when we allow for accounts that safe-guard his author intention, which may or may not have been to be ambiguous, *and* divergent readings. Brenner (w. Marshall 2014) suggests:

> Up to the second half of 18th century some of Machiavelli's most intelligent readers – philosophers like Francis Bacon and Spinoza and Rousseau – read him as a thinker who wanted to uphold high moral standards. They thought he wrote ironically to expose the cynical methods politicians use to seize power, while only seeming to recommend them. Which doesn't mean they thought he was writing pure satire, a send-up of political corruption. He had constructive aims too: to train people to see through plausible-sounding excuses and good appearances in politics, and think harder about the spiralling consequences of actions that seem good at the time. Even his worst critics doubted that Machiavelli could be taken at face value.

A main tenet of my teaching philosophy, something I try to make students understand and suggest as a healthy attitude, is that they should "use concepts with confidence, but not take them at face value". It is precisely the kind of "thinking harder" (not: thinking "hard", in the same way as there *can* only be "better" but never *are* there "best practices") that I would claim Gordon, Brenner('s Machiavelli), and I might be able to have a productive conversation over. As you have seen so far into this book – and perhaps previous works of mine – you will have realized that I like to deploy two standard tactics to make people think a bit more differentially about certain ideas. These two acts are "linguistic" in nature: One is, of course, the comparative use of non-English words, especially my native German, and explicating problems of translation to reveal – for example through problems of extension and intension – aspects of meaning that can open new possibilities or critique existing restrictions; the other tactic is the etymological approach, discussing the Greek or Latin "origin". Both tactics, but especially the latter, are not without problems – to some degree, the etymological tactic is inherently pretty colonial. That is not to say that it cannot be productive and helpful, but whenever it is deployed, both the author and the reader need to be aware of its restrictions – i.e. of how it restricts – and of its coloniality: Going back to Latin/Greek roots has, even in its critical mode, a power to sacralize, affirm, and/or reify concept(ual content)s. The procedural and ethical ambiguities of the etymological approach must always be reflected in its deployment; nonetheless it is an effective approach. Lewis Gordon manages this tight-rope walk like few others do, and it is worth quoting him at length here (2014: 83/84):

> The word "science," although also meaning knowledge, reveals much in its etymology. It is a transformation of the Latin infinitive *scire* (to know), which, let us now add, suggests a connection to the verb *scindere* (to divide – think, today of "schism"), which, like many Latin

> words, also shares origins with ancient Greek words, which, in this case would be *skhizein* (to split, to cleave). Oddly enough, this exercise in etymology is indication of a dimension of epistemological colonisation, for most etymological exercises report a history of words as though language itself is rooted in Greek and Roman classicism. The tendency is to find the sources of meaning from either the European side of the Mediterranean or from the north. There is an occasional stop off in Western Asia, but for the most part, the history of important terms suggests a geographical movement that is oddly similar to the movement of *Geist* in Hegel's *Lectures on the Philosophy of History*. Some further inquiry reveals, however, the relationship of the Latin and Greek words to more ancient Egyptian/Km.tian words *Crethi* and *kotket* by way of the Hebrew *Crethi*, which was derived from the root *carath*, which means "to cut". The word *Crethi* referred to the ancient Egyptian/Km.tian royal armies, which were split into two classes. We thus see here a transition from one form of ancient centre to various others on a course to European modern times. Oddly enough, there is an etymological link during the Latin transition with another Latin infinitive, *secare* (which also means "to cut"), through which is more transparently connected the Hebrew *carath* (if one imagines *cara* as a possible spoken form). *Secare* is the source of the English word *sex*. A link between science and sex brings biology to the fore and the question of life sciences. Such a consideration indicates the importance of life reflections on the unfolding developing of systematic inquiry: As the question of a supreme deity motivated theological reflections and metaphysical inquiry, so, too, did concerns over the generation of life initiate scientific inquiry, although life was loaded with metaphysical content as anxieties and fear over the salvation of the soul without the theological guarantees attested to this day The subsequent unfolding story is familiar to most of us who study colonisation.

The reference to the context biology in this account of productions of social scientific knowledge, a context I also insist on, is connected to the "problem" of the production of Others and the critical question of their voice: Here, we see not only the colonially of power and Being at work in *the political* of the (social science) disciplines, but also and furthermore the intertwined processes of *silencing* and *Othering* of radical Others.

Gordon, in channeling Fanon, opens this line of inquiry to the methodological conversation – it is, for me, a brilliant coincidence given that I am very concerned about, more importantly that my *way of caring "sociologically"* is precisely in terms of methodology (for example through what I call SAT serving as a methodological evaluation framework for methods and theories). Fanon allows us to ask the question, if colonization happens on the most primary of levels and infiltrates even grammar – Walter Mignolo has repeatedly reminded us of the problems beginning with the I-You structure of the European (the colonizers') languages – does it not mean that methodology itself is already colonized and that, therefore, "methods devour themselves"?

The issue here isn't so far from a concern that Western social scientists themselves had repeatedly throughout their own history, and which was a great concern above all for Talcott Parsons. I speak of *reification* – by the way, I think that fear of equivocation and avoidance of equivocation breed reification – even if Parsons himself, too, could not avoid it, for example when in his middle phase he introduced "evolutionary universals".

Gordon unfolds and exposes the paradox that methodology finds itself trapped in, a paradox that requires the distinction between rationality and reason to be solved. It is a distinction that many social scientists and even some philosophers have trouble with – even if, for example, Kant did not:

> To evaluate method, the best "method" is the suspension of method. This paradox leads to a demand for radical anti-colonial critique. But for such a reflection to be radical, it must also make even logic itself suspect. Such a demand leads to a distinction between rationality and reason. The former cannot suspend logic, for to be what it is, it must, at minimum, demand consistency. The demand for consistency eventually collapses into maximum consistency, in order to be consistent. In effect, this means that rationality must presume its method, and it must resist straying from its generating grammar. Reason, however, offers a different story. To be maximally consistent, although logically commendable, is not always reasonable. Reasonability can embrace contradictions. Even more, it *must* be able to do so in order to evaluate even itself. This means that the scope of reason exceeds rationality.
>
> Science is more at home with rationality than it is with reason. Departure from consistency-maximisation would disintegrate an important foundation of modern science, namely, the notion of a law of nature. A law in this sense cannot have exceptions. Since reason at times demands exceptions, a marriage between science and reason would be shortlived.
>
> (Gordon 2014: 85)

The inability to differentiate between reason and rationality now stems from this state of affairs leading to science/rationality colonizing reason, leading to the differentiation and institutionalization of disciplines and disciplinary divides even in the face of failure of these efforts. But these efforts do fail, as Gordon states with Karl Jaspers, because reality does not obey consciousness as there is "more to and of reality". This is precisely the point that Rudolf Hermann Lotze (1817–1881) was making, when speaking about "the Absolute". For Lotze, who taught both medicine and philosophy (Jaspers, too, gained his doctorate in medicine but eventually became a professor in philosophy), "the Absolute" was more of a *regulative ideal* that could be used in place of this notion "reality is always more than we (can) know"; herein he differed greatly from Hegel for whom the Absolute was constitutive and – in the account of the *Rechshegelianismus* (see Löwith) – religiously founded. Hegel/Hegelianism and the dominant Neo-Kantianism (Marburg and Southwest), what I have called (Stingl 2008/2010) the *biological account* of knowledge of the social (or the social sciences) epitomized in Lotze, nearly disappeared at the end of the 19th century and beginning of the 20th century, with only a small string of scholarship remaining active that "thought" in this epistemological vernacular. In this string of scholars, I have identified Jaspers as an influential figure. Thus, while we come from very different directions, Gordon and I seem to converge on central issues. In this way, when Gordon identifies in the "failure to appreciate reality" as being much greater and the act of "recoiling" from this greater reality an "inward path of disciplinary solitude" and "disciplinary decadence", then this covers a crucial aspect of what I refer to when I speak of *White Collar academia/sociology*.

I would also like to add here that the tendency to *monoculturality of mind* in sociology is partially rooted in the (unnecessary) solitude of producing theory. In part, the German context of social theory is to blame in its relation to how writing and reading are done, aka poetics, which Lepenies (2006: 247) describes as "lonely acts". The poetics of theory is a lonely act because it is conceived and practiced so, not because it has to be so. But in being so practiced, as a lonely affair, it lends itself to *Theoriepolitik* (the politics of theory), in making theory a competitive and, thus, deeply antagonistic enterprise. To precisely avoid this, Peeter Selg (2013) has suggested that debates in sociological theory should follow Chantal Mouffe's proposal and, instead of swinging between antagonism and (false) consensus, should work in terms of agonistics. I agree that this is one important way to change the practice to doing theory; but an equally if not even more important change would be to *do theory together with Others*, to have actual conversations. This would require to rethink theory in the social sciences as a mode of belonging and (political) participation in new and different, in *radically Other* ways. One of those ways could have been "interdisciplinarity", but as Gordon shows, this has been conducted in a way leading to decadence as well – what I call merely associative interdisciplinarity. Gordon proposes a kind of transdisciplinarity that deploys a "teleological suspension of disciplinarily" in "going beyond" the modes of knowledge production offered within one discipline, and to do so fearlessly and unconcerned with these acts producing new disciplines or revitalizing older ones, etc. This is not a self-directed practice or *ethos* – it is not a form of *paideia* – it is, if it is an ethics, a way of knowing and doing that is very different from the *scientific ethos* which is monologue and technologies of the self: "We see here the sacrificial irony of all commitments to liberation: It is always a practice *for others*" (Gordon 2014: 90).

The *Oikos* in Sociology and the Information Order Problem

While I am far from agreeing with Talcott Parsons's (1964) notion of "universals"[23] in his attempt to remind social scientists of the importance of biological ways of thinking, especially with regard to evolution – which we should remember was made before important contributions such for example by Lynn Margulis – I would, however, argue that revisiting Parsons, despite certain shortcomings owed to his historical situation, can be a productive adventure. Parsons's many contributions to sociology and to political life in the United States and post-war Germany (as Uta Gerhardt has shown throughout her career) – nowadays largely ignored or, following the 1970s/1980s denigration of Parsonian system theory, hated – are noteworthy and often enough offer insights that current sociology has often come to overlook, such as the role of "motivation" in health care interactions, or the difference between power, influence, and money, or how to think power not as a zero-sum game, to name only a few. I find two aspects of Parsons's work from his middle-period, however, particularly interesting: The Power Bank (an unpublished variation on his general media theory, which anticipates and antedates present NGO-theories [see Stingl 2008/2010], also: NGO stands traditionally for non-governmental organization, of course) and his revision of land factors in economic theory. The latter, I consider, can be productively linked to an aspect from his work on universals (which Parsons considers the elements that increase or have increased a system's adaptive capabilities, but without which any further development would be no longer possible, save in the volatility of niche-existences), as present in his 1964 paper on "evolutionary universals in society".

Parsons, working and wrangling his stream of thought through Weberian channels, considers "bureaucratic organization" as one of the candidates for universals, an element that precedes money and markets in his account. He clearly had a "Western" kind of bureaucracy in mind, and at best we could say that the idea of bureaucratic organization can be used in terms of a *general theory* that can be helpful *iff* precisely the reificatory force of "universalization" can be avoided through provincialization of the concept – while Parsons had moments of such insight throughout his career, the normative way he and, more importantly, many of his followers went about does not reflect this kind of modesty when it comes to the aspect of his later theory on evolution and universals.

But once we do provincialize his thoughts, they tell us an interesting story about the roots of coloniality's political algorithms. It makes sense to remember that Parsons had a unique understanding of power, which did not reduce power to mean "suppression/repression"[24] but was already more complicated than that – even if it did not reach the level of Foucault, whose views on power I share[25]:

> I shall call this capacity to act, or more broadly, that to make and promulgate binding decisions, power in a strict analytical sense." Although backed by coercive sanctions, up to and including the use of physical force, at the same time power rests on the consensual solidarity of a system that includes both the users of power and the "objects" of its use. [...] Power in this sense is the capacity of a unit in the social system, collective or individual, to establish or activate commitments to performance that contributes to, or is in the interest of, attainment of the goals of a collectivity. It is not itself a "factor" in effectiveness, nor a "real" output of the process, but a medium of mobilization and acquisition of factors and outputs.
>
> (Parsons 1964: 347)

Power is, then, a limiting factor that plays a significant role in what bureaucracies do (external relations) and how they function (internal relations). But most importantly, it is Parsons's evaluation of the historical emergence of these aspects and factors, when we see them through a lens of co-evolution or co-production, namely that the notion of the *oikos* seems to stand at the very beginning of these developments – and I would claim, going further than Parsons – that the *oikos*, even through various transformations and additions (such as, for example, the colonial notion of "disposability"), remains a crucial discursive formation across Western and Global Northern history. It is worth quoting Parsons here:

> Even with rather clear differentiation, however, position in a stratification system is likely to be highly correlated with position in a hierarchy of authority.
>
> [...]
>
> In the very important cases, like the *polis* of antiquity, where power is widely dispersed, private units of organization are not likely either to be large enough or to command sufficient resources to become highly bureaucratized. Perhaps the *oikos* organization of the interests of important aristocratic lineages in late antiquity constitutes one of the most important relatively early examples approximating private bureaucracy. The Western Church is clearly another, as are modern business firms. The basis on which I classify bureaucracy as an evolutionary universal is very simple. As Weber said, it is the most effective largescale administrative organization that man has invented, and there is no direct substitute for it. Where capacity to carry out large-scale organized operations is important, e.g., military operations with mass forces, water control, tax administration, policing of large and heterogeneous populations, and productive enterprise requiring large capital investment and much manpower, the unit that commands effective bureaucratic organization is inherently superior to the one that does not. It is by no means the only structural factor in the adaptive

> capacity of social systems, but no one can deny that it is an important one. Above all, it is built on further specializations ensuing from the broad emancipation from ascription that stratification and specialized legitimation make possible.
>
> (Parsons 1964: 348/349)

Turning the exemplification of the *oikos* downside-up (from the archeological to the genealogical direction), we come to entertain the thought that *oikos* is what later bureaucratic, administrative, and *accountative/auditative* systems are informed or inspired by; in short, the *oikos* is the actual root-operating-system of Western society, without which Europe could never have (see Oyewumi 1994; Quijano 2000; Mignolo 2011; Grosfoguel 2013; Santos 2018) in the 15th and 16th century *invented* (a) itself as Europe, (b) (the society of Northern) "America" as its great experiment of what was to be "Western society", (c) the "rest" of the American continent as "Others", and (d) the populations of the American continent and of the African continent as "disposable". With these *inventions*, with the colonial matrix of power, the epistimicides against *other knowledges* occur in co-production of the "modern" Western university. The operating system for the Western academy, then, would also be in the spirit of *oikos* and subject to the supplements that the *oikos* underwent such as rendering disposable one class of teachers/researchers on the one hand while elevating and maintaining an aristocracy on the other hand. The aristocracy is, of course, not only found in the "higher echelons", such as US-American Ivy League universities, nor are disposable knowledge laborers only teaching at community colleges. The aristocracy referred to here, which I dub *White Collar academicians*, is largely found in an attitude or style. Even at a community college, those can be found, who worship at the altars of *White Collaredness*, while one can find exceptionally productive and open laborers with *a BluesCollars's* attitude at an Ivy League or upscale university. But a sure sign of a *White Collared* attitude is the worship of Ivy League universities. If a community college professor holds researchers s/he cites in high esteem or propagates certain ideas and results based on a kind of warrant that can be called entitlement rather justification (following Tyler Burge's work on warrant) – i.e. by claiming that result x is true, person y is a great scholar, or idea z is innovative, *because* it originates from an Ivy League-ish place – then we can suspect that there may well be a certain *White Collared* mindset in play. Not to be misunderstood: This is not supposed to mean that all research coming from such a university must be seen as suspect, nor that immediately every scholar's work, regardless where they are from, must be regarded as ultimately equal. My point of critique is about fending off an attitude that promotes a preference warranted in the form of entitlement while *denigrating* others; the resistance to that cannot be to just: Turn the tables! In the same way, as Walter Mignolo as shown, if the response to coloniality is simply and/or only De-Westernization, the result is usually also not very progressive or helpful. Or, with Chantal Mouffe and Peeter Selg, by furthering one type of antagonism over the other, little is gained. This is why *BluesCollarship* should not be about "fighting" *White Collar academia*; it should establish itself for example in agonistics, epistemic delinking, and border dwelling. Instead of fighting *White Collared academicians*, their epistemic entitlements should be made explicit and the right for justification and the right to be justified to should be proliferated. Not in order to "win" arguments or to find a consensus, but to keep open conversations and to continuously create new opportunities. Ultimately, as I will argue in the final chapter of this book, this is about transforming the infrastructure of societies and of academia.

And this leads us to the connection between *care, power, and information*. One way of expressing that connection, reiterating the issues brought up earlier in my discussion of Bellacasa, could be by connecting these three dimensions as such: Care *matters* through Power which *is* a *mattering* of concerns, which is condensated into and sediments as Information through a *mattering* of facts. While this conceptual approach would not be my first choice of account, it would not be a bad choice either. *Thinking with* this kind of relations will create interesting and useful results. The crux of my methodological approach (Stingl 2011, also: 2008/2010) is precisely to say that there is not this "one account" which is correct or true; instead, we can create numerous accounts, which are *plausibly* only more or less likely, but not a single one is "*the* truth". Because the actual reality is or was (or will be) always larger than any account we can give, different accounts sometimes can allow us to grasp different aspects of that reality; so even accounts that are incommensurable may complement each other to a more comprehensive picture. (In a most simple example, this strategy helps with accounts of why people can have intransitive preference orders, i.e. why someone likes A better than B, B better than C, but C better than A.) Of course, there is the accusation that this was a relativism of the nihilistic kind. So "anything is possible"? Of course. But: Everything is not possible. Of course, for some people the slightest smack of "relativism" is evil and all relationalism is already to be denied. Too often their rejection is based on the idea that "anything means everything". My approach, however, is clearly different. While there are different accounts possible, and any possible account should be taken seriously, that doesn't mean that every account is possible. For example, Adorno, in his dissertation, painted Husserl as a bald positivist. This should be a possible account, although in the way Adorno created his account, it is not very defensible. Another question altogether is, however, this one: Was Husserl a Poststructuralist? How could he have been when Poststructuralism was created after Husserl and to some degree also because of Husserl. So that account is not possible. We cannot "wish" away, for example, that Talcott Parsons trained intensively in biology, but we can have different accounts of how that training played a role for various aspects of his life and work. But we cannot create an account where that training didn't exist. So any account that cannot be reconciled at all with Parsons's biography is not possible. An account that argues that his biological training played very little role – and most accounts of Parsons, even when they mention it "as" important, do not account its importance for Parsons's conceptual development– is possible and we can inquire how they justify this minimality. In contrast, my account (2008/2010) makes explicit ways in which biology as an epistemological vernacular was central to Parsons's, and how that resonates with early sociologists throughout the 19th into early 20th century. That does not mean that my account is "*the* truth"; it only means that I have good reasons as to why my account is very reasonable and, thus, more likely than many others. Where, for example, Bruce Wearne's (1989) religious account of Parsons is both incoherent and often unreasonable (for example when branding one of Parsons's most influential teachers, Karl Jaspers, a Marxist because of the use of the word *Weltanschauung* ["worldview"] in the title of a book on the psychology behind worldviews), where Harald Wenzel's account is very coherent but contains some conceptual problems for example in the way he reads some influential sources for Parsons such as for example Alfred North Whitehead (in part, because some archived correspondence in the Parsons's archive belongs and/or refers to Whitehead's son at the Harvard Business School instead; Wenzel was forced to rely on a flawed German translation of Whitehead's works), Uta Gerhardt's (2002) political biographical approach is very reasonable and coherent while she also decided to limit her scope to specific aspects based on the actual material basis that archives provided her with. Thus, Gerhardt's and my account complement each other even though they are not fully commensurable; Wenzel's and my account are – despite our mutual recognition of the importance of biological researchers – decidedly different and can only co-exist even though I would claim mine is more probable, whereas Wearne's "story" can be safely discarded (at the time it appeared, it was, at least, still an interesting source, because he cited archive materials few others had previously accessed).

This brings us to the question of what information matters and how. Science and scholarship, when it engages the world, operates on the idea that the world gives us data and/or information. Sometimes, the two words are used interchangeably. Furthermore, there is this idea that the world consists basically of "raw" data and that the more data we have (the buzzword is, of course, "Big Data"), the more accurate our "picture" of the world, which is t(r)ending towards complete correspondence with how the world is. Now, one problem is that even in Big Data approaches, we cannot have all the data – in part, because data itself is in flux because, well, of time. Still, the idea is that there is an amount of data that is enough to make a safe judgment or prediction on how things really are or will be (or used to be). In quantitative approaches, it is accepted that, if we ask at least *this* many people out of a larger population, we will get a result that will resemble the population. That it is, of course, not that easy – and therefore open to all kinds of manipulations, as well – can be seen in pre-election polling. So "data"-favoring scholars tend to make more and more complex their approaches, until they think they safely got it. But higher levels of technical complexity, in my view, deviate further from the messiness of the actual reality. The reason for this is the problem with the *information order* – to use a term of historian Cristopher Bayly. First of all, I do not think that "the world" *exists* as data or *gives* us data; "we" are in very complex relations with different levels of environments and ecologies, which provide affordances, which lend themselves to multi- and metastabilities. Actualizable affordances can become data-points, but various processes and agencies are in play when it comes to making affordances *actualizable* at all and then lead to the *actualization* of affordances.[26] What we *call*(!) data emerges precisely as the in-between of *actualizability* and *actualization*. The tension between what we are thus "afforded" as data and the ways we actually perceive data is resolved by the production of order among the data, and this order is what we then trade/treat as information. On the one hand, we have the messiness of how data comes to be, and on the other hand we have the no less messy production of an order among the data, while the only thing we can practice on and with (discursively, whether textually or not, whether consciously or not[27]) is what has been ordered. This I call the information order problem, and, respectively, there is no "raw" data because we only have ordered information to practice on, which gives us no clues about that level, which people tend to think of when they say "raw" data, which is the level before *actualizability*. But my point is that what we consider data does not exist before *actualizability*. Data is an epistemological unit that we construct to make sense of something that happens on ontologically in between *actualizabilities* and *actualization*. So, based on the information, we guess or infer on data, but that inference is always already ordered. And therein it is always inherently already political (not: social; that dimension comes later). So the data afforded and the data we like to speak of and then "do stuff with", stuff such as "science" or "sociology", are not false but dependent on processes and on orderings over which we have very little control but depend on ourselves if we are to have any agency.

An overly simplified example, perhaps, is "sound" which we can hear but also "feel". At a concert or when clubbing, we can hear sound, or feel sound, or both, but of course, the sound affects far more, not only in our organism but in the space we occupy: The sound can, perhaps, do something to our gut bacteria which we will have to deal with later, or it can loosen a screw in the ceiling of the club. Now, in reading that description, most people will already distribute between things we consciously perceive and things we just do not perceive consciously but will be able to infer on and from later. For example, if the ceiling comes down, we could infer that heavy bass loosened some screws and that made the ceiling come down. But that is already depending on various orders and data-points: For example, that we conceive of a screw in certain ways: It is separate from the beam it held; it is made of certain materials, which may be subject to corrosion; but then, we could also consider the beam and the screw in a relation, maybe because they were of different materials that led to corrosion; we could also think of the beam and screw as one object – in fact, we often understand or practice with composite objects as one object; sometimes a practice even depends on "oneness" as the order of an object to function at all. Now, if you read the example *and* my subsequent elaboration again (or reflect on it), you will likely realize that while there is hardly anything that can be effectively disputed there are also a lot of ifs and possibilities and unclear aspects. Now, a first reaction is, of course, to say my depiction is just not satisfying, that if more "data" was brought in, we would get a clear picture, and so on. My account, however, is that every situation we think about is like this, full of ifs, possibilities, and unclear aspects, and that any practices – including knowledge practices to "clear up" the situation – in order to be *meaningful* (as "meaningful" I understand any practice that is "able to connect to other [meaningful] practices") must use available information, which is, however, already ordered. So knowledge practices about the ceiling coming down are already ordered: The sound, for example, is understood as "sound that can be heard and felt", but we cannot know how gut bacteria experienced the sound. Gut bacteria may deal with affordances in the sound that resonates with agencies that have to do with the ceiling coming down but they are just not in our *information order*, because our information about sound and about bacteria must be ordered in ways that denies us a different order that would enable us practices of building beams in the way gut bacteria *could* (but *wouldn't* because gut bacteria, at the same time, have no need to build ceilings and do not even have a concept of "ceiling" – I am not sure if gut bacteria can dance or not, though).

Care Power Information as in-formed by the *oikos* is a Western/Global Northern assembly, and in the moments and the spaces of the *expertise* in this knowledge assembly, we encounter *White Collar Academia* as it presents in its current (neo[post])neoliberal and neocolonial form of *oikein* (management): "We might even say that all traditions of political thought that assume rulership or sovereignty as the essence of government and politics find their origins in practices of household rule" (Owens 2015: 3).

As a resistance to and liberation from White Collar Academia, I propose a different kind of action, gesture, attitude, and being together in *learning together with* one another. I call this BluesCollarship.

Ghost Notes: *The Blues* and *BluesCollarship*

This proposition, intentionally vague, equivoque, and unconcrete, for I consider it a necessary *anti-accent*, notable and yet a softer and therein fluid note, I call *BluesCollarship*. It is inspired, above all, by decolonial authors like Catherine Walsh, Maria Lugones, Sylvia Wynter, Walter Mignolo, Madina Tlostanova, and Lewis Gordon. The latter's inspiration is of particular importance here, because BluesCollarship is directly inspired by his essay "Is philosophy blue?" (2014) and tries to heed the warnings in his paper "Black Aesthetics, Black Value" (2017). Let me try and explain.

We need to recognize that, first of all and at best, we can only provincialize Western/Global Northern science, social science, academia, and the university, and that this is the necessary but never sufficient first step to decolonize and to taking up seriously decolonial options. A truly decolonized university, for example, would not be a university anymore but, perhaps, something more akin to what Ramon Grosfoguel and others (such as Gurminder Bhambra) have called a *pluriversity* (see, for example, the special issue of *Human Architecture* co-edited by Grosfoguel with Capucine Boidin and James Cohen in 2012). In the previous book, *The Digital Coloniality of Power*, I have subsequently argued that while there could be a postcolonial sociology, for example, and while sociology can be provincialized and engage in productive conversation with decolonial options, there can be no such thing as a decolonial sociology. Walter Mignolo and others who work on decoloniality have argued against the appropriation of decoloniality in this way as well. Decoloniality is not an "academic" endeavor; it is not a school, or anything of this kind. Engaging together with others decoloniality can change the academy, of course, but that means to leave behind such distinctions such as changing it from within or from the outside. Decoloniality means to leave behind, or to go "before and beyond" (as I propose, in going before, but without going back to some imagined and imputed precolonial past, and then go from before beyond the [being still] colonial present) such distinctions that were created (or de-noted) in the name of Euromodernity or modernity/coloniality. It is noteworthy that there is a "trace" of this possibility for "going before and beyond" in the sciences, when read through their own *diffractions*, as Karen Barad shows: It is as follows:

> an embodied practice of re-membering – which is not about going back to what was, but rather about the material reconfiguring of spacetimemattering in ways that attempt to do justice to account for the devastation wrought as well as top produce openings, new possible histories by which time-beings might find ways to endure.
>
> (2017: 63)

For the social sciences and sociology, provincializing means to understand its co-productive role in the creation of Empire and racism – a role shared, of course, by the university/academia at large, but with social scientific keynotes, as authors such as Zine Magubane or Julian Go have emphasized (more towards the end of this book). In identifying White Collar Academia – which includes many references such as Whiteness, white privilege, as well as white collar – and addressing the many "problems" of decolonizing the university, of black people and other minorities attending college, and finding good jobs including in the academy which offers ever more precarious ones rather than real jobs and in many fields and disciplines has an abysmal record of hiring women or people of color and minorities, and of addressing the "problem" of the knowledges of non-Western people, of black and brown people, of women, i.e. the "problem" of epistemic (in)justice, in all of these aspects, I am reminded of a point Lewis Gordon makes, evoking Richard Wright:

> To illustrate, Jean-Paul Sarte had spent some time with Richard Wright in Harlem during his visit to the United States in 1945. Sartre purportedly asked Wright to tell him about "the Negro problem".
>
> "What Negro problem?" Wright responded. "There is no 'Negro problem'. There is a white problem".
>
> (Gordon 2014b: 15)

Following Lewis Gordon's essay into the questions of the blues, including the question whether white people can, may, or should listen to or even play the blues, I hope to draw lessons for the *blues* in BluesCollarship:

> We could carry Wright's insight to the presuppositions of thought in relation to people and things black. The presumption is that black music must be a particular, even more—specific—kind of music bereft of universal significance. As black music, the blues encumbers the burdens of a presumed particularity awaiting the illumination of universal analysis. The result of this presupposition is at least two misconceptions: the first is that because it is black, blues music is trapped in its own particularity, which means it can only offer that which refers back to itself—which is invariably *experience*. This makes the music in effect unreflective in that, simply as experience, it suffers from trying to figure itself out. Its metanarrative is thus not a reflective one, until, of course, it receives the universal light of theory and analysis and receives meaning. The music thus suffers from a peculiar crisis of legitimacy at the level of thought: it could only ex-press and attempt to but never justify itself. The second result is at the heart of the logic of particularity and universality. Du Bois, in his discussion of double consciousness, noticed that what is often called "particular" tends to include the contradictions of society. Thus, the effort to construct the dominant and the normative "pure" often involves disassociation, rejection, disavowal, and denial of so-called "dark" elements of

> society. In effect, the dominant claim to universality is often premised on a false assertion; namely, its universal scope through simply ignoring what transcends it. The presumed particular, however, requires admission of just that, which means attunement both to its limitations and transcending them. In effect, the former requires an artificially limited reality and the latter demands admitting the artificiality of that limitation. In effect, it means the so-called particular is at times more universal in scope than the proclaimed universal. Let us call this *potentiated double consciousness.* The message to be learned should now be obvious. That blues music, as black music, may actually be more universal than avowed or proffered classical and mod-ern music, though in truth such comparison is like contrasting apples and oranges without remembering the nutritional value of both. It may be so not only through an incorporation of the elements of avowed universal music but also through reaching into territories the latter dare not venture.
>
> (Ibid.)

There is, Lewis Gordon notes with Frantz Fanon as his interlocutor, an understanding of the relation between blackness and *the blues* to be found: Blackness or, rather, the very experience of it contextualizes *the blues.* Fanon, of course, through his writing captures this experience in an "understandable" way – and, of course, he experienced it in that famous exclamation that was made by a French boy to his mother. But please allow me here to emphasize what in *The Digital Coloniality of Power* I also insisted on, referring to Walter Mignolo: Can we, can I (a white, labor class-born, European male) understand what it means to be black (and/or a woman, etc.)? The key-point that Walter Mignolo made clear to me is that, while I can begin to understand what it means, I can never know what is to be black (etc.). Thus, a kind of understanding is possible, but it has its limits (hence, there can be no "universality" of knowledge) – and grasping this, the existence of these limits, which is an integral part of provincialization.

Lewis Gordon points out what I think of as a related issue, when he calls out Fanon on his initial thoughts on the blues. The blues emerged out of the experience of oppression, but Fanon proposes that if the blues is played for and listened to by a person who has not experienced this oppression or is not a card-carrying member of the group of people whose history does entail this experience, then that would mean the blues were played for the "admiration of the oppressors" (Fanon). In other words, a white person listening to the blues is "entertained by the suffering his political location has created" (ibid.: 18). Gordon suggests that Fanon falls for what he calls the fallacy of causal permanence: This would, in Gordon's understanding, mean that when the blues is played or listened to, the very conditions – i.e. the oppression – that made the blues are part of and even "maintained" in every instance it is played. Gordon says here that there are non-white, non-black people who also count as listeners of the blues, and here we run into the very trouble of how that fits, truly, into Fanon's account. Bringing back into the discussion the notions of the *potentiated double consciousness* and the "particularity of misconceived universals" allows Gordon the conclusion that "Thus, while born of black suffering, the blues speaks *to modern suffering itself.* It thus speaks to anyone confronting the entrails of modern existence, and since that also relates to postmodern existence, transcends its specificity" (ibid.).

Gordon shows in the argument that black people "were produced by the modern world", that matters are entangled and relational, and that, what I would describe, perhaps, somewhat awkwardly as a *theory of the blues* and/or *ethics of the blues* must make us aware of that. Enrique Dussel, for example, makes a similar point in connecting the production of what he calls the *goodness claim* to the production of the victim. The production of "black people" out of a group of people happened in the same breath as, for example, the co-production of America and Europe as entities, and the co-production idea that there was a context where the latter had "discovered" the former: this is what Anibal Quijano identified as *the colonial matrix* or *the coloniality of power. The blues*, and the many genres that derived from it, the blues as "expression of suffering *and*(!) as (its) reflection" (Gordon), thus cannot speak without the conditions of production that made it; however it is not reduced to the moment it was made, because the conditions of production are more entangled and complexly related than the moment of its production *and* in light this have become so much more. I have tried to capture a similar point in my call for making lateral movements of "going before and beyond" (B'n'B). This move should never be done in a way that loses sight of or respect for the event/phenomenon, the suffering it caused, and the people that were made to suffer. The very problem is that in navigating these waters, we find contradictions, conflicts, and questions of injustice that are often felt and experienced in the present and do not open themselves up to easy resolutions and sometimes to no resolutions at all. How to live with that and, perhaps, more importantly how to live together *with* one another *in and with* (and not simply *despite*), this is the ongoing and never stopping issue at the heart of what I call *BluesCollarship.*

BluesCollarship must also take seriously non-Western knowledges, non-human agencies, silenced and Othered beings, knowledges, and practices. That entails awareness of epistemic injustices. It also, importantly, entails "decolonizing *BluesCollarship* as a set of methodologies". Here, I am referring, of course, to what is so wonderfully and creatively articulated by Linda Tuhiwai Smith in her inspiring book *Decolonizing Methodologies* (2012). The fact that for many indigenous people, as she points out, the word "research" is one of the dirtiest words is something that *BluesCollarship* must take to heart. It is, in short, about how to play and/or enjoy *the blues* without appropriating it. This involves a lot of *learning and doing together with one another*; it bears not just the risk of, but will and must experience many moments of failure, and may be a never ending process. But without it, we would forever be stuck in doing, and doing carelessly, White Collar (social) science that conducts (*oikein*) itself and Others – as well as integrates/encloses the world it builds from many worlds into one from (*oikos*) under the myth of calculability and control – by the ways of the *coloniality of power and Being*. Lewis Gordon raises this issue in interesting ways in a recent essay on "Black Aesthetics, Black Value" (2017): "A problem with constructing black aesthetics is whether aesthetics has been so colonized that its production would be a form of colonizing instead of decolonial practice" (19).

Embedding or, rather, relating his discussion to a wider point of "a theory of dimensionality" helps understand the multiplicity of worlds including the plurality of human worlds, and connects aesthetics and value(s), for he states that

> colonial and racial theory necessitated at least three intellectual responses form those it degrades: (1) critical reflection of what it means to be human, (2) meditations on freedom, and (3) a metacritique of reason. The human world is not, in my view, reducible to a single element but it is instead capable of multiple elements manifested in complex, interweaving relationships of meaning.
>
> [...]
>
> A manifestation of reality disclosed through the communicative and symbolic richness of language is culture, which, too, is a dimension [*Comment A.S.: I think emphasis should be on the article, "a" manifestation and "a" dimension*]. As should be evident, no dimension is self-contained but instead functions as a key to the manifestations, simultaneously, of other aspects of reality. Aesthetic reality is such. It emerges as how human beings live as *deserving value* in it s full range [....].
>
> [...]
>
> This means not delinking[28] aesthetics from freedom and metacritical reflections on reason.
>
> (24)

An important point Lewis Gordon helps us see by introducing "dimensionality", and he helps us see it by not addressing it directly, is that "value", embedded in its Euromodern hegemonic apparatus/dispositif, has been used in monoscopic, monadic, monocultural, and monodimensional ways that obfuscate relationalities. In my *Between Discursivity and Sensus Communis* (2008/2010), I have pointed out that there is (and today I would add the line: "even, especially, and auto-violently within Euromodernity's geopolitical epistemic territory the Global North and its era-political enlightenment/post-enlightenment Modernity") even within the institutions of knowledge (aka science/academia) an obfuscation of what "value" is permitted mean, split into two dominant strains that are inherent in the two dominant schools of Neo-Kantianism in the 19th century, the Marburg and the Southwestern School: After Rudolf Hermann Lotze, who was the last to hold a – no less Euromodern and problematic but potentially more open/relational – concept of value, the Neo-Kantian split (with all its consequences, including the partial and thus Thomas-theorematic fiction of the continental/analytic[Anglo-American] divide in philosophy) cemented in *value* as a mathematical-physical and thus quantifiable/measurable unit on the one hand, and, on the other hand, a specific definition of value as part of moral philosophy, and more or less applicable as a politics of morality. Outside of politics and in the arena of "the social" aka of "society",[29] the idea of value as discrete, measurable units has usurped very dominant meaning, namely that of – to speak with Bataille – a specific economy: an economy that measures everything in money (which it[30] needs, in turn, to be included in growth, profit, etc.). This value has begun to colonize the world, but its most important effect is not just exclusion but, as I claim *the precarization of everything*, because exclusion and destruction themselves do not produce rent and the current neoliberal, postneoliberal, and neocolonial capitalism in its age of Plantationocene is interested in rent – thus, we see a neo-feudalization, which goes, of course, hand in hand with *precarization*. In my recent work on the agenda of *the* BioEconomy, I deal with this in more detail. Most importantly, I emphasize that this leads to the conclusion that the deciding feature of the current capitalism is an embedded logic and rationale that characterizes the "political and moral economy", if you will, of (historical) colonialism *and* of coloniality in general: The logic of extractivism (which requires a concept of [material as well as intellectual] property and ownership, which, in Athenian law is named *oikos*), which ultimately requires that all(!) *the Other(s)* – "other" to those who receive the "rent" from extraction – are (constructed as) *ultimately disposable*. Why? Because they are viewed as not having any *rent-able* value. Here, it becomes clear that we can think of an almost binary opposition between *rent-ability* and *response-ability* (Haraway). We need to "interrogate and trouble" the concept of value. Even in the Global North, we need to provincialize and have the concept travel – to visit each *B'n'B* it can travel to. Value is a plurality, values are relational. And, I think, this is something Lewis Gordon helps us see and understand better – so yes, this is an invitation to read his insightful text, which he concludes as follows: "It is not about who 'owns' thought and has the 'right' to aesthetic experience but about what needs to be thought, the kind of thinking and experiencing worth fighting for in the quest for aesthetic self-respect" (32).

With regard to the earlier paragraphs, one can certainly ask why, in describing some of what *BluesCollarship* can be, I seem to focus on Black people in the "American" context (by which it may mean North America in the present and North and South America in the colonial past) when there are so many groups of people and knowledges that are clearly part of both post- and decolonial discourses. Indeed, I could have talked about contemporary India, South America, Maori people, indigenous people in Australia, and black people in Africa; I could have also made more out of decolonial feminism, addressed ableism, included non-humans more elaborately, and so on. But I simply chose one entry point of many possible ones, because I have been inspired by Lewis Gordon's work and in the papers I refer to, there is just that kind of focus. But this can be extended. However, I also think it allows me to point out a few things, starting here: To begin with, much of the theoretical discussion of racism and the conceptual tools for critique offered are limited to intra-European histories of race, as Alexander Weheliye has pointed out in his book *Habeas Viscus* (2014). Hannah Arendt, Michel Foucault, and their successors have primarily addressed "racism" in light of the event of the Holocaust, creating concepts and theories wherein Black and Brown bodies and experiences are not included in the moment of conception, even though scholars later applied these ideas to injustice suffered by Black and Brown people. Lewis Gordon makes that very point about the issue of aesthetics in Black aesthetics, that literature on Black aesthetics is decidedly sparse on including actual "theory of aesthetics" formulated from the point of inclusion of what he calls "the aesthetics of everyday life in liberation" (2017: 19/20); instead he identifies in the literature a "form of epistemic imperialism was at work in which, by way of practice, black life was ultimately devalued as a source of theoretical reflection on itself".

Second, this approach also allows the possibility to interrogate and trouble the ways in which academic discourses often focus only on one side of injustice. In their famous paper "Decolonizing Antiracism"[31] Bonita Lawrence and Enakshi Dua ask some very uncomfortable questions, exemplified in what culminates in a provocative statement, they raise: Whose land do people think the 40 acres (and the mule) are taken from? The provocation they make is that, precisely, in many efforts to fight racism and its inequalities for example in North America, other groups, such as First Nation and other indigenous people, are forgotten or even made to suffer even greater inequalities as a result. This is as important as it is controversial. In addition, a number of readers will immediately and correctly point out that in bringing this issue up, I am also conducting an act that may deserve some critical questioning, because of my positionality as a white, European male. Do I have the right to do so? On what authority? That of being an academic, and the authority it gives me, e.g. qua a PhD? Or by acting all righteous by proposing *BluesCollarship*? But then again, am I kidding myself that I am not usurping an epistemic authority that is just, again, imperialistic? I think it is right to ask these questions, and not only do I ask them to myself, but moreover, I am not sure there can be a satisfying answer – but I will insist that all of the above, the questions addressed to me as well as the issue of decolonizing antiracism are all extremely important. To begin the (maybe unending) search for answers, perhaps I need to ask another question to myself: Am I, Alexander I. Stingl, a racist? This is, of course, a horrible and horribly difficult question. To answer with "Yes" would put me in company of people like, well, truly horrible people, no? But to answer with "No" would also beg the question, how can I say so? Can I guarantee that I have really never said or done anything that is even remotely racist in any way? Am I really convinced I haven't? Or would saying "no" make me the same as those saying "I am not a racist, but….."? I will drive this provocation further, but before giving you my answer: Let's add a different question into the mix, namely "Can Black people be racists?"

This question is hotly debated; it's a hornet's nest, and if there is a question that I, in light of my positionality, should perhaps never ask, let alone try to answer, it is that question. But in thinking about this question for example by having worked with students with whom questions of this nature come up frequently, I also do not seem to have the luxury to evade it. I think if I did evade it, when students bring it up, I'd be no less guilty. So, what if we start with saying that phrased as such, these kinds of questions are making matters too simple, and that asking for a "Yes" or "No" is the wrong kind of answer. And yet, even a "Yes" or a "No", if asked by people aware of the context and part of a longer running discourse, can sometimes codify many elements that needn't be unfolded but are understood by the present company. Sometimes, one must start with a "Yes" or a "No" and then hope for the patience of the other people in the conversation to let them go on with giving warrant in the form of justification. So, my response to "Am I racist?" might be "Yes, and how could I not be?" This response would have to be taken in the light of, for example, Robin Di Angelo's work on *white fragility* (2011), because if I simply responded "No", I might be very oblivious to the positions of white (and male) privilege that I do certainly occupy, and I might try and preemptively absolve myself from even the possibility of error or ignorance. But my thought is that, in light of work such as by Bonilla-Silva on colorblind racism or DiAngelo's on white fragility, and many, many others, the reflection that I think lots of people have when they are asked "Are you (a) racist?" is "I don't really know, and it's kind of hard to know if I am being or acting in ways that are racist if only sometimes". [By the way, there are people who are very openly following a racist ideology such as White supremacy, and they'd probably be proud of saying "Yes", but they are not even part of this discussion here, obviously.] Some people will react to this reflection defensively, because of *white fragility*. But, I claim, anyone who takes the reflection seriously will come to the conclusion that at one point or the other in their lives, thoughts, words, actions, ideas about society, etc. that were theirs may probably have been imbued with some form of racism, color-blind or not. So, an honest answer would be that "it's very probable that I may have been guilty here or there of racism, without me realizing it". Second, one may just understand that we are living in societies where (color-blind) racism is often a structural component and living in such a society implicates one in this racism. Taken together, one may accept that the answer should be "Yes". What matters is what one does with that knowledge. To be implicated is one thing, to take steps not to further intensify racism – in one's own action or in society – is another. And the latter process cannot begin before the prior is acknowledged and the acknowledged processed and taken seriously.

Now, what does this imply for the other question: Can there be Black racism?

This is a question that I cannot answer, and – to some degree – it would be wrong of me to answer it, if by "answer" one would mean a *definite* answer. But perhaps, even from my positionality, one can have a sort of a "response" to it. However, that would yet presuppose that I can or I should be allowed to even (pro)pose the question in the first place. Is there a problem with me posing or even asking this question? By implying that there can be a difference between proposing, posing, and asking a question, I already hint at an important issue. Instead of exercising an "analytic philosophical exercise" – readers may well think up their version of what I am alluding to, e.g. which of those ways of putting a question "out there" requires an answer, etc. – the point is that context matters. For example, if one raises the question in a discussion about white privilege, it just be a "whataboutism"-diversion. And clearly, the sentence "but Black people can be racist, too" has been used by too many people who defend white privilege or even White supremacy. And that means that this question can never be posed innocently. But, I think, it can be asked in a response-able way, which means, among other things, that someone – especially a European, white male, like, well, me – posing it, must be aware of the difficulties of the question, the pain it can cause, and so on. And that even with all of that kept in mind, it may not be enough. And to assume that, just by being careful and attentive and responsible, one is absolved in any case is just utter hubris. Just as simply acknowledging "of course, I am, in a sense, racist, because I know that in today's society as a white person I will have access to privileges" is not absolving anyone, nor does it make them a better person than others (perhaps thinking of some people being "better than others" is part of the actual problem. To think a word said on one's own positionality was any absolution is just pure and utter hubris: Nobody is "the better white person" or "an ally" just because they are so "brave" to recognize their own privilege, their membership in the group of profiteers of structural racism, or past ignorant statements in the eye. In this way, for example take the analogy of Germans and the holocaust: If all that the current generation of Germans does were to follow a "cult of apologetics", this would amount to nothing but narcissism: Apolog*ism*, a culture of *Entschuldigung*, is self-centered. The German phrase for "asking for forgiveness", i.e. to apologize is *sich entschuldigen*, which taken literally means to "to get rid of or take away one's debt/guilt". The word *Schuld* means both "guilt" and "debt". Thus, if all we do is *entschuldigen*, what we want from the other (in accepting our apology) is forgiveness – to forgive us our guilt and our debt. But the *Entschuldigung* is never about those who suffered – nor about their descendants – or about learning from the crime committed. It is about one self: ***I*** apologize. And even in apologizing *to* you, I seek an answer *from* you – possibly and preferably, your forgiveness (of my guilt/debt). This is, however, oddly narcissistic. By making a culture out of apology, we accomplish at once a trifold split into stereotypes: Those who do not *want* to apologize (we can call them, quite rightly, Nazis, fascists, racists), those who are tired of apologizing and want to move on (often asking when is it enough), and those who are celebrating the culture of apology. All of these are subject to a narcissism, some more and some less malignant. I am not saying that apologies are unnecessary. But there are things that come before an apology (e.g. understanding and accepting one's responsibility as the *guilt* for what has happened) and beyond the apology: The apology is not the end of a process, nor should it be self-serving. What response-abilities follow after the apology? Ideally, what comes before and beyond the apology doesn't even require the apology itself. But the apology can express the acceptance and a growing(!) understanding of one's guilt and accepting that *together with* one another, we need to undertake steps to learn (and/or recompensate, socially transform, etc.) from what happened. An apology that does not include what comes before and after is, well, really just bullshit. In other words, the apology may even be necessary, but it is not the point. Those who make the apology itself the point, they miss the point – among those we can include those who think "there is no apology necessary", "it's been made often enough", or that "time for it has passed". The issue of the history of slavery in colonialism *and* in the United States is as important an example as is the treatment of indigenous peoples in historical colonialism *and* after. And here is my point: Have the respective apologies ever happened? Not really. Have things that need to come before and have to happen beyond an apology ever happened? No.

The few "political apologies" that exist in this context are easily identifiable as bullshit – even if the individuals who made them may indeed have felt strongly and truthful about them. But apology, like democracy, is a process – a process that does not have a fixed, foreseeable end, if it ever has one; it needs to be regained by every generation – involving those who inherited the responsibility from the perpetrators as well as those who inherited a legacy and memory from those who suffered, and, perhaps, also those who were and are just "standing by" – that are the inheritors of the twisted legacy of the event that necessitated the apology, just as every generation living in a democracy needs to reconstitute the foundations of their democracy (because these are, in a sense, post-foundational) and not take it for granted. Perhaps, the need for the apology will have disappeared, but it will possibly at the earliest have done so when people have reached a point when the way they are living together has made the original event and its historic consequences unthinkable in the first place. But that is not the case with racism, colonialism, and coloniality. We still live in racist times, in neocolonial (neo-extractive) times, and in an age where coloniality/modernity holds a hegemonic sway. And there have been no real apologies for slavery and colonialism – and in saying "there are no apologies for slavery" points to the many complexities of language that I tried to allude to by using the German above: For there are those kinds of "apologies" that can be used as "excuses" such as in saying "there are reasons, why it was ok to do a bad thing", in other words, if there were "apologies" as reasons external to the perpetrator, it would mean they would be used to absolve them of guilt and responsibility. This is part of the malignant narcissism of "inauthentic" apologies that eschew the *before and beyond* of the apology – take for example any "fucked up" excuses for colonialism that take the form "but it brought them civilization". Such an excuse (despite its idiocy and giving warrant to violence) does not reflect on, for example, the ways that turning mere difference between peoples into a dichotomy of "civilized/primitive" did epistemic work and ended in ontological harm; it obscures the suffering and the ways in which the concepts were used to obscure and reify themselves when "resistance to invasion" was encoded as "primitive response to civilization" and then (re)written and learned (in school) and studied (in *uni*-versity) as "History".

An honest, truthful kind of apology towards, for example, Black people in the United States or the abduction of millions of people from the African continent in colonial history and what was done to them has never happened. And that is why a white person asking "Can there be (a kind of) Black racism?" is always posed in way that carries this context of an apology missing, but missing because there never happened what has to come before and beyond an apology. It continues the (colonial) wound. But the problem is that it is not, therefore, a question without legitimacy nor many contexts and nuances, since the idea of a Black racism can refer to many things: A recent episode of the television show *Black-ish* (Season 5, Episode 10, "Black Like Us", 2019) addressed the question of "colorism" in terms of differences of skin tone being enacted among the African-American community itself, a phenomenon not restricted to the United States but also found in many African societies, as well as in India and Bangladesh as was addressed in the novels by Salman Rushdie, for example; a novelist may imagine a counter-historical, speculative narrative in a novel (if Africans had "discovered" Europe and colonized it, and developed racism). And, of course, there a people who are black who hate white people, etc. But here is an important question: Would any of these practices/ideas exist without the prior existence of "racism" and "coloniality" which *are* the creation of Euromodern white people? I say, they would not exist otherwise. And, to speculate that the answer was "yes" would require to think that the making of racist distinctions and subsequent practices of racism were a universally human trait – but that would make all humans racist and the qualifier "Black" in "Black racism" would be meaningless. But then again, I don't believe in universalisms, anyway. In short, can there be "Black racism"? Maybe, but the issue in this question is not the "blackness" of any racism, but the fact that racism exists because of coloniality/Euromodernity. That doesn't make it wrong in itself to ask about the possibility of Black racism, and it may be important to ask the question. And, yes, perhaps it may sometimes be posed by a white person, for it has emerged as an important question in a context, but was not asked by anyone else. But even then, that white person is, I suggest, never the "right" person to ask the question. And for them to consider asking it means to become aware and accept that in asking it one is bearing a kind of responsibility. And who can really do that without falling hubris? So, what is the question here other than a question I can't really answer, namely: How to ask a question that may or may not be wrong to ask, but I am not ever the right person to ask it? I really don't have an answer. And my only response – although it is a response – to proposing the question is to point explicitly out that I am never the right person to ask it. What I have been doing by proposing the question is something problematic, uncomfortable for all involved, and it leads me to say, I don't really know how to deal with it. In other words: It's a failure. But the point here is precisely that: BluesCollarship is in many ways about doing such things, namely fail. It's about being and becoming uncomfortable. And in ways for which there is no excuse, but also in ways where there is no more narcissism possible, even though one can never be sure and only wary that one didn't fall into narcissism. It is about taking responsibility for asking it, and the history in it, and for accepting that one is complicit in this history. This is why BluesCollarship is *not* something one can do alone. It is – against many White Collar academic practices, against any hubris-filled idea of the politics of "academic career" – a practice of *doing and thinking and learning together with AnOther/Others.* Thus, if I ask questions such as "Am I a racist" or "can there be Black racism" on my own – even if I ask an audience, a classroom, etc. – or if I ask for forgiveness and apologize, doing so on the basis of an isolated event and me as an individual speaker removed from those who I ask, not only will I fail but nothing will happen except in the worst case that I am doing more harm. But if asked "together with one another", the situation is different. Now, what does it mean to ask for forgiveness, to apologize together with one anOther? I will not answer this question here, because doing so would just be another monologue. I know that such a question can, again, be horribly misconstrued – e.g. to "shift blame to the victims". Or, some would use it to absolve them of the work of the "before" of an apology – but without making an effort, what can be accomplished. That's all a risk, but I think it is worth taking. Of course, BluesCollarship is about more than that. Any development and movement that is scholarly, intellectual, valuable (in many dimensions) and that defies the confines of White Collar Academia involves failure, making mistakes, being and becoming uncomfortable, and doing so in ways that do not mean that accepting failure leads to hubris or serves as an empty apology.

BluesCollarship is about getting, or rather feeling uncomfortable, sharing in an uncomfortability, a sadness we all may feel; it can and should sometimes be enjoyed. It is like the blues. But it originates from a place, originally, that entails suffering. Also, I will re-emphasize this: *BluesCollarship* is something that is done as *(learning) together with one another(s)*, but it also needs or, rather, is itself when integrated as artisanal practices or craft as solidarity (as suggested by Sabrina Weiss and me in *Worlds of ScienceCraft* [w/ Restivo 2014]), what we could call "an infrastructure". It is, however, important to understand that I do not make a concrete proposal for "how things need to be done" or for "institutions" to be built. What I am trying is to talk about opportunities that were closed or were opened but remain obscured. I am not simply "opening" a debate in some patriarchal gesture, but try and make explicit, in going "before and beyond", certain things that are already there, processes already underway or were just cut off and silenced. It is not my place to simply "start" something on my own and from above, nor should we miss the point that the matters at stake are anything but relational and procedural, becoming and ever finished.

Finally, on some past occasions, I spelled BluesCollarship with a space between Blue and sCollarship, but I will use the spelling without the space, because BluesCollarship (and Blue sCollarship and, perhaps, BluesCholarship) can be used interchangeably. It is about the idea of blue collars and scholarship (and therefore linking craft and science/technology as Sabrina Weiss and I along with Sal Restivo [2014] have considered, or linking artisanal practice and scholarship as Boaventura de Sousa Santos proposes [see his latest and most complete articulation of this: 2018]), as much as it is about the blues. It was, however, very important for me to make this line of thinking with and without the space between words, the criticality of the space, i.e. the critique entitled in it and the criticism that is always potentially to be deployed to inquire after my use of the spelling without space as an inquiry into whether it is a genuine, charitable, and humility-aware effort of bridging that space or if it is a (colonial) erasure – a critical inquiry that must always remain possible to be made. The possibility to always re-make the space and to re-unmake it, the fact that it is never clear – that is like a *ghost note* – which it is while always hinting at different orientations, directions, and opportunities being retained, points to the importance of equivocation. As I have said, I perceive *White Collar sociologists* as being highly afraid of equivocation; in turn, equivocation in BluesCollarship is an important source of liberation productivity, creative delinking, and epistemic resistance. If sociology were music, you could say that main difference between *White Collar Academia* (including, in particular, *White Collar sociology*) and *BluesCollarship* is the latter's ability to play *ghost notes*.

Notes

1 Contra sociologists' fears and especially those of the (Schuetz-ian) social phenomenologist cliques, and other kinds of approaches are possible within the Western episteme. Merleau-Ponty (2002: 7), for example, hints at the fact that he is not averse to equivocation, that equivocation may just be a fact of (perceptive) life: "We must recognized the indeterminate as a positive phenomenon. It is in this atmosphere that quality arises. Its meaning is an equivocal meaning." (This is not a fluke by the translator Colin Smith, but confirmed by the original: "Il nous faut reconnaître l'indéterminé comme un phénomène positif. C'est dans cette atmosphère que se présente la qualité. Le sens qu'elle renferme est un sens équivoque, il s'agit d'une valeur expressive plutôt que d'une signification logique.")

2 In a fragment comparing the epistemology, ethos, and philosophy of science of Goethe and Kant with one another; wherein, implicitly, their mythical Promethean potential is uncovered and the Promethean Goethe seems juxtaposed to the Epimethean Kant (cf. Hans Blumenberg).

3 Sociology itself, particularly *White Collar sociology*, has served and continues to serve as such a colonial inclusion-and-difference engine.

4 There is also a rehashing in Latour's distinction of fact and concern of Radcliffe-Brown's account of the (older) distinction of *idiographic* and *nomothetic* approaches, as Ingold (2008: 70) describes very well.

5 It pains me greatly that it is not possible to continue to work further on these ideas in collaboration.

6 I tend to agree that "critique" even in the parsimonious approach I endorse, namely as the general (not[!]: universal) inquiry for the conditions of possibility of something, is certainly not "objective" and is most certainly political. But by making that explicit, possibly in terms of provincialization, and by being aware of this and of the vulnerabilities it affects, one can retain its very pragmatic usefulness, for example through *critical realism*.

7 Don't get set into one form, adapt it and build your own, and let it grow, be like water. Empty your mind, be formless, shapeless – like water. Now you put water in a cup, it becomes the cup; you put water into a bottle it becomes the bottle; you put it in a teapot it becomes the teapot. Now water can flow or it can crash. Be water, my friend (Bruce Lee).

8 Reading Kant from the point of view of biology and psychosomatic medicine, which were lines of inquiry very important to him, presents the "matters" in play for Kant himself very differently. However, the divergences into the various Post- and Neo-Kantianisms have rendered this nearly obscure – with a thin line within the discourse, including notably Lotze and Jaspers, being the exception. At the time I was working on my PhD thesis, remarkably few philosophers had published on the depth of Kant's engagement with biology, historians of science being the notable exception (see Stingl 2008). Currently, there is more available and even sociologists (who were extremely opposed to the idea, when I was researching for my PhD) are willing to entertain that idea, while it is still almost impossible to discuss the comprehensiveness of the biological and biomedical roots of sociology. Sociologists still fear the shadow of Social Darwinism, which they confuse with 19th-century biology and medicine.

9 Here, with the rejection of Wittgenstein, as in many other places, it becomes clear how important Deleuze is for Latour, even though he is less frequently cited, and Latour seems to me to swing back and forth about the Deleuzean influence when asked about it (see Latour in Ihde/Selinger 2003).

10 A not entirely unproblematic translation for *agencement*. My own use of *assemblage* in *SAT*, while related to and influenced by Deleuzian thought, e.g. by Brian Massumi or Jasbir Puar, but also deviating versions that hark back to more institutional theoretical approaches such as Saskia Sassen's, differs somewhat.

11 For example in my iterative deployment of concepts by Nicole Karafylis's and Stefan Beck's, I go further a bit than Latour, too, because I see the issue from the point of view of what I call the "narrative dialectics of technoscientific practice" – wherein I consider human practices in general technoscientific in nature, and my concept of narrative is perhaps more parsimonious than, e.g. Greimas's: techno(f)acting and bio(f)acting (see here, too, further below in this book, section 8. On Green Precarity).

12 And "coming first" should also not be understood in terms of perennial philosophy either.

13 Following a proposal by Barbara Tversky (2004).

14 But, it should be said, this is conducted in a way that considers agents not as individuals but as constantly "in individuation", wherein I am following ideas by Gilbert Simondon.

15 See here also the brilliant work by Ruha Benjamin on *biodefectors*.

16 The kind of *akrasia* bureaucracy induces (Stingl/Weiss in Weiss/Restivo/Stingl 2014: 161).

17 And by deploying this term, I may still retain some chance of connecting with a Whiteheadian account of the matter, too.

18 This could, by the way, also refer to care that *silent Others* give. Care is, of course, not a one way street. In other words, what offers of care are we not even ignoring but not even hearing at all?

19 An example for such silent Others are other sexes and genders that existed before colonization. European colonization rendered sexes and genders into binaries (and on occasion into hybrid forms built from these binaries). Therefore, forms of living that could not be conceived simply disappeared; in another and harsher form, if they could not be conceptualized (as to be included or as subject to exclusion), they were rendered "killable"/"disappearable".

20 Although both have positive force in rationally supporting a propositional attitude or cognitive practice, and in constituting an epistemic right to it entitlements are epistemic rights or warrants that need not be understood or even accessible to the subject. We are entitled to rely, other things equal, on perception memory, deductive and inductive reasoning, and on – I will claim – the word of others. The unsophisticated are entitled to rely on their perceptual beliefs. [...B]eing entitled does not require being able to justify reliance on these resources, or even to conceive such a justification. Justifications, in the narrow sense, involve reasons that people have and have access to (Burge 1993: 458/459).

21 There is, of course, some similarity to Tyler Burge's proposal on warrant to be considered here, as long as a poetics of relationality is reflected on.

22 It should be noted that I follow and endorse Mignolo's use of "praxis"/"practical" here, reserving the use of the concept "practice" to actual "practices" and "practice regimes" such as typically used in the works of Michel Foucault.

23 That Parsons would treat "universals" in his middle-period (not in the beginning nor so much at the end of his career), in a way that became itself somewhat reificational is rather unfortunate. Originally, Parsons followed, after all, Whitehead methodologically and would not only have agreed that but also taken quite seriously and without resorting to ambiguities that universals are only abstractions from our particular point of view and its framework.

Like with Radcliffe-Brown (see Ingold 2008), Parsons has much more to offer to think *with* than the current depiction of his work's place in sociology lets on. However, many of his disciples and critics focus pretty much on this middle-period as the anchor-point from where one must read Parsons – not to forget the peculiar reception history his 1937 book *The Structure of Social Action* enjoyed – so that it has become nearly impossible to talk about the slumbering contributions that a serious engagement with Parsons could still make. One does not have to become a Parsonian in order to profit from such an engagement.

24 Even social network theories of power have not changed much and conceptualize power as hierarchical (e.g. the generally rather impressive and insightful work of Hafner-Burton/Montgomery 2006).

25 Unfortunately, many sociologists, especially in the United States, do not deploy Foucault's power concept really well. Even when they make the "obligatory" citations, the actual deployment in empirical research or conceptual discussions (again, I would say the lack of a genuine methodological discourse is partially at fault here) reduces power again to function as mere suppression/repression. Similarly, Parsons's concept of power and the important differences between influence, power, money, and value commitments are often overlooked (on different concepts of "power", see Stingl 2015: 25–30). However, for example the past two or three decades of failed US "Middle East policies" can be explained much better using Parsons's analytical clarity on power, influence, and money than even with Joseph Nye's (2005) quite sophisticated and interesting *soft power/hard power* approach. US foreign policy effectivity in the so-called Middle East can be partially explained by its active epistemological reduction of power, influence, and money (and, to some degree value commitments), whereas the social and political ontological relations were more complex than that. As a matter of fact, even Nye's concept lends itself too easily to reflect merely Western ideas of what elements count a *soft power* – the "Coca Cola diplomacy". Parsons's concepts could be put to a more productive use.

26 While not wanting to go into too much depth here, this string of argument is derived from both (and thus reconcilable with) Whitehead and Viveiros de Castro. There are also similarities to the ideas of Karen Barad and Manuel Delanda.

27 Here it becomes clear that my understanding of practice is much more comprehensive than the usual sociological ones, which often tend to reduce practices to mean textualizable discursive units.

28 Obviously, the conceptual-grammatical use of "delinking" here is not to be confused with the theory-conceptual use by Walter Mignolo, Madina Tlostanova, and others.

29 See in the final sections of this book my opening of a discussion of the "state without society".

30 Taken here only in the context of its sentence, the connection seems like a circular argument. The connection is intended by me to make sense, though only in the context of this entire book and what you, dear reader, do with it from there. Sapere Aude! This illustrates what I mean, evoking Barthes, by this – like most of what I write – is a writerly not a readerly book.

31 I am deeply indebted to my students at Leuphana University for emphasizing that this paper deserved more attention than I had given it.

3 *Oikos*, Property, Urban Normativity

The Source Code for Class and Psyche

In this book, the central tenet of my critique is directed towards *oikos*, the Ancient Greek, aka Athenian concept. The reason can be found and summarized in a single statement by Jean-François Lyotard:

> And yet the *oikos* in the Greek tradition (...), is not, and I insist on this, the place of safety. The *oikos* is above all the place of tragedy. I recall that one of the conditions of the tragic enumerated by Aristotle is precisely the domestic condition: relationships are tragic because they occur in the family; it is within the family that incest, patricide, and matricide occur.
>
> Tragedy is not possible outside this ecologic or ecotragic framework.
>
> (Lyotard 1993: 97)

Where Is Greece When You Ask "A German Empire or a Latin one?" A Rhapsody of Concepts, Words, and Intranslatable Terms as Nearly Useless Facts

In my native Germany, which prides itself with a nimbus of a post-war economic miracle and prosperity on a national level and – compared to many so-called "developing countries" or an internally socially "uneven" country as the United States – social welfare level, we have coined an interesting word (among so many others, from *Schadenfreude* to *Fremdschämen*, or *Treppenwitz*): This word that I would like to spend a moment with is *Exportschlager.*

Germany's "economic miracle" (*Wirtschaftswunder*) after World War II has, to deploy another interesting German phrase, *viele Väter* aka "many fathers" – of course, if they were persons in a novel that I was writing, they'd all carry the name Urbanus. It's an interesting contrast: A very common account of the post-war economic situation goes like this: During the war, the Allies send men to the front, while women took on the (factory) jobs only to be set aside again after the war, when the men returned. Germany was destroyed and rebuilt; when the economy picked up again and workers were needed to replenish the diminished workforce, Germans looked to the male population in Northern Italy, Greece, and rural Turkey.

Of course, this is a glossing and somewhat stereotypical account, but it includes two important aspects: (a) No matter how the stories are told, women were often neglected both as equal participants in the economy and labor market as well as in many socio-political and economic historical accounts concerned with this time. (b) Germany "imported" labor in the form of male workers from other countries with no plans for their long-term residence or re-describing itself as a country that had to socially transform due to significant immigration – this strategy would back-fire, but from the 1970s to the beginning of the 21st century the term "immigration society" was what the Germans like to call a *Kampfbegriff*[4] between the parties in Germany's own version of the *Culture Wars* between the conservative parties Christian Democratic Union (CDU) and solely Bavarian-based Christian Social Union (CSU) and the progressive wing of the Social Democrats (SPD) and the German Green Party. But the reality Germans had to face was that while their miracle was stable, the imported labor force was needed over an extended and continuous period of time, leading to *de facto* – even though not, at first, *de jure* – immigration of these workers, and later their families. Eventually, German politics and legislation caught up to this reality – largely, when a SPD/Green coalition took power in 1998, although this came for some conservatives as a lasting *culture shock*.

It is sort of funny, since it also makes sense even in a Marxist frame of reference that Germany would "import" labor and then create its profit from it by turning into an export-oriented economy. But it is an observation many international and more realist German economists have made, that Germany has a problematic focus and dependency on export. It's not so much the success in exporting what's "Made in Germany"; it's the danger of not having a sustainable enough internal market once "shit really hits the (global economic) fan". With Germany's current implicit definitional power in terms of the collective cognitive policies of the institutions of the European Union (EU) and the terms of their embeddedness in national contexts through banks – a power, which it shares with Great Britain and interconnections of the global financial markets – this problem has disseminated into the EU makeup, as can be seen on the example of Greece. Germany has effectively profited due to its lower unemployment rate over the rest of Europe during the economic crises of the past years, because of the Euro and the improvement of export conditions for Germany within the Euro, as for example former US-American Federal Reserve chairman Ben Bernanke illustrates (2015a,b). However, I would say, a bit cynically, that overall Germany's *Exportschlager* is a "technocratic mindset".

Having unwrapped the first half of this cluster-word, let's pay attention to the second half – *Schlager*. A blank translation of the word reveals it to mean nothing other but "hit" as in a "hit-song". But there is a bit of a twist to it, because Germans have imported the word "hit" as well, and there is a difference between the two. Schlager are, to some degree, even more of an *easy listening* category; and to be particularly easy to listen (and sing along) to, they are most often in German – or if referred to as something that would be considered a *Schlager* from another country, assumed to be in that national language. The (in)famous Eurovision Song Contest used to be the prime example of different national *Schlager* songs. A *Schlager* is easy to listen to and sing, clap, and/or sway (rather than dance) along to – and as such (more or less) successful in air-time and sales. The mood is positive, at times romantic (although some aren't far from being as creepy as 80s stalker songs), but to a large degree *Schlager* songs can be considered to be *Kitsch* and are a mass-produced item.

Now, it should be sort of clear what is meant with a product or idea that is an *Exportschlager*: It is something that is exported easily and meets a massive demand. Interestingly, some US-Americans think that "democracy" either is or should be their *Exportschlager*, which is ironic given the current state of US-American democracy. Their actual *Exportschlager*, only disguising itself under the name "democracy", is something else entirely, something closer to what MacPherson would have called *property* or *possessive individualism* paired with a British-"Continental European" idea of technocracy. Thus, you have a quagmire ready to go off in every direction: This mix could be called neo, near, or late Liberalism which gave birth to a unique form of (anocratic, post-democratic) governance. I have unfolded this a bit further elsewhere (Stingl 2015).

How come that something that is not democratic can exist and be disseminated under the label "democracy" and how can someone claim ownership over it in order to both claim to exhibit democracy in its purest and best form and go on to export it and suggest others should mimic this form of democracy…or else (be called communists)?

Of course, the fact that the United States is not the world's "greatest democracy" or "greatest country" – regardless if it ever was that – has been subject of academic and journalistic discussions as well as fiction. Famously, the protagonist of the Aaron Sorkin TV series *Newsroom*, portrayed by Jeff Daniels, in the first scene of the pilot episode is placed on a stage discussing politics. A student asks a question of the panelists, asking them to "say why America is the greatest country in the world", echoing the very typical sentiment of what is known as *American exceptionalism*.[2] Daniel's character's response is a long explanation of why – based on empirical facts – the United States is not (anymore). This opening scene alone has been watched countless times on sites such as *YouTube*.[3]

Not that we would find a fully realized democracy[4] or republicanism in the EU, as the recent negotiations between Greece and the EU show. What Yanis Varoufakis reports (2015) is telling:

> There was a moment when the President of the Eurogroup decided to move against us and effectively shut us out, and made it known

> that Greece was essentially on its way out of the Eurozone. … There is a convention that communiqués must be unanimous, and the President can't just convene a meeting of the Eurozone and exclude a member state. And he said, "Oh I'm sure I can do that." So I asked for a legal opinion. It created a bit of a kerfuffle. For about 5–10 minutes the meeting stopped, clerks, officials were talking to one another, on their phone, and eventually some official, some legal expert addressed me, and said the following words, that "Well, the Eurogroup does not exist in law, there is no treaty which has convened this group."
>
> So what we have is a non-existent group that has the greatest power to determine the lives of Europeans. It's not answerable to anyone, given it doesn't exist in law; no minutes are kept; and it's confidential. So no citizen ever knows what is said within. … These are decisions of almost life and death, and no member has to answer to anybody.

But this paradox of a US-American (obsession with) democracy, and democracy in the societies of the Global North, has deeper historical roots. We should remember, of course, that "[h]istorically, the Athenians 'own the copyright' for 'democracy'" (Raaflaub in Raaflaub et al. 2008: 12).

Source Code (of the Euromodern Psyche): *Oikos/Polis*

Possession, property, and distribution are the core themes of Western political and social development, identified in the classic Greek writings. Society was organized around the *oikos* and the property it stood for: the possibility to participate politically depending on the ownership of property. To have a psyche, one must not only possess senses (and later with Locke: "have a body"), but also rationality. To be owner of property and to be owner of one('s)self can be seen as mutually constitutive, negotiated by powerful imaginaries (*phantasmata*). The possibility of the lasting power of these phantasmata is the object of the discussion.

There is one cautionary comment that I must make, though. When taking into consideration the question of *oikos* in "ancient history" (starting somewhere around Mesopotamia and traveling up the ladder of chronological history to Classic Greece and Rome) from an economic point of view, matters are notoriously difficult in terms of academic material available. A decade ago in 2006, the author of the seminal *The Roman Market Economy* (2012; see, however, the critical review by Erdkamp [2014]) Peter Temin (2006: 133) summed the problem up as

> Ancient economic history is in its infancy, both because few economists have learned much about the ancient world and because ancient historians have typically not incorporated economics into their analysis. This essay provides an economist's view of the Roman economy that emphasizes the role of markets. Some ancient historians and

> archaeologists have been receptive to this economic point of view, while others disagree strongly with it. [Footnote1]
>
> Footnote 1: Previous generations of ancient historians divided into "modernists," who followed Marx as applied to ancient history by Rostovtzeff (...), and "primitivists," who followed Polanyi as applied to ancient history by Finley (...). Ancient historians today universally argue that these positions are outmoded and counterproductive, but they frequently lapse into one position or the other when pushed.

This situation, in the decade hence, has somewhat, but not yet profoundly enough, changed. Following the recent years of financial and political crises, there is a resurgence of Marxism across the board of academic discursivities, and few – such as the notable and brilliant exception of Jason W. Moore – manage a genuinely inventive and effective engagement. And while there is a neoliberal backlash, "Finley-ism" is not their way to go. Some have, instead, gone as far as trying to occupy Foucault (whom some Hegel-Marxists, while far off the mark, have always been all-too happy to brand as a closet conservative/reactionist). Discussing the *oikos* from an economic point of view as a *coherent*[5] historical element is difficult and, understandably, opens up some avenues to slew criticism at me – though I think that proposals such as George E. McCarthy's (2002) work on the strong influences of classic Greek writers on Marx, Weber, and Durkheim should be enough to suggest to critics that what I am suggesting here could at the very least be an interesting exercise. A second point of criticism is, of course, that there seems to be gaps on "classic Greek" thought in the middle ages until the generation of Aquinas got their hands on translations that were originally circulating in the Muslim world.[6] But these translations did not fall into a vacuum: The worlds of ideas of the *oikos* were partially preserved in Roman (urban normative) legal traditions, which were preserved in the Catholic Church, and also fell in line with the discourse on *oikos-ecclesiology* (Synek 2005; Conradie 2007; Zamfir 2013; Sakupap 2018). Preservation of the structure of the *oikos* meant that the evolving imaginary of European societies since the Athenians contained the notion of *class*-ification as a *coherence*: The political notion of a class society emerges with forms of governing society in Athenian society, and transforms over time into the structure and (political) economy of labor. On the other hand, what it means to be a political being and the structure and (moral) economy of self derive their meaning in a co-production. The main concepts and major steps of the co-productive development are recounted as the story of the source code of what is going to be identified in (post-)modern times as class society.

I do understand my account in terms articulated by decolonial scholars and postcolonial sociologists, specifically by Catherine Walsh, Walter Mignolo, Maria Lugones, Julian Go, and Claire Decoteau on the one hand, and on the other hand, regarding the genealogy of the present by philologist Brooke Holmes, by Josiah Ober, by Michel Foucault, by Hans Blumenberg, by Luce Irigaray, and by feminist science studies: With regard to the prior, specifically in terms of the critique provided in Walter Mignolo's "Spirit out of bounds returns from the East" (2014) and Julian Go's "For a postcolonial sociology" (2013b), I am attempting to unravel a Western episteme (its possessive individualism and its social theory) from within itself – think: *provincialization*–, I stay within that frame but also expose the assemblages and arrangements deployed to assemble its coloniality of Being and Power. Therein, for the present purpose, I do draw far more on Northern Theories far more than I do on Southern Theories, but I am also not talking about the Global South but about the powers the Global North wages to colonize the Global South and, of course, to (continuously re-)colonize itself. With regard to the prior Northern Theories and their genealogical heritage, I consider a move by Brooke Holmes[7] productive, which understands that there simultaneously exist similarities, intelligible and unintelligible differences between ancient Greece and Rome and our present – although my directions of fit and scholarly interests are quite reverse, as she is a philologist/philosopher focused on Ancient Greece whereas I am a sociologist/politologist and by design more focused on the present. Hence, I do not claim at any point that my references to ancient Greek or Roman authors are in any way interpretations of these authors' work pertaining to their own time and authorship nor are they historic accounts of these times, their socio-political economic contexts, etc. My interest is in the effects of these authors in terms of metaphors, myths, phantasms, etc. that have had and continue to have effects in shaping the world as it is today – even and especially when these are misunderstandings or series and heaps of misunderstandings (see also Stingl 2011). Any criticism or critique that my arguments may illicit should refer to this interest. Little is served by claiming that "in 362 BC Xenophon really could have meant only this or that", when what I am saying is that "Xenophon's discourse on the *oikos* had *analogy effects* by means of myth and metaphor that can be found in the present among US-American so-called conservatives understanding of what they think a marriage is".

To put it differently and avoid at least some unnecessary (and uncreative[8]) misunderstandings: I am not so much interested in what the source code originally really was (in terms of a correspondence) than I am in what it *enabled* and how it *enabled* (and therein by means of creative misunderstanding may also have changed) that. This means acknowledging that there necessarily and pervasively are equivocations and creative misunderstandings, on which histor(icit)y itself is built.

So, let me ask: "Who's afraid of equivocations?"

Territory or "To manage the *polis*, to manage the land": An Essay on *Oikein*, Disposability, and Settler Colonialism's *Terra Nullius*

Stuart Elden, in an insightful book on "territory", makes the important point that

> [a]lthough it is a central term within political theory, geography, and international relations, the concept of territory has been underexamined. Where it is defined, territory is either assumed to be a relation that can be understood as an outcome of territoriality, or as a bounded space, in the way that Giddens described the state as a "bordered power-container." In the first, the historical dimension is neglected; in the second, the conditions of possibility of such a configuration are assumed rather than examined. Both take the thing that needs explaining as the explanation.
>
> (2013: 3)

There are various consequences attached to both the lack of understanding the depth behind the concept of *territory* and the focus on territory as the main/sole dimension of the boundary-making of statehood – the latter understood as either the *territorial trap* or as the reduction of "boundary" in boundary-making to the dimension of spatiality.

Statehood's boundaries, I would argue, should not be reducible to territoriality in the same way that "the state" should not be reduced to the "actor"-perspective nor should it be identified with government (see the seminal Rowland/Passoth 2010; Fishel 2017). At the root of these problems is to a large degree, however, tactical *arrangements* that require collapsing space and ownership into territory: the idea that space – maybe even all space – is actual or, at least, potential property[9] (i.e. belongs to someone). On philosophical grounds, "space" appears closer to ontology, whereas "territory" requires us to think about epistemology; but a sociologist might be tempted to consider "territory" in terms of social ontology, i.e. the study of the "nature of the social world" as Brian Epstein (2015) understands it. We *assemble* territory in our practices largely by way of *constraints*. Territory is also deeply entrenched in the (historic) exercise of coloniality; it is not only about the terms of how we think the world should be, but it is how we think the world is pre-*arranged*. We assume that actors have space, a space which is "their own". In thinking about other human beings, we do acknowledge that they have a personal space and that this space can be invaded, penetrated, and disrespected. When we encounter people who do not seem to understand this idea of space, they are often thought of in Western terms of pathological cases (for example, by calling them autistic).

In terms of political rather than personal (psychological) institutional levels of social relations, Elden (2013: 9f.) argues that we must distinguish between territory, "land" (which can be owned), and "terrain" (which is the domain of political entities such as states). But in my view, "land" and "terrain" are concepts or *assemblages* that *enable* certain practices, whereas "territory" obtains in specific *arrangements*. With land, for example, we can assume that it comes with the potential to be a factor that will produce some kind of "rent" or other, whereas "terrain" may describe the spatial extension or reach of government.

And here, for "territory", the *coloniality of power and Being* applies. "Land" and "terrain" *intensify* relations, whereas territory *is* what these relations are made of; therein territory is not identical with a space but with how people *do* space which is *technique:*

> These techniques, or arts, of governance have an important bearing on the development being examined here. These techniques include legal systems and arguments; political debates, theories, concepts, and practices; colonization and military excursions; works of literature and dictionaries; historical studies, myths, and— the technical in the narrower sense—geometrical instruments, statistical handbooks, maps, land-surveying instruments, and population controls. Territory is not simply an object: the outcome of actions conducted toward it or some previously supposedly neutral area. Territory is itself a process, made and remade, shaped and shaping, active and reactive.
>
> (Elden 2013: 16/17)

> The idea of a territory as a bounded space under the control of a group of people, with fixed boundaries, exclusive internal sovereignty, and equal external status is historically produced. (…) There is, of course, a danger of presupposing the thing we are looking for, which we then find.
>
> (Elden 2013: 18)

From these Western discursive epistemological economic (*oikos-nomos*) notions of land/property and generalizable land factors as they are present in and have shaped our contemporary imaginary into a historical and political ontology flow also, [*aqueduct-ed*], for example, the likes of Locke, the conjoining ideas of property, methodological individualism, methodological nationalism, Liberalism, and, eventually, capitalism and democracy – wherein the nature of this "and" connecting capitalism and democracy is fluent between a non-connective neutral, a connective, and an aporetic "and". Famously, democracy has its roots in Athens as has the notion of property/household united as *oikos*. Democracy has, from its inception, been a contested political form, in part because of the problems that property/household/family being united in *oikos* posed for and since Athenian politics.

Take a look at Josiah Ober's book *Political Dissent in Democratic Athens* (2001) as an example; this book deals largely with the critics of Athenian democracy and reminds us of the very fact that it is almost only in democracy that critics of the political system can publicly articulate criticism of the political system. We should also be reminded that while Ancient Greece and Athenian (political) culture in particular did and still do have lasting and generative influence over the Western world and through "Global Northern"-centric globalization exert an (colonial) influence epistemologically beyond "the West" and geopolitically beyond the "Global North", this influence is not reducible to any "happy" idea of Athenian democracy alone – suffice it to be said, the Athenian model of democracy is also not the only one in history either. Moreover it needs to be said that the "democratic" phase of Athens is not only complicated in its emergence and, as Ober shows, was deeply contested by its contemporaries with arguments that continue to shape opposition to democracy today, but it was also somewhat "short-lived" in that it existed basically throughout the 5th century BCE (before the common era). Solon famously laid some, equally short-lived, foundations for democracy a century earlier – foundations meaning above all the idea of an instituted code of law, which in my reading can be considered proto-republicanism. But the establishment of an actual "democracy" is often credited with reforms of Cleisthenes in 510/508 BCE. However, it still restricted democratic participation in many ways, e.g. to male citizens. But democracy eroded, was abolished during the Peloponnesian Wars in favor of first the *Four Hundred* and a second time for the rule of *The Thirty*, but each subsequent revival had to give way for the later Hellenistic Empires and the rise of the Roman Empire. And, of course, Plato and Aristotle – two of the West's/Global North's most influential "network addresses" in the discourse of knowledge in general and political thought in particular – were no friends of democracy. Not to forget that Aristotle (and to some degree Plato, although this reading is contested) defended an idea that proposed that slavery was "natural" for some people (Ojakangas 2016: 101ff.). In Western intellectual/political history, subsequently, the reception of Athenian democracy as something to be valued positively occurred rather late, i.e. sometime in the 19th century. Yet opposition, both in politics and philosophy as well as in their conjunction, remained strong: One just needs to think of Heidegger and Schmitt in conjunction with the end of the Weimar Republic and the political and ideological power grabs of Fascists across Europe in the early 19th century, and again in the recent 21st-century rise of authoritarianism and proto-fascism within both the populace and the administration and government in Hungary, Poland, Brazil, the Philippines, Italy, and Austria as well as the significant support that right wing extremist movements have in other countries, such as particularly in the Eastern and former communist states of Germany. Through the lens of Right and Left political populism, the issue is rendered as a form of capitalist critique, whether through rejection of the capitalism/multiculturalism(migration) constellation or by rehashing the trope of class struggle.

With the emergence of discourses and sciences that make explicit "social issues", such as *die soziale Frage* ("the social question"), both "class" and "psyche" became intertwined as political economic problems that were explicitly discussed, studied, considered, and attempted "to be solved". Competing visions of an American Dream and a European Dream that were the result of these attempts both rested on notions of "upward mobility" and helped along, if not in the creation but by the institutionalization of disciplines such as psychology and sociology – but sociology, too, is rooted conceptually in an imaginary that is deeply *ancient Greco-Roman*.

However, "upward mobility" and modern democracy are, I claim, just Euromodern applications of the "source code" of *oikos*, and thus of its *primacy of property*. Since the logic of *oikos* remains at the heart of the democratic psyche and "free" market society, class does not disappear but becomes constantly reified. Moreover, the very sciences/arts that study these processes and were intended to help "understand and solve" the social issues, for example political psychology and sociology, are implicated in the process of reification. The diagnosis of a contemporary post-democratic, class-structured society of people whose political psyche is based around imaginaries of (production/)consumption is a logical outcome of the *oikos* and of the *coloniality of Being and power.* Upward mobility, the (US) *American McDreamy* strife to "own" your house, is both broken and, paradoxically, therefore stronger than ever before – in part because of the cultural hegemony exerted by the (US) baby boomer generation. Therein, however, "upward mobility" may prove to be, in other words, a vanishing mediator for the next level of class society. Those who cling to such hopes, such as the secret hope of being rich one day despite the odds, or something more sinister such as "making others their bitches",[10] and so on and so forth, are not particularly "perilous and courageous" in the Nietzschean understanding of these terms. We are faced with what I think the (increasingly problematic) scholarship of Slavoj Žižek (2014[11]) describes quite adequately:

> When we have to deal with a specific struggle, the key question is: how will our engagement in it or disengagement from it affect other struggles?
>
> […]
>
> The courage of hopelessness is crucial at this point.

And, he also cites a memorable phrase from Deleuze:

> *Si vous êtes pris dans le rêve de l'autre, vous êtes foutus,*
>
> ("You, who are you caught in the daydreams of the other, are done for.")

It is those dreams, those phantasms that I am concerned with here, that have colonized our imaginaries and continue to colonize the worlds we assemble to dwell in.

Bringing Down the House

How do we dwell? How, where, and in what temporal structures do we reside?

> The history of ancient houses is strongly affected by how historians think about ancient cities and social relationships, and important new ideas for Greek cities have been those of Jean-Pierre Vernant and his associates and followers. These scholars have continued to analyze mental processes, social habits, and the deep kinship structures of myth familiar from structural anthropology, but they have reorganized the agenda to address historical situations, in this case Greek history. A leading idea has been that all political and social events have an essentially spatial structure in Greece and elsewhere; the idea arrays the analysis of houses with that of cities, and bridges the gap between how a citizen might live in a polis and how a philosopher might think about the cosmos. As Pierre Vidal-Naquet and Pierre Leveque have shown, the individual who could reorganize mental and physical conceptions of space in the city (philosophically and politically) and make them akin to the space of houses was destined for political success. This was because, in antiquity, any act of apportioning the *moira* or "lot" (which can mean anything from a house lot or a plate of food at dinner to an individual's ultimate fate in life and luck)w as an act of majestic authority.
>
> (Métraux 1999: 393)

There was a general concern within academia of the 1990s and forward, as Métraux points out, with the city, *la cité* and with houses but less with their essentials than more with their form as something that could, I would propose, be understood in terms of the *epistemics of planning* ("city planning" [Métraux]), which are inherently colonial, if one understands that coloniality is as algorithmic as the (idea of the) algorithm is colonial: It is not accident that Deleuze, authoring a book titled *Le Pli* (*The Fold* [1988/1992]), deploys the notion of *multiplicity* so frequently: *Multi-Pli-Cité* – the city as the cite of many folds, or, the city is the Manifold. Perhaps, we will have to consider what this means within an imaginary confined in urban normativity and within an imaginary that is radically different. There is, it seems, an inherent contradiction: The city, as the place of the *polis* and public, is the stark contrast to the *oikos* and the place of private affairs. Hannah Arendt, among others, had her share to say about these matters. And yet, we have to accept that this is precisely the point, since one cannot be without the Other. My sociological interest, in other words, is in how "the social" under investigation has been folded – temporally and spatially; and it must be emphasized that folding is a political process, meaning that the political comes before "the social" (as "the social" is not what explains but needs to be explained).[12] Herein, I do follow post-foundational views such as Chantal Mouffe's (2013) or Oliver Marchart's (2008). Foldings, while political, are meaningful in terms of practices we come to understand explicitly only in nested terms of – as Kockelman (2006) argues – their representations *and* their residences, i.e. as embodied and enacted which we can unlock through an "ontology of dwelling" (Ramirez Barreto 2010). But the "ontology of Global Northern dwelling" is infected by the Western arché-tectural epistemology (*ἀρχή τέχνη* [*arkhḗ tékhnē*]), which is as a narrative dialectic that presents as an archive[13]:

> In conclusion, the recent historiography of Greek and Roman housing indicates a new and vivid awareness of how domestic space sheltered the parallel histories of antiquity: ordinary folk and the elites, quotidian habits and the glittering etiquettes, homely piety and the great systems of belief and thought, men and women, and owners and slaves. The parallel histories converged in houses, in villas, in palaces, and in the way each dwelling related to its setting, urban or rural. Houses are the archive of antiquity, and in this respect, the history of housing provides an alternative to the histories of political and military events, of great men, of institutions, and of great philosophical and literary discourses. For Greece and Rome, the great archival collection of written documents-wills, legal papers, marriage contracts, letters, title deeds, architectural plans, and all the other materials upon which modem history and civilization rely-is mostly gone. For that reason, the history of Greek and Roman houses gives us the archive: it shows us people making great or humble decisions about how to live, how to allocate their space, how to maintain themselves as social entities, and, ultimately, how to achieve a balance between necessity and desire.
>
> (Métraux 1999: 400)

Here, I want to point to Xenophon's *Oikonomikos*,[14] written after about 362 BC, variously translated under titles such as "The Estate Manager" or "The Economist" (see Xenophon as referenced in Foucault 1988 [1984]; Detel 2005), and much analyzed. Xenophon is also understood as a notorious critic of democracy and proponent of oligarchy, in part due to a long-running attribution to him of a text on the *Constitution of Athens*,[15] which was, however, written by another unknown yet older author who is called, for lack of better name, Pseudo-Xenophon (see Ober 2001: 14–51). I wish to point out here that the social imaginary's source code of our times as much as it inherits folds from the Greco-Roman *oikos/domus*, while transformed through notions of alternative practical notions "property" and "individual", it inherits simultaneous – whether antagonistic, dialectic, paradoxical, or aporetic that remains to be seen – social vectors of democracy and oligarchy. The result, as I have argued elsewhere (o), is not only the re-emergence of a quasi- or pseudo-feudalism but also a spinning wheel of *timocracy* to *techno-/aristocracy* to *plutocracy*, which in fact arranges a state of *anocracy*, a state constantly in-between autocracy and democracy, constantly in a crisis mode: Here, the distinction between *fiction*, *narratives*, and *assemblies* may be helpful on theoretical grounds: The world[16] we live in is *assembled* quite *anocratically*, but we like to tell each other rich *fictions* and *stories* of particular freedoms, of what occurs in voting booths, as well as of conspiracies of the rich and powerful, while we actually act in practices that can become enunciated by *assemblages* and in *arrangements* that we think fit the stories of democracy or oligarchy/plutocracy (or that present as strives or transitions between them), which shape narratives, i.e. sequences in time and space, which we decide upon empathically and then decorate endlessly. The trick for any sociologist, political scientists, philosopher, or otherwise is not to become lost in the decoration, while understanding that it is impossible to completely "unsee" it either.[17]

I cannot nor would I want to reconstruct here the development of the argument from Pseudo-Xenophon to Xenophon, Tucydides, and Aristotle, but summarize that the Athenian defenders of oligarchy in the face of Athenian democracy – despite their individual differences – unite in their distrust of the demos (which is by no means, we must remember, all-inclusive but poses quite a lot of restrictions on who can be considered a citizen to be part of this demos) and its capability to work for the interest of the *polis* at large. Not only were the non-oligarchic citizens, the *demos*, considered to be inferior (since, and maybe because, they are poor), but, unlike the members of the oligarchy, politically they are also presented as governed by a self-interest. This self-interest can make itself known in individual as well as in class terms. Even if, as Pseudo-Xenophon stressed, that

> [s]ome Athenian elites simply failed to see democracy for what (in Ps.-Xenophon's view) it really was, that is, government by a specific social class in its own selfish interest. Ignorant of democracy's actual nature, some elites cooperated with the democratic government because they assumed a compatibility between their own true interests as elite individuals and the interests of the individual members of the demos. They assumed this compatibility, we may suppose, because they thought of the polis and its citizenry as a unitary whole. They failed to recognize the polis' status as an environment for the competition for spoils among mutually hostile social subgroups, engaged in a zero-sum game.
>
> (Ober 2001: 22)

The true interests of the elite's individuals can, of course, be considered along with the interests of the *polis*, of the *city* itself.

It is easy – way too easy, actually to be true – to use this as a pattern of corresponding interpretation for contemporary American politics and its intertwinement with interests of corporate/financial leaders. But even though not corresponding/equivalent, the comparison is usefully analogue/equivoque to describe the behaviors and motifs we can observe (though not, perhaps, the motivations).[18] What is more important here is that the "individual" entity in question is a(n almost socio-political-"transcendental-empirical") double: an individual person of the demos and a demos that has self-interests as a group. Neither Pseudo-Xenophon nor his successors have a problem with that self-interest for the group by itself. On the contrary, they can, at times, only praise the consequence of both recognizing and pursuing it (almost naturally); but they also note that the group interest is not in the interest of the larger collective and that, therefore, only oligarchic rule can work out in favor of the larger collective.

But self-interest for the individual person *himself* is problematic, yes even running the danger of becoming pathological for Thucydides (Ober 2001: 68). This is, of course, one of the elements that will transform towards the "hyper-individualism" that we witness in present Western society.

The "maleness" of the person is important in this context, its relation to himself and Others that plays an additional role here as we learn in Xenophon's work, which plays an interesting and somewhat contested (Detel 2005) little part the second volume in Michel Foucault's *History of Sexuality* (1990). As Nelson-Maldonado Torres reminds us (2007), there is a feminization in the colonial construction of the Other, because the Western/colonial construction of women as weaker becomes superimposed on Otherness itself.

In Nelsestuen (2017: 74f.) we find a retelling of today's myth, that those who manage (*oikonomos*) – meaning White Collar MBAs, scientific managers, and captains of industry – supposedly should lead us in everything else (conflating timocracy [the rule of those who hold property], plutocracy [the rule of those who are wealthy] into a unique aristo-technocratic oligarchy). It is a myth that is turned into a reality of ongoing precarization of everything:

> [Even Aristotle, whose *unequivocal* rejection of the categorical similarity between rule in the *oikos* and rule in the *polis* (e.g. 1252a7–16) underwrites his analysis of the city as a natural and unique entity, nonetheless views the *oikos* as both a foundational element of the polis (e.g. 1280b33–34) and, like the polis, even its own "association" (*koinōnia*) of speech, reason, and morality (e.g. 1253a7–18). Whether the comparison is made in terms of the agent (i.e. *oikonomos*, *politikos*) or institution (*oikos*, *polis*), all three examples attest to the power of

analogical reasoning between the household (as the "source domain") and the city-state (as the "target domain") in Greek political thought.

Ultimately, Nelsestuen's goal is to show that Xenophon's writings are an example of what he calls an "analogical shift" from *oikos* to *polis* and subsequently from *oikos* to *arche* (ἀρχή), the latter in his account stands for "empire":

> a type of polity that, transcending the spatial bounds of the polis, takes the form of "rule over very large territory and many peoples without consent"; assiduously cultivates the terms of imposed hierarchy between the dominant "metropole" and its subaltern peoples through a rationalized system of difference; and generally entails exploitation of that subject "periphery" by that metropolitan "core."[19]
>
> (Ibid.: 77)

In a similar vein, connecting *oikos* and urbanism throughout history, Paola Pellegrini stated in an interview that

> [...] urbanism is a dispositif in itself is not a novelty. What is more interesting is precisely that it is a set of collated and coordinated devices, linked in some sort of organization or as Foucault's termed 'apparatuses'.
>
> [...]
>
> The fear of the other, the poor and the stranger has often fostered the formulation of specific policies, while the history of the European city can be described as a succession of systems of intolerance, removal of the difference and normalisation efforts. The adoption of devices to prevent permeability and accessibility (such as walls, infrastructural and environmental barriers) in the past, has been replaced today by multiple and complex forms of segregation.
>
> (www.planum.net/the-new-urban-question-br-a-conversation-on-the-legacy-of-bernardo-secchi-with-paola-pellegrini-1, last accessed March 15, 2019)

Channeling the notion of porosity by architect Bernardo Secchi (1934–2014) into her conversation, Pellegrini goes on:

> The notion of porosity, borrowed from physics but also from literature, i.e. Benjamin, is as well analytical as a design tool, and refers to the percentage of open spaces in relation to built spaces and to the possibility to have different flows (of people, public transport, water, activities, practices, differences and vegetation). Porosity does not only include green areas and agricultural land, or abandoned, vacant

and under-used lots; it rather implies the possibility to re-signify non-built areas as a whole, especially the space for mobility. Furthermore, porosity is strongly related to permeability, represented by the single connections between the 'pores'. A porous city is widely accessible thanks to a new structure of public transport (a network described by the metaphor of a sponge) and highly sustainable new biological corridors, as well as, more space for the water network/wet lands.

(Ibid.)

We are, here, quickly aware that we come face to face with the narrative of two alternative conceptions of communal living: The *gated community*, which works as a socio-spatializing inclusion-and-difference paradigm, and *la ville poreuse*, which is also the analogy to the conceptions of the body as either that of the *immunopolitical* one with designated borders and an immunity-system (Cohen 2009) that fends of all "foreign bodies" in a clear distinction between self and non-self,[20] versus that of a body that is porous and leaky – the latter an idea that in Euromodernity and through its coloniality (of power and Being) is "gendered" through the construction of the female body as leaky, porous, wet, etc. (e.g. Benbow 2003; Manning 2009a); similarly, the Euromodern territorial nation is constructed as that which fends off migration through the tropes of immunity (see Cohen 2009; Thacker 2011), an imaginary that reactionaries, fascists, and the like still cling to. Fascists and the neoliberal *riches* share in this fantasy of "gating machines" (Vesselinov et al. 2007) whether on the national or the communal level.

In the Euromodern imaginary and its predecessors, the "congregation" of bodies, cities, and police-able gates, even the Agrarian Myth, the rural landscape, the outside of the city turns into a product of urban normativity: The rural, agricultural landscape becomes an outside that is, however, not radically Other but still foundational part of the territory (*khora*), *qua inclusive exclusion* from the urban metropole yet still subject to productive labor; but there are lands (land factors) that are not considered to be labored on, that can be occupied, colonized, because they are, effectively, Othered in a way that the labor it holds (which can include non-human and extra-human agents, besides "merely" indigenous populations) is silenced. The city and its inclusively excluded outside are, thus, an "attentional anchor" (Hutto et al. 2015: 7) and a point of folly.

In his reading of Jacques Derrida's "Point de folie" (1986), in the context of unpublished seminars on Plato, Francesco Vitale (2013) identifies Derrida's considerations on the dwelling and about a related "law of the *oikos*", a *phantasm of the origin*: A phantasm rooted in the *chôra* (χώρα), i.e. the very territory that was demarcated by the Ancient Greek *polis* as that which was outside the city proper but still owned by it, "the countryside" (of the country/city-state) but it could also describe "one's place in life".

As Stuart Elden (see 2013: 21ff.) opines, there is a close and organic link between the people, the land (khora), and the *polis* (23).

Badredine Arfi (2012), by the way, will go as far as identifying in the *chôra* (χώρα) the (Lacanian) notion of the *Real. Chôra* (χώρα) is a "phantasm" of origin; thus, we need to understand that there is no and cannot be an (actual) origin,[21] but – functioning as an anchoring metaphor – the *"landed" origin*, even if only phantasmatic, gives (necessary) meaning to make experience possible. "The experience of meaning must be dwelling, the law of the *oikos*, the economy of men or gods" (Derrida 1986: 69).

> Finally, for Plato, it is not important to establish what *khōra* is, but that there is a proper place, an unique origin which is identical with respect to itself, stable and permanent, an origin which, remaining indeterminate in itself, institutes the criterion of every possible ontological determination: the identity understood as stability and permanence. In this way, the law of the *oikos* secretly founds ontology and, thus, our experience and conception of the dwelling. A law but also an archaic phantasm to which no reality corresponds, as no reality can correspond to the myth of Erichtonius, the myth of autochthony, which founds the political identity of Athens. And which still haunts our experience and conception of the dwelling and of the architecture.
>
> (Vitale 2013: 70)

The law of the *oikos* as our secretly foundational ontology is, thus, what I refer to in terms of the "operating system" of Euromodern society – however many patches and updates we have "downloaded" into it. This "ancient" operating system still persists and creates the mythical (and tragic) *phantasm* (Boesch 2002) of "stability and permanence" that claims to prevent the chaos. That claims to save us from the chaos that we ourselves are in this ontology. The chaos we were brought up to believe in through having early in our lives learned to live in the "myth of lurking chaos" (Boesch 2002). *Oikos*, as Douglas M. McDowell (1989) reminds us, has three interrelated and interdependent meanings: House, a household's goods and substance as its *property*, and family. This gives, further, the (ontological) foundation for Athenian society and, subsequently, *qua* the (historically progressing) "work on this myth" the (ontological) foundation of Euromodernity (entangled geopolitically in [Settler] colonialism and epistemologically in *coloniality*):

> it means the sum total of goods belonging to an individual man, including land, buildings, crops, animals, slaves, furniture, clothes, money, credits owed to him by debtors, and anything else. Legal disputes about an oikos are disputes about the ownership of this totality, not about some particular item.
>
> [...]

> A man who already has some property and then inherits someone else's is not said to have two oikoi; rather, his oikos has become larger. It has sometimes been thought that an oikos belonged not to one man but to a family. But, at least in classical Athens, that is not correct. The error arises from confusion of two senses of the word. Oikos may itself mean 'family', but when it means 'property' it refers to the property owned by one individual.
>
> (McDowell 1989: 11)

We can follow this "myth" and the work on it as well as the work it does (as operating system) – despite some historic i(nte)rruptions – from Athens to Catholicism, from Aristotle to Aquinas. The Athenian/Aristotelian proposal gives us an interrelated unity of the wider political community, its constitutive domestic communities of family and property, and the regime of practices for producing and acquiring (by means of exchange and barter): These three interrelated elements are named as *polis*, *oikos*, and *chrématistike*. Such a unity – which for Aristotle has its teleology in maintaining both life and the means of life (which is why the agricultural production on the land/soil, as a bioeconomy, can be included in this discussion[22]) – in a *natural teleology* reappears in the Aquinasian merger of natural law, divine law, and human activities such as economic activities (however irrational).

Disentangling Aquinasian views on "usury" from the Aquinasian position on private property, today's *neoliberal* and neocolonial order is thus given warrant by an economic theory that defines as natural an economic order in which ownership, wealth, and economic success occupy an ontological foundational position. It can do so either in the German neoliberal order known as Ordoliberalism (which thinks in terms of economic activities serving towards a *social whole*, that is – unlike in Liberalism and neoliberalism [see Markwardt 2019] – not simply reducible to individuals) or in terms of the juridico-economic Chicago School, which views markets (as quasi-*khora*) separated from corporations that count as persons (*oikos* as family), and "price" as a natural-functional and foundational unit of activities that were once called *chrematistike*, wherein the wealthy and financially successful individuals are those that are leaders of their respective *oikos*.

Similarly, for the latter group, it is a "natural" conclusion that economic success/power is the foundation of political power, i.e. that financially and/or economically successful people – as those who own/manage (*oikein*) land factors – should be the ones who comprise the *polis* and govern, aka *manage* (*oikein*) affairs (the government seen like a household) in the interest of the majority (a social whole comprised of individuals). *Refeudalization* and precarization fall into this context:

> Subprime loans demonstrate that to grow and make profits finance needs to involve, other than the middle class, the poor too. To work,

> this capitalism must invest in the raw lives of people that cannot guarantee anything, that offer nothing in not themselves. It is a capitalism that makes raw life a source of profit.
>
> [...]
>
> In such a way, the financial logic produces a commune (of goods) that then divides and privatizes through expelling "residents of the commune" by means of the artificial creation of scarcity of all kinds—scarcity of financial means, liquidity, rights, desire, and power. This is a process that reminds us of the time of the 17th century enclosures where the peasants—living on and off the land as a common good—were expunged by the processes of privatization and division of the common land, the processes that gave rise to the modern proletariat and its bare life.
>
> [...]
>
> The originary or primitive accumulation, as was shown by Sandro Mezzandra, i.e., the salarization and proletarization of millions of people like expulsion from the land, is thus a process that historically reemerges every time the expansion of capital clashes with the commune produced by social relations and cooperations free from the laws of capitalist exploitation. The commune produced by free social relations thus precedes the capitalist appropriation of this very commune.
>
> (Marazzi 2010: 33/34)

That the Aquinasian position on private property (and thus *oikos* and *oikos*-ecclesia as ontology) is held on to into the 20th century as a driver in these developments can be evidenced in the writings of Pope Pius XI (1857–1939; see Spinello 2009: 190). Pius XI affirms (the Settler Colonial position) that property – which he deems natural – is made "first" *qua* occupation and by labor. But then our line of inquiry becomes: What about those whose lands (factors) are occupied or are deemed occupiable, and whose labor counts as such?

The response is derived from urban normativity and colonial/racist binaries that it sets up accordingly, including the "technological condition" that extends from Ancient Greece to Digital Culture, as Gilbert Simondon concludes:

> The technical operation that form imposes on passive and undefined matter is not only the operation abstractly envisaged by a spectator who only sees what goes into the workshop and what leaves it again without understanding the process as such. It is essentially an operation that is ordered by someone free and carried out by slaves.... [...and] reflects a city consisting of citizens and slaves.
>
> (Simondon cited in Hörl 2015)

Oikos Controversies: Settler Colonialism and Racism: The Problem of Disposability

Disposability, which is in my view the main issue – especially in the context of slavery – is something that seems to me more apparent in Ancient Greek/Athenian thought than in the spirit of Roman Law. And, indeed, in scholarly discussion of the status of slaves in Roman Law, there has been a debate on the question of the double status of slaves as both property and human beings, which bestows slaves right to obtain through emancipation freedom as well as obtain safety from bodily harm, at least in theory. For slaves as being part of the property (*oikos*) in Ancient Greece, the same cannot be said. Some authors have also remarked on the fact that in Rome, the slave discourse was not similarly a matter of racism (Verstraete 1980; Bradley 1987, 1990, 1994, 2011; Vlassopoulos 2009; Buckland 2010; Silver 2016). It is a provocative thought; however an argument can be made that not all forms of slavery are equally moral repugnant[23]: That we – at the very least in the Global North – seem to condemn all forms of slavery today in equal measure and, respectively, understand societies that would contain slavery as a social structure as "primitive" has to do with the Euromodern understanding of freedom and individuality. But there is an apparent danger here, which is why I think we need to differentiate between different forms of slavery and to make clear that some practices of slavery have been more brutal, terrifying, and violent than others, and that their distinctive feature may be persistent. If we consider the history of slavery – and we can and should add the contexts of indentured labor and feudalism here as well – and if we understand, that historically some forms of slavery and servitude had legal protections for the enslaved, as well as acknowledge the existence of parties that enforce these legal protections, whereas in other contexts such legal protections either exist but are never practiced or enacted only on behalf of specific groups and ignored with regard to others, or cases where such protections do not exist, then we come to ask what are the distinct features of each of these forms. If we consider for a moment the idea that there were legal provisions in Roman Law that could, for example, guarantee the bodily sanctity of a slave, e.g. someone who for financial reasons sold himself or herself into slavery (as that was legally possible), we must question a number of issues. For instance, while some scholars hold that the rape of slave was punishable because it was considered a "damage of property", we must also think about the ambiguous status of slaves as both property and human beings that not all historians of law have sufficiently included in their considerations. In Roman society, women of standing and slaves had legal protections when it came to rape – women with no standing (actresses, prostitutes) were in a different situation – as slaves had possibilities of becoming free citizens – a complex situation that appears somewhat different from the situation of slaves in Ancient Greece. If we think of the history of slavery in the United States in this context, we can find that the defining element is that in culture, practice, and structure there is a defining element of "disposability", an element that can be found in the treatment of indigenous people as well but there it has taken on its own forms. But that is precisely an important point: While slavery was abolished in the United States, the construction of the Other as "disposable" in the collective minds was not. But filtered through the Euromodern lenses of "freedom" and "individuality", those who enjoy white privilege construct the excuse that with abolishing slavery in society and with granting civil rights to Black people, justice and equality were already ensured, because from their privileged positionality a primary measure for equality is "being free" and "being treated as an individual". But if we accept that there are (and then review) different, complex, and difficult histories of laws and social practices entangling slavery, feudalism, and indentured servitude, then our gaze should fall on the elements that demarcate different forms of slavery, feudalism, and indentured labor and that may not be intrinsic to them. Thus, I propose that we must understand the no less complex history of *disposability*, which is primarily rooted in colonialism, and the still persisting colonial matrix of power or coloniality of power and Being. The point is that when we talk about structural racism today, a term many people enjoying white privilege refuse to accept, there is a persistence of a group of people existing in a society that has nominally shed slavery and racism but that still constructs the formerly enslaved through aesthetics and politics of *disposability*. This plays a role, for example, in the economies of relevance: In Global Northern societies, in the United States in particular, it is difficult to make relevant the problems faced by Black people. Take Black Lives Matter, for example, as the response by many White people – including many who would not consider themselves as being racist – was to say All Lives Matter. And, of course, it seems like indeed all lives should matter, but what "all lives" proponents fail to see is that not that "all lives don't matter equally" but that what matters to them is that on the surface to matter means to have the political right to enjoy the freedom of being an individual, regardless of color of skin, sex, etc. That some people still are socially and culturally construed as "disposable" does not come about or them as a *relevant* thought. But here is precisely the point: Paradoxically, in saying "your being Black should not/does not matter" to your already having equal rights as a person both assumes as well as assigns "no relevance" to "being Black" and exactly thereby rendering Black people disposable, because any problems stemming from anyone being Black are given no relevance. It is, in the "traditional" logic, truly hard to grasp but the reality of it is inescapable: In stating "All Lives Matter!" it follows that Black lives don't matter – as so many Black people in the United States know by their *lived experience*. The relevance of the experience of being Black is negated in favor of both a false universal and a fiction of individuality; in similar ways, women and their embodied experience have been made *not to matter* in philosophy, as Christine Battersby (2011) shows. There are, in short, fundamental processes of constructing the Other, that are destructive and persist not so much *despite* but almost *because* of the way in which certain social forms and structures – such as slavery or overt feudalism – were abandoned. Through the coloniality of power and Being, disposability is such and its roots can be traced back to the notion of the *oikos* which rendered slaves into objectified parts of the household and property that could be disposed of. One should see here the work of Benjamin Issac (2006, 2017) on the "invention of racism" especially by the ancient Greeks, and the Aristotelian idea of "natural slaves" as the source code for the coloniality of power, which further implicates the sociological canon as practiced, maintained, and revered in White Collar social science, which – as shown by McCarthy (2002) – has its origins in Ancient Greece.

As for the connection to this discourse *qua* reverence to the idea of "canon" in White Collar social science, McCarthy (2002) illustrates the depth of the debt that sociological founders/classics – Marx, Weber, and Durkheim – owe to the ancient classics. But the inherently unfolding logic, specifically of Marx (as his later musings in the *Ethnological Notebooks* show, which McCarthy briefly references towards the end of his Marx chapter), is still deeply entrenched in the coloniality of power and Being; this entrenchment is, however, not in opposition to but in alignment with the colonial logic. Contrary to what Marx's (or Weber's or Durkheim's) apologists would like to have us believe, Marx was just as implicated in racism, misogyny, and coloniality.

McCarthy's book discusses many interesting cue cards for my engagement with the connection between the sociological classics and the ancient classics – precisely to take them from behind in the Deleuzean/Foucauldian sense –, and I would almost be in the danger here of committing mental plagiarism, if it wasn't for a few discontents and disagreements, specifically with regard to Max Weber (66):

> The classical world motivated his thought, opened a debate with the ghost of Marx, and helped him clarify his major thesis about the rationalization of Western society and the development of modernity. It is interesting to note that as he developed his theory of rationalization and modernization, he moved closer and closer to the perspective of Marx in Capital. Second, through the secondary interpretations of sociologists such as Georg Simmel, Ferdinand Tönnies, and Alois Riehl, the cultural world of classical antiquity had an impact upon Weber through the philosophy of Friedrich Nietzsche.

To begin with, I had a hard time seeing Riehl primarily as a sociologist, since he was a well-known neo-Kantian philosopher. Sure, Solms-Laubach (2007) in his investigation of Nietzsche's role in the genesis of early German and Austrian sociology points out that Riehl's (1897) book on Nietzsche was a significant influence on early sociological thinkers, and Szakolczai (1998/2014) points out that Riehl was a "forgotten figure" in Max Weber's life who was instrumental in introducing the Webers to society in Freiburg, while also being the first philosopher to "take Nietzsche seriously". But Riehl's work was largely conceptual and Kantian on the one hand, and the degree of an actual reception of Nietzsche's works by Weber, let alone the influence on Weber's oeuvre is not without dispute.

Where others, such as Bataille, can be seen as cleft between Marx and Nietzsche (Landa 2015a,b), the famous quote – the basis for strong claims about Nietzsche and Weber introduced by the likes of Raymond Aron or Wilhelm Hennis – attributed to Weber by his nephew Eduard Baumgarten, that *Zeitgeist* was cleft as such, should also not be overstated, as Wolfgang Schluchter (1996: 166–185) has argued with some reasonable arguments to the effect that Weber had some reservations about Nietzsche – as Weber related to Else Jaffé in letters regarding the "moralism" of Otto Gross and Nietzsche (see ibid.). Recently Bryan S. Turner (2011) revived his own earlier notion from 1982 that Weber was strongly influenced by Nietzsche, to take issue with Joachim Radkau's recent, comprehensive Weber biography (2009).[24] However, even Turner had to concede that the direct influence is hard to prove and appeared to settle for a discussion of some mere "similarities".

"Some similarities" is something that even Schluchter can find little fault with, but a statement such as the earlier one by McCarthy that (ancient) classical authors had their impact in Weber "*through* Nietzsche" seems like (at least) a problematic overstatement of Nietzsche's influence.

The ancient classics had, I would think, an influence on Weber largely because he read them (rather than Nietzsche) and deemed them as contemporaries of their own era, allowing for contemporaneous insights for his own.

The statement "With this view of sociology, Weber integrates the thought of Kant, Nietzsche, and Aristotle." (McCarthy 77) does have merit; in other words, I would underwrite this statement in so far that it represents what I call elsewhere the *intellectual climate*, i.e. the *dispositions* and *thought-scapes*, but that does not precisely mean that Weber directly translates from Nietzsche's *vernacular* and *conceptual frame* into his own (negatively, i.e. to explicitly refute Nietzsche or positively, i.e. in the affirmative). In following an idea of a "translational frame", both Turner and McCarthy overshoot.[25]

Oikos, Neoliberalism, and Back Again

There has been much talk about the *entrepreneurial self*, but I want to point out another aspect of our shareholder economy, which is, indeed, a tent of neoliberalism@capitalism.oikos [DO NOT CLICK THIS LINK]. Talking about shareholders, I realized that part of the problem does not lie in the "shares", their size and distribution, or sign-economy (what is/represents a share), but in the issue of "holding": If you are holding, you are gripping (too) tightly, and have trouble letting go. Here, the English word for *oikos* is interesting, too, since it is the house-*hold*: a hold is a fixed place, intended for holding stuff.

It should also be noted in this context, too, that while Michel Foucault was certainly very interested in the problematization of sexuality, his work on the classic Greek sources should not be reduced to this problem. He was generally interested in the question of health – one could speak of the metabolism or politico-physiological economy – that is found in the concept of dietetics. Quite contrary to what, for example, Wolfgang Detel (2005) proposes when he argues that

> Foucault regards it [classical dietetics; A.S.] as a comprehensive art of existence and a thoroughly regulated practice of the self, but it is soon apparent that, at the most, this view can only be applied to an excessive, prophylactic form of dietetics that was in no way representative of the classical world. [...] This restricted view of ancient teleology and dietetics had the effect that Foucault arrived at an incorrect model of the regulation and practice of sexual desire: [...].
>
> (Detel 2005: 3)

Instead, we must consider Foucault's earliest writings on Kant's anthropology from 1959/1960, wherein his analysis focuses precisely on questions of Kant's psychosomatic philosophy and dietetics. This is exactly the issue: *that* Foucault is aware of continuities and discontinuities, precisely those continuities that he himself aims to follow and think through, that inform how he thinks about dietetics. Detel, thus, does not see why and how Foucault accounts for dietetics not simply as comprehensive but as a psychosomatic ideal; he thinks we can still find in Kant's philosophy: Knowing the world (*Weltkenntnis)*, in Kant's anthropology is a cosmological view of "Nature and Man" (Foucault et al. 2008: 32); and not for nothing, as Foucault would have known, Kant developed his anthropology lectures back to back using his lectures on physical geography (see also Elden/Mendieta 2001). More importantly and in relation to this, however, we find this awareness in Foucault's late genealogical work on the topic that the Greeks not only considered the dietetical aspects in a medical and an economical perspective, but also in what we would call, today, an ecological perspective:

> Q. So the Greeks were more concerned with health than with pleasure?
>
> M.F. Yes, about what the Greeks had to eat in order to be in good health, we have thousands of pages. And there are comparatively few things about what to do when you have sex with someone. Concerning food, it was the relation between the climate, the seasons,

> the humidity or dryness of the air and the dryness of the food, and so on. There are very few things about the way they had to cook it; much more about these qualities. It's not a cooking art; it's a matter of choosing.
>
> (Foucault 1982)

The kinds of choices and decisions, for the Greeks, happen in a way that is translatable to what we will now find in the term *okeios*.

In this chapter and to this point, I have been ruminating about various aspects regarding *oikos* – accepting against the general (and universalistic) attitude expressed in many corners of contemporary sociology, the power of equivocation –, as a rough sort of source code. Look at the issue from two angles that concern it as an assemblage in the context of *care, power, information*: These angles are specifically a discussion of the distinction between *oikonomia* and *chrematistike*, wherein I follow on the one hand a discussion created by Dierksmeier and Pirson (2009) in business ethics and George E. McCarthy (2009) in history of sociology, and, on the other hand, a discussion of the metabolic rift and the concept of *okeios* by Jason W. Moore (2011).

Dierksmeier and Pirson contrast Aristotle's theory of life, the academic reception of Aristotle through the lens provided by Thomas Aquinas, and the organization of market society with one another; whereas Moore's interest is in the fact that capitalism is an ecological regime presented in what he calls world-ecology, both lead to a question, how, at the very source code, the "life" in human life is thought of as constituted and qualified as related to other entities of all kinds, which I claim means that life in human life is qualified as "human" and the human is the being that is constituted through oikos. Thus, it's not homo oeconomicus but, if anything, *homo oikos*. Aside from this being a bit tongue-in-cheek, it's supposed to emphasize, once again, that any notion of economy does not exhaust the meaning of *oikos*, that *oikos* has a greater extension but also in various ways further intensifies as well as generates a longer running historic depth and evokes the long-standing colonial legacy in our imaginaries of our ways of caring, doing, and knowing. It's important to remember here, too, that leading political philosophers, such as Hannah Arendt, discussed the *oikos* alongside the notion of the *polis*: Together these concepts demarcate important political and, subsequently, social relations. While it has been mentioned that in ancient Greece the *polis* was a "males only" place, *oikos* as the household included women (as family) and slaves (as property) with little distinction. Far from simply excluding women and slaves from political considerations, this *assemblage* of *oikos-polis* deliberately regulates exclusion as a double mobility of inclusive-exclusion and silencing of Others. It is this first move of inclusive-exclusion of the oikos that sets up the inclusion-and-difference engine (described in *The Digital Coloniality of Power*) of contemporary Western society and its sociology (as the academic discipline). Sociology itself is (as I have begun to argue in *The Digital Coloniality of Power* (2015), following transformatively the redeployment of Steven Epstein's (2007) concept of inclusion-and-difference paradigm queered by critical ideas put forth by Walter Mignolo (2014), Syed Farid Alatas (see Chapter 10), and others) functioning as an *inclusive-exclusion-difference engine*, and its source code or operating system is based, I claim, on the complex history of derivation of the *assemblage* that is originally contained in the notion *oikos*.

Claude Levi-Strauss, in his famous *Triste Tropique*, observed the following – as Boaventura de Sousa Santos (2018: 61) reminds us – with regard to the emergence and purpose of writing:

> The only phenomenon with which writing has always been concomitant is the creation of cities and empires, that is, the integration of large numbers of individuals into a political system, and their grading into castes or classes. Such, at any rate, is the typical pattern of development [...]: It seems to have favored the exploitation of human beings rather than their enlightenment [....]. My hypothesis, if correct, would oblige us to recognize the fact the primary function of written communication is to facilitate slavery. The use of writing for disinterested purposes, and as a source of intellectual and aesthetic pleasure, is a secondary result, and more often than not it may even be turned into a means of strengthening, justifying or concealing the other.
>
> (Levi-Strauss 1973: 298/299)

It is striking how this passages clearly demarcates, what I elsewhere have identified as the deployment of a *Kulturtechnik* (cultural technique) as a *Herrschaftstechnik* (technique of domination). Furthermore, the Western geopolitical and Northern epistemological interlacing of *writing* both with domination and with the Enlightenment reverberates in the Northern (categorical) construction/norm/"epistemic limitation" of seeing as political agents only those capable of "civil agency" – an agency equated with possession of a cultural capital in the form of the cultural technique of writing. This is precisely the critical point that can be derived in the wake of the work of Partha Chatterjee, namely that civil society and political society are no identical or co-extensive, i.e. that, for example, illiterate peasants such as in India are indeed political agents and have proven themselves to be so.

> The question becomes one of change and power.

However, we must be careful not to revert to a simplified "causal model of power". Detel (2005: 11ff.) identifies the problem of the realist discussion on power that Foucault attempted to avoid in a 1950s conflation of "power to" with "power over", spurred on by behaviourist thinking. While it spawned interesting alternatives, which do not necessarily have to fall into the trap of conceptualizing power as repressive and/or coercive, it is Foucault's relationality in his concept of power that lifts him above the realist conceptions. Foucault's concept of power incorporates his understanding of how people place themselves in the world and develop a relation of how they experience their relations to the world and themselves. For this, the word "human" has become such an important point of reference in terms of Western epistemology and flowing and shaping – geopolitically (Mignolo) – from the Global North, along those lines, it is precisely not universalistic or holistic but still comprehensive and thus underscoring what Moore calls world-ecology. While Foucault did not go as far as developing an ecological program that was environmentalistic, i.e. the notions of *envrionmentalité*, *ecogovernmentality*, or – *pace* Lemke (2015) – *government of things* were not seriously developed by him[26]; this is precisely why we can consider that the term "ecology" is not otherwise too conceptually contaminated in Foucault's conceptual realm and we can happily see what it does if brought in on a conversation between his work and that of Moore and others.

A relation between *oikos* and *oikesis* has been remarked upon, among others, by Derrida. But only Derrida does prominently refer to *oikesis* as "tomb" – instead of using the word *tymbos*. Referring to the letter "a" in *différance*, Derrida writes,

> let me recall that this discreet graphic intervention, which neither primarily nor simply aims to shock the reader or the grammarian, came to be formulated in the course of a written investigation of a question about writing. Now it happens, I would say in effect, that this graphic difference (a instead of e), this marked difference between two apparently vocal notations, between two vowels, remains purely graphic: it is read, or it is written, but it cannot be heard. It cannot be apprehended in speech, and we will see why it also bypasses the order of apprehension in general. It is offered by a mute mark, by a tacit monument, I would even say by a pyramid, thinking not only of the form of the letter when it is printed. as a capital, but also of the text in Hegel's Encyclopedia in which the body of the sign is compared to the Egyptian Pyramid. The a of différance, thus, is not heard; it remains silent, secret and discreet as a tomb: oikesis. And thereby let us anticipate the delineation of a site, the familial residence and tomb of the proper' in which is produced, by différance, the economy of death. This stone (provided that one knows how to decipher its inscription) is not far from announcing the death of the tyrant. And it is a tomb that cannot even be made to resonate.

But others refer to *oikesis* in a different manner. One can easily find etymological lexica and dictionaries that define oikesis as "The establishment of an organism in a new habitat; accidental dispersal and establishment in a new area." or refer to it as "the act of dwelling", even find it referred to as "(e)migration" as well as find it in its adjective form as a pertaining or relating to the nature of a place; in this latter aspect, it enters the territory of the word *oikeiosis* (see on different accounts: Depew 2016). The comments to the context of the German translation of Aristotle's *oikonomia* refer to *oikesis* as *Hauswesen* as it appears in Plato's context: The *Hauswesen* here does not mean so much the "essence [*Wesen*]" of a house, as it does its coverage, range, affairs, and busy-ness: The household in its enaction, which in the mentioned text is claimed to be seen, with Plato, as an analogue in terms of its governance of both a large household and small city; but therein then Plato – as the comment suggests – sees an "analogy of essence" (*Wesensgleichheit*) in the administration of house and city that will *become the target of Aristotle's polemics* (ibid.)

Eric Nelson (2015: 134) will refer to Hippocrates' household (*okesis*) to extend to include the actively practicing members of his "medical community":

> Hippocrates, his sons, and his students work together to combat the plague as members of an *oikesis*, or household that Hippocrates maintained [...] Hippocrates' Thessalian *oikesis* is not simply a residence: it serves as a kind of *locus Hippocraticus*, a place where medical knowledge is received, processed, and dispensed. This broader functionality may be implied by the use of the term *oikesis*, which can imply the conditions of residency in general (...), a household company where administrative functions take place (...) or even indicate the activities of administration itself(...). Hippocrates household, perhaps like that of Herophilus, is a medical community.
>
> (Ibid.)

Hans Blumenberg assigns to it the double-meaning meaning of *Gehäuse* in reference to the Platonian cave: an enclosure but also a "possible dwelling among others" (1989: 187ff.). The choice of the word *Gehäuse* is interesting: Max Weber's famous evocation or "traveling idea" (Baehr 2001) of a *stahlhartes Gehäuse*,[27] which means a steel-hard enclosure but which under the pressure of the publishing house Unwin in the 1930s had been translated into English as "iron cage" (see Baehr's 2001 summary of discontents and with regard to the genealogy of the translation corrected by Scaff 2005, 2012; Gerhardt 2007; Stingl 2008/2010). But the Weberian formula echoes on in Heidegger's notion of the *Gestell* ("enframing").[28] The jump to Weber's friend Simmel from here is a short one, too, given Simmel's rarely acknowledged influence on Heidegger (Gawoll 1993; Jalbert 2003; Levine/Silver Introduction to Simmel 2015). Simmel used the term *Gehäuse* in reference to "culture" in relation to fragile, yet interactive, social relations that produce the *crystalline concretizations of social interaction* (with reference to David Frisby's erudite work on Simmel see Mieg et al. 281). The term pops up in the work of Karl Jaspers, and it is likely that both Simmel and Weber – with whom Jaspers was acquainted each – are the source. The "concretization" of oikesis in the *Gehäuse*, the "enclosing", and "enframing" par excellence are decidedly urban and/or metropolitan and, perhaps, harking also back to Derrida, if we follow Blumenberg's (1989: 76ff.) discourse on a speleological anthropology: "Two cultural phenomena give expression to the biological-physiological need for regression: The tomb and the city. The city is the repetition of the cave by other means" (76; my translation).

Here, urban normativity is reconstructed by Blumenberg in explicit reference to Simmel, because the *work on capturing reality in the life form that is the city allows for higher degrees of abstraction, a work requiring symbolic representation of wares and values* (ibid.). The means for this is, unsurprisingly in the evocation of Simmel's discourse, money. The "steel-hardness" of the city, in comparison to the cave, is that it not only encloses but also closes off creative imagination: the now urban-normative imaginary is restrictive and constraining, where the cave was enabling and empowering:

> [B]ut in the cave one could dream of what was not reality within it; the city does not allow this anymore. [...] In that fashion the city no longer requires magic; quite on the contrary it is the place of its depotentiation, a prospect harbored since the Enlightenment. For in it, realities are accounted [*verrechnet:* algorithmically processed]; illusions not suffered, presumptive profits not rendered into discounting. Abstraction is the surrogate for magic. In yet other words again: The cave was the anticipation of the city. In it, one would train for the other, regardless of the number of steps that had to be taken in between. Money is the symptom for the notion that the drawings on the cave walls are no longer needed.
>
> (77; my translation)

Besides "markets", the city offers, instead, "archives and land-surveying offices" (ibid.) that secure possession of property: "Eventually", Blumenberg concludes, "the administrative function of the city would supersede and replace the market function".

Furthermore, Blumenberg remarks on the city's relation to "luck" and refers to the famous sayings of "making your luck in the city" and the required abandoning of one's rural home, which reifies the rural-urban divide. But more importantly, and with direct reference to Simmel and the functional differences he identifies in the city, he calls the city an "acoustic cave": It does not require, he states (80), walls to shield itself; it has other means of doing so and what it shields itself against, according to Blumenberg, is of ultimate interest – he describes the city as the ultimate, autopoietic (yet, with a nod to Heinz von Foerster, quite trivial) machinery: The city shields itself against all that which "it does not produce itself or recognizes as its reality." The city's unbearable "noise" is the shield that encloses everyone into her own acoustic cave, thereby dulling the sense and privileges that can be seen: urban-normativity in terms of Blumenberg's Simmel is ocular-centrism par excellence.

Notes

1 Intranslatable: it describes a word/concept/metaphor deployed to provoke political opponents; it makes the bifurcation into antagonistic camps immediately explicit; and it elicits an affective reaction of immediate resistance to or approval of what it stands for.

2 There are, of course, different kinds of exceptionalism possible. One kind of exceptionalism is open-minded, accepting that every culture is different and thus exceptional, and each one of this can be an example for others in a positive way, but that doesn't necessarily imply a hierarchy. Another form of exceptionalism considers one's own culture automatically the best and greatest – in religious terms, perhaps even chosen and privileged by a higher entity – which other cultures should admire, aspire to be, perhaps even copy as best they can, and maybe even accept as their natural leaders. The latter is not for nothing called *imperialism* (or worse).

3 The opening scene is, of course, worth watching. And in his explanation, why America is not the greatest country, the character continues to argue – as written by Aaron Sorkin – that it once was, citing a great number of social, political, scientific, and other achievements. But what is also striking in this account of America's past achievements is the obvious male privilege and male-centrism in Sorkin's text:

> [...We] explored the universe, cured diseases, and we cultivated the worlds greatest artists *and* the world's greatest economy. We reached for the stars, acted like men [...]. We were able to be all things and do all these things, because we were informed...by great men, men who were revered [...].
> (2012 HBO/Warner)

In American exceptionalism as a lived ideology or as a nostalgic sentiment (all intentions, even good intentions aside), it seems that there is an inherently masculine-privileging inherent to it.

But Sorkin is in "good company", so to speak, as many philosophers of the Left also "suffer" a kind of amnesia or selective blindness when it comes to the suffering or the contribution of women, as for example Rosi Braidotti and others remind us, when emphasizing that this forgetfulness that one finds as common among many of the intellectual fore-thinkers of the present Left – the Left is, thus, thinking with its fore-skin, so to speak – has a "grand tradition". We find this, as Nicole Miller (2006: 20) states, quite exemplary in Agamben.

4 With Karl Jaspers (arguing, with Max Weber, against Hobbes), I understand democracy as an ongoing process, not as an existing state (of freedom).

5 Following Foucault, the point is that it is not continuity, consistency, or adequacy that we seek in tracing the patterning effects of *oikos* as an *assemblage* in our political imaginaries, but that we are looking for the analogical *coherences*.

6 Of course, I am simplifying the timeline and historical crossings here.

7 An accessible version of the following can be found in Holmes's interview with Richard Marshall, originally for *3AM Magazine*, now to be found here: https://316am.site123.me/articles/philosophical-frontiers-of-ancient-science?

c=end-times-archive [the story surrounding Marshall taking all the interviews of his long-running series from *3AM*, following 3AM's editor retracting one interview deemed controversial, is documented on various philosophy news, including here: http://dailynous.com/2019/03/20/richard-marshall-resigns-3am-magazine/].

8 I presently consider as uncreative those misunderstandings that are usually the result of deliberately wishing not to understand nor even engage seriously with an argument, as well as mindlessness. Uncreative misunderstandings, in science and scholarship, are often a source of "undone science", as well as a typical tactic deployed by White Collar Sociologists, which will surprise no one at this point.

9 An interesting discussion of "dialectics" could be made here out of the meanings of "property". Property can be space or a thing that someone owns, but properties can also be the qualities that define someone. It is not without importance to consider how these two meanings are related.

10 As the colloquial expression goes, see: www.urbandictionary.com/define.php?term=make%20you%20my%20bitch (last accessed: March 14, 2019)

11 Or several other bibliographical years, since this quote is reused by him.

12 I have previously and frequently complained about a tendency in sociology to confuse *explanans* and *explanadum*.

13 On the archive today, see Chapter 3 in Stingl (2015).

14 Historians, especially those specializing in "classicists" regarding texts, tend to read Xenophon's work *seriously*; in the wake of particularly Leo Strauss, an *ironic* account of Xenophon has emerged, however, as well. Nelsestuen (2017: 76f. Footnote 7) presents a quite informative bibliography on these different accounts and mutual responses.

15 A rather common title at the time.

16 Some countries especially so: We should, thus, speak not of the United States but the Anocratic States of (North)America (ASA).

17 A good "exercise" of what is meant by "unseeing" can be found in China Mieville's novel *The City and the City* (2008), equally well-addressed as the issue of knowledge bias in Neal Asher's *The Technician* (2011)

18 Let us stay, for a while, "usefully vague". Something is, in my *Nü-Pragmatist* view, "usefully vague" when vagueness trumps the danger to commit the fallacy of misplaced concreteness (or reification).

19 Nelsestuen refers to the (differential) comparison in Xenophon's works to the Persian Empire, noting that the Greek *oikos* goes a step further in installing a kind of military-industry complex: "Unlike the Persian Empire, then, production and security are common concerns of both spheres of function for the oikos" (85).

20 An issue that I have learned the most about from Sabrina M. Weiss during our collaboration (2011–2016).

21 Says Elden (2013: 21):

> Foucault warns us that genealogists will never confuse themselves with a search for origins. It is for this reason that we cannot simply find a birth of territory, a singular moment, which could be outlined and its lineages traced backward.

For a more metaphysical reading of *khora* in Plato, see Brisson (2016).

22 With reference to agriculture and land extending to land factors, see Ault (2007).

23 To emphasize this point I am opposed to any and all forms of slavery. I am making the case that there are historically different forms of slavery, and that it is important to distinguish these forms, precisely to understand forms of violence and suffering that prevail after slavery is officially abolished.

24 I disagree with Turner (2011) in calling Radkau's dismissive position on Nietzsche and Weber "revisionist, perhaps in the extreme" against what he thinks is a consensus in Weber scholarship. Again, given Schluchter's long-held skepticism (1996), I would not consider this issue as clear a matter as Turner makes it out to be, given that some of Turner's most valuable "witnesses" are (Fleischmann and) Aron in the 1960s and Stephen Kent in the 1980s. Schluchter, with whom I do not fully agree on a variety of issues, is one of the co-editors of the Max Weber-*Gesamtausgabe* (Complete Edition) and must be considered a leading expert on Weber. Regardless if one agrees with him or not, he deserves to be taken seriously, when he says that Weber was *not a Nietzschean social scientist* (1996: 185, my translation). [Schluchter is not cited in the 2011 paper, and the sole German publications are by W. Mommsen and E. Baumgarten's].

25 Not untypical for translational approaches.

26 Of course, others have developed it, subscribing like Arjun Agrawal (2005), to Foucault's idea of "permanent provocation". Others such as Cepek (2011) question critically how useful such a concept would actually be at all.

27 I am grateful to Donald Levine for having pointed this out to me several years ago.

28 Ibid.

4 Remanence of the *Oikos* in the FQu Economy and Digital Coloniality

Dīvide et Īmpera? Divide, Introduce "Scarcity", and Then You Are Soooooooo Governed.

In the conclusion to *Worlds of ScienceCraft* (2014), titled "Harder Better Faster Stronger" inspired by a song by *Daft Punk*, Sabrina M. Weiss and I identified four key issues that Science, Technology, and Society Studies (STS), Sociology, and Philosophy of Science have to tackle in order to generate real progress. One issue is that innovation is not the same as invention, which we argued is related to the fact that efficiency is not the same as effectivity. Similar trajectories could be found since we wrote our piece, of course. Summarizing the likes of "Bifo" Berardi's and Maurizio Lazzarato's diagnoses of the crisis of subjectivity under capitalism, McKenzie Wark writes:

> Capitalism launches new subjectivities like new model iPhones, only these days the subjectivities are all basically just bloatware versions of the same model – just like the iPhones. We are all supposed to be entrepreneurs of the self, an impossible task, leading to depression – as Franco Berardi might concur. This version of capitalism – if that is still what it is – no longer promises to be the 'knowledge society.' All it offers is just debt servitude and lottery tickets. Contrary to the slogans repeated over and over, there's not much 'innovative' or 'creative' about it.
>
> (Wark 2015)

This kind of sentiment that Wark deconstructs here is indicative of the operating system of contemporary societies and global economy, wherein "innovation" and the "economic productivity" and "growth" it leads to, along with an inherent methodological individualism, are established in a global paradigm of *extractivism* (Ye et al. 2019), which, ultimately, leads to the further (if not ultimate) precarization of the "units" methodologically individualized: regardless, even, of whether the unit in question is alive, for example a human individual, a cow, a bacterium, or an ecological niche. Everything is a unit that can be individually subject to *value capture*, "captured", and used "more efficiently".

Representation: From Physics to the Body Politic

A major tenet or touchstone of the political organization, political economy and imaginary of the Global North, its colonizing past, contemporary neocolonialism, and digital coloniality of power resides in the way it is based on an underlying *operating system* that both "implicates and intensifies"[1] the body as *technics*, as poetics and politics[2] – especially in terms of representation. Representation is itself, of course, a loaded and ambiguous concept – a concept that in its various deployments has both caused and experienced a lot of crises: From the *linguistic turn* in philosophy, the crisis of representation in ethnology/anthropology and subsequent *ontological turns* to the crisis of democracy, and the current state of post-democracy, these crises are, at the same time, distinct uses of the concept *and yet* interconnected.[3]

The body and representation have come to operate as a *technics* that can be understood as a kind of *immunopolitics* from the works of Roberto Esposito (2011), Ed Cohen (2009), and Vanessa Lemm (2013). Lemm further elaborates this discourse while engaging with the works of Friedrich Nietzsche as an archeology of *Einverleibung* (incorporation/embodiment) and *Aneignung* (appropriation) as re-presentation.

To situate this discussion in the context of sociology it makes sense to recall that along with the philosophical imaginary, the social sciences "proper", *coloniality* (*of power and Being*), and imperialism are mutually imbricated and thus implicated in the neoliberal, capitalist enterprise, and extractivism as a mindset, a psyche formed in urban normativity as *Euromodernity*: In other words, representation allows comparison and vice versa. Hence, representation is both situated in *and* situates discourses of identity and difference on the macro-, micro-, and meso-levels, determining *or* regulating how the relevant unit boundaries are negotiated. With regard to sociology, our most important Euromodern interlocutor in this context may be, unsurprisingly, Max Weber. But whose Weber is it? Or rather what type of accounts of Weber should be considered as effectually present in the discourse on representation? The question of determination *or* regulation of boundaries emerges, I would think, in terms of the question of a Nietzschean versus a Kantian account of Weber, and a Nietzschean Weber would then likely be a strongly *decisionist*, almost Schmittian Weber. This is a complicated proposition: Since I appreciate the difficulty of actually finding Nietzsche impressed on Weber, I consider the proposition – made for example by McCormick (1997: 83ff., also: 1995; Cerella 2016) - that there was a genealogical connection between Nietzsche and Schmitt which comes from Schmitt's debt to Weber, difficult or, rather, untenable.[4] Again, Wolfgang Schluchter proposes to slip between the horns of the Nietzschean and the Kantian account – although sliding down the curvature of the Kantian horn – in favor of a methodological individualism that he identifies in Weber. It is this methodological individualism that is at the heart of the problem here: On the one hand, I would like to agree with Schluchter (2000), who recognizes that there are versions of individualism and methodological individualism that are problematic at best, corrupt and destructive at worst, which leads him to propose that there is a version of methodological individualism, which he suggests is Weber's actual methodological (pro)position, that can not only escape the various "dark sides" of individualistic approaches but is actually – in terms of a *Verantwortungsethik* (*ethics of responsibility*) – a positive contribution from a social-moral perspective. On the other hand, I don't see how even a "weak" individualism once it is deployed as a methodological individualism is not ultimately bound to end up in antagonism (see Stingl 2015). Methodological individualism cannot but draw boundaries between units that reify differences and identity as mutual exclusive: Even if identities – of individual persons, groups, communities, nations – are never fully achieved, they become the constitutive, determinative goal people strive for and compare others by. In methodological terms for Weber this would refer, of course, to the *ideal-type*; as can be seen, while this is not adopted from Nietzsche into Weber, it still resonates with a possible account of incorporation in Nietzsche, wherein "*Einverleibung* is understood as a process of life through which ever more powerful wholes (*Ganzheiten*) are constituted and preserved by the annihilating and excluding incorporation of the other" (Lemm 2013: 4).

This is a point that Lisa Lowe (2005) and following her argument Hong and Ferguson (introduction to Hong/Ferguson 2011) have made against Weberian sociology while reading it through Foucault's heterotopia:

> ideal-typical construction of 'pure rational action' presumed the individual within the context of modern western industrial society, and measured the different degrees of rationalization as 'deviations', by comparing concrete social instances to this normative regulatory type.
>
> [...]
>
> To study both the history of the modern apparatus of unitary formal theoretical discourse, and the 'entanglements' or 'imbrications' that the discourse has classified and suppressed, would entail excavating cultural geographies of encounter and mixture, not as 'origins' of singular subjects, but as epistemes, as ways of inhabiting and knowing various divisions of the social at given times.
>
> (Lowe 2005: 410/411, 412)

As Ann Stoler or Peter Redfield did, an argument I follow in my previous book (Stingl 2015), Hong and Ferguson (2010: 5) find Foucault's work commendable in general and yet lacking in identifying the colonial mechanisms that construct, for example, the materialities of race and racism. In an effort that I would describe as deeply decolonial or as delinking at the very least in spirit if not already in practice Hong and Ferguson argue that the deployment of *difference* cannot be executed in terms of boundary-making that merely establishes "pluralist" modes wherein individuals and cultures co-exist in parallel spheres that can then be potentially suspended or entirely negated through processes that Achille Mbembe has analyzed in the term *necropolitics*. Multiculturalism – although I would add here that this refers specifically to Jeffrey Alexander's *multiculturalism mode I* (Alexander 2001, 2006; Stingl in Weiss et al. 2014) – can thus be identified in their (and their collaborators') analysis as a "mobilization by the neoliberal formation":

> Contemporary regimes of power naturalize brutal racialized, gendered, and sexualized violence, labor exploitation, and the rendering of subjects as redundant and disposable by creating new, nominally nonracialized categories of privilege and stigma, or, in other words, valuation and devaluation.
>
> (Hong/Ferguson 2011: 17)

In my previous book (2015) and elsewhere, I argued in similar terms that mobilization by neoliberal formation presents as the Janus-face of what Steven Epstein (2007) calls *inclusion-and-difference-paradigms*. In short, there are categories of inclusion created that produce new types of difference (Hong/Ferguson: "new ideal types"), wherein only certain forms of difference are allowed, certain forms of exclusion are acknowledged, and those who don't fit are more than "just" excluded but are, as *radical* Others, silenced, i.e. they are denied even any voice to make explicit that they are (not) being included.[5]

While I consider, therefore, that the *ideal-type* describes epistemic regimes through which Western identity-formation and Global Northern 20th-century social sciences have operated, once we attempt to delink from this view, without, however, ignoring the historical (and hysteretic) presence of *the West* and factual existence of *the Global North*, we should consider something like a *liquid type* (see also Stingl/Weiss 2014, 2015): Thus, we should consider what a *liquid type* could mean, by allowing, for example, for the possibility of equivocations[6]; we are contrasting ours, the liquid type, with the idea that the *ideal-type* describes a type that is a kind of specific abstraction, as Weber notes:

> An ideal type is formed by the one-sided accentuation of one or more points of view and by the synthesis of a great many diffuse, discrete, more or less present and occasionally absent concrete individual phenomena, which are arranged according to those one-sidedly emphasized viewpoints into a unified analytical construct.

With *liquid type* Weiss and I had originally intended (2014) an ontological fluency rather than an epistemological one: something that exists but can take on more than one form in many worlds. It is a (hopeful) complement to the ideal-type. I think there exist real tensions that can be clearly explained because in a given context, we encounter a though-image (Weber: *Gedankenbild*) people have in the form of ideal-type and unmatched *liquid types*. The political economy of the neoliberal labor market – with its extreme individualization of people and skill sets – and the real and actual people that look for and occupy jobs are interesting examples.[7] This has led to a political economy of constant firing and quitting.

The Political Economy of the FQu Labor Market (Fire and Quit)

The political economy considered in this section, we can call the FQu labor market – FQu is the short form of Fire and Quit. FQu-dness bears a deep irony, for it connects, at the same time, to the New Spirit of Capitalism that demands commitment (Boltanski/Chiapello; here, though: Chiapello/Fairclough 2002):

> The 'spirit of capitalism' is the ideology that justifies people's commitment to capitalism, and which renders this commitment attractive. It is a necessary construct because in many ways, capitalism is an absurd system: wage-earners have lost ownership of the fruits of their labour as well as any hope of ever working other than as someone else's subordinate. As for capitalists, they find themselves chained to a never-ending and insatiable process. For both of these protagonists, being part of the process of capitalism is remarkably lacking in justification. Capitalistic accumulation requires commitment from many people, although few have any real chances of making a substantial profit. [...]
> (Chiapello/Fairclough 2002: 186/187)

Thus, the "ironic" aspect of the interconnected intensification between FQu (firing/quitting), aka the intensifying mutual disposability of occupancy of a job on the side of both employer and employee, and the commitment to the new capitalist economy constitutes yet another case of what I call *implification* (the paradoxical situation when the implication of x is accompanied by an intensification of x). While we are, especially from within academia, *implicating* the neoliberal stress on individualization, we see it increasingly *intensifying*, including in academia (pretty much like an ideology in the Althusserian sense): Implication produces all kinds of intensification of "individual" by means of the pedagogy of the individual (all individuals, yet all the same) via a possessive individualism where "possession" as drawing the lines around the *oikos* as "oikos of one" is the demarcating factor. What we do not, however, encounter is a *psychagogé* of the individual (Foucault), where the individual is truly both individually and together with others (on failures, see Sykes 2017; McComack 2012). Similarly, Christina Scharff (2016) reconstructs the pedagogy of the individual in the psychic life of the entrepreneurial self through its exclusionary "assemblages and arrangements" of producing subjectivities. She notes further about the inherent temporalities (see also Adams et al. 2009) that

> [a]ccording to du Gay (1996: 193), there is a real sense in the world of enterprise that 'one is always at it'. This sense featured in accounts that emphasized the importance of being active [...].
>
> [...]
>
> The emphasis on being active ties in with a neoliberal philosophy of time where being idle is to be avoided [...].
>
> [...]
>
> The entrepreneurial self orients to time with a view to making the best use of it. The resulting constant activity means that there is also a feeling of a lack of time.
>
> (112)

I think this activity characterizes not entrepreneurial selves per se, but those that live under the regime of the *precariate*, and that it is an imaginary/phantasmatic coping mechanism: In other words, I think that people in precarious positions, especially in academia, send out CVs and engage in a large number of activities, even if they do not realistically invest as much time as they may claim to do, but as long as they can maintain the narrative (not only the fiction!) that they spend all this time, they can compensate even their lack of success or lack of reaching a goal – such as job security and full time income – to actually become in fact a member of the "middle class". Only this way can they phantasmatically cope with the *myth of lurking chaos* (Boesch 2002), i.e. the idea that merit in a supposedly merit-based society does not matter as long as they behave in a way that the narrative structure they do live can be made to correspond to the narrative of society and its mode of subjectivation (including in terms, not of time but temporalities, i.e. it doesn't matter how much time one spends on applications, qualifying activities, etc., as long as one allocates time-slots for them in a proper way that *arranges* them with other activities).

If it is true that there is a shift from the liberal notion of subjectivity as owning oneself, as Scharff (112/113) quotes Gershon and Rose, to the neoliberal one where one does "see oneself" as business, then I would consider this as a regressive step: A business, where the individual subject herself/himself *is* that business, is nothing but an *oikos* that owns property (material, living, or otherwise) – either property as other businesses, as owned or as "feudalized" peasants/animals in the sense of "I own you. You're my bitch!", to phrase it in the words of Kevin Spacey's character Dave Harken to Nick Hendricks (played by Jason Bateman) in *Horrible Bosses* (Gordon 2011). A bit unlike Bröckling, whom Scharff also cites, I would further clarify that the notion of the *entrepreneurial self* not only means that this subject is always in a "constant mode of becoming" (ibid.) but must be refined further under the risky subject concept (Patel 2006), as it applies to Late Liberalism and Near Liberalism, and more to the "possibility of risk" (Amoore 2013). The point about risk is, of course, not that something *will* happen, but that something *could* happen (with some probability) as Ulrich Beck famously outlined in his book *Risk Society* (1986). Business and governments have integrated the idea of risk, and – more importantly – quantified and financialized risk as well. The *politics of risk itself* have, however, become much more refined and escalated, as Amoore illustrates, and established a "politics of possibility" wherein state power and corporate power – through technologies of anticipation and preemption (inspired by insurance and casino algorithms) – have combined together to provide a new *security infrastructure*, which is yet another expression of coloniality as the data-ification necessary to enable this infrastructure spreads into other areas of life and re-forms our psychic life.

My point here is not to refute, criticize, or profoundly modify Christina Scharff's account of her empirical material or theoretical approach; instead, I merely seek to complement it, in an area I think it can be complemented, by including the notion of *precarity* (which is mentioned but not unfolded by Scharff). I do so precisely because Scharff rests her account on her understanding of Judith Butler's approach to "psychic life" to which I propose we must add the notion of precarity because Judith Butler, since her work on "psychic life", has not only unfolded "precarity" much further but also became part of the much wider discourse on precarity, for example in relation to affect theory among others.

There are several aspects to Scharff's argumentation, which, I think, emphasize the antagonistic nature of *the Differend* that the neoliberalism she describes is part of, and which infuses individuals as well as macro-levels. She writes,

> In the interviews, the research participants also established boundaries, between different kinds of individuals, particularly between 'lazy' and 'hard-working' people
>
> (Ibid.: 12)

But isn't it that rhetoric is also strangely familiar in view of the Greek/EU-crisis of 2015? Here, entire nations – in tactics reminiscent of the late 19th and early 20th century – were re-imagined as collective personalities: The tabloid portrayal was that Greeks were "lazy" and Germans were the "hard-working" people who were asked to pay up for the "lazy" ones, while simultaneously among the so-called "institutions" and committees, the topologies of economic critical realism (e.g. Yanis Varoufakis) clashed with the geometric imaginaries and maps of technocratic path-dependencies (e.g. Jeroen Dijsselbloem).

Both positions (realist and elite technocrat), however, are strangely connected in a shared, and perhaps mistaken, premise. Not so much because – paradoxically and despite their supranationality in the EU and Eurozone – these different interlocutors still consider states as nationally sovereign states. What is really the point is that they still consider these states as actors.

But therein lies a problem, as states are not so much actors (let alone are, in my view, only nation-states "states") but are more complex. This can take different forms conceptually such as in Passoth/Rowland's (2010) analysis of *actor-network states.*

If we consider the notion of "seeing like a state" (Scott 1999), then both the positions of our two vignettes Varoufakis and Dijsselbloem are attempting to see like a state that is imagined to be an actor, an individual actor that is constructed to have a *collective psychic life* of an entrepreneurial self of Neo-, Late, and/or Near-Liberalism. The state is, if imagined as an actor, indeed a risky subject. It becomes precarious because imagining and performing the state as an actor do not correspond to the more complex nature of statehood. While Passoth/Rowland and other theorists and researchers expand the concept of the state in terms of new theory developments such as actor-network theory (ANT), I go in a different direction, perhaps even one step further, arguing that statehood is a very old phenomenon that has never been merely national or an actor; on the contrary the techno-imaginary of the national actor-state was a late (Euromodern) phantasm that could only work as the sole "for of statehood" on the surface level for a short amount of time, an intermediate period. As both older and newer forms of statehood other than nation-states and mere actors express themselves quite forcefully again, we must learn (and relearn) how "the political" works. Neoliberal technocrats – whether their governance regime is economic, accounting, or administrative – are quite ill-equipped to do so. They are, actually, an institutional relic of the 19th century and a discourse relic of the colonial era, embodied exemplary in the East India Company, where

> racialized and classed subjects tend to be positioned as the 'other'. Thus, the exclusionary dynamics of entrepreneurial subjectivity do not only extend to the positionalities required to become an entrepreneurial subject, but also to the kinds of subjects that are othered in this process of becoming.
>
> [...]
>
> [...E]xclusions under neoliberalism do not only relate to the kinds of subjects that do not have the resources to become entrepreneurial. If the constitution of entrepreneurial subjectivities involves othering, exclusionary processes may lie at the heart of neoliberalism.
>
> (Scharff 2015 online first: 13)

Transferred to the university and academia, the same logic applies:

> So what's so bad about leadership, service, and creativity? What's bad about them is that, as they're understood on campus and beyond, they are all encased in neoliberal assumptions. Neoliberalism, which dovetails perfectly with meritocracy, has generated a caste system: "winners and losers," "makers and takers," "the best and the brightest," the whole gospel of Ayn Rand and her Übermenschen. That's what "leadership" is finally about. There are leaders, and then there is everyone else: the led, presumably — the followers, the little people. Leaders get things done; leaders take command. When colleges promise to make their students leaders, they're telling them they're going to be in charge.
>
> [...]
>
> Which is really too bad, because the biggest challenges we face — climate change, resource depletion, the disappearance of work in the face of automation — will require nothing less than fundamental change, a new organization of society. If there was ever a time that we needed young people to imagine a different world, that time is now.
>
> (Deresiewicz 2015)

But people are, of course, imagining themselves to be political people, and that is especially true for many young people that attend university in order to "be educated", "to learn", and to be "made fit to join the labor market", etc. all of which is, inherently, pedagogy. This is, of course, a part of the *FQu politics*: It is not surprising, then, that the first reaction student activists, from Left to Right, have with an academic person (whether tenured professor, precarious adjunct, or academic administrator) is to demand this person be fired. It is also no accident that many White Collar professors who get tenure and are fairly safe from firing (unless, of course, their offense is genuinely as grave as bordering a crime) have no better response to these student other than to "quit" on them and on political arguments,[8] and to call them "infantile". The game played here is one of politics, but it is hardly political.

It is, however, striking that people, from all "political" and "ideological" fronts, will call for somebody being fired; they will act in a similar way when having an unsatisfactory shopping experience, or just caught a bad day when eating out in a restaurant. Not only do people, in asking for someone to be fired, say: I want to take away your (and, possibly your dependents) means to make a living. In the same stroke, they reify the importance of "having a job", asserting the worth of an entrepreneurial self that sells its services on a labor market, by calling for someone being fired; in essence, they play into the ideological premise of neoliberal psychic life.

Making "Freedom" and Participation qua Property: On "*Polis* is *Oikos* and Vice Versa", and Other Implicit Historical a priori of White Collared Social Science

Political scientist Jack Turner in 2008 concluded that there is an "elective affinity between [US] American individualism and structural injustice", which he sought to illustrate through an analysis of Tocqueville's descriptions of individualism, white supremacy, and gender domination in the "classic" *Democracy in America* (1835–1840). Such a take on Tocqueville is exceedingly rare across the history of the social sciences. In fact, Tocqueville's work on slavery and racism, especially the chapter from *Democracy in America* discussing the "three races" in [US] America, and the accompanying volume – *Marie Ou L'Esclavage Aux Etats-Unis* (1835) – written by his "traveling brother" Gustave de Beaumont, received close to no reception among social scientists before the 1960s and the US Civil Rights Movement, and even then the constitutive and complementary role of Beaumont's work for Tocqueville was largely disregarded, and Tocqueville's position was often interpreted and dismissed to suit particular ideological positions.[9] Having received my sociological training in Germany and been for a long time part of European and US-American "mainstream" sociological discourses, I neither actively knew anything about Beaumont's importance nor had I been under the impression that Tocqueville could play any role for my work aside from perhaps the odd marginal reference in a footnote. This perspective changed remarkably in the weeks before the final deadline for finishing the manuscript for this book upon attending a talk given by the always insightful Gurminder Bhambra, titled "Dispossession, Justice, and Social Science".[10] Bhambra's critical discourse on the social sciences verges in a similar direction as mine; admittedly, her work is both more eloquent and clear.[11] Bhambra reiterated her position for her audience (including me) that the subject matters and conceptualizations of sociology such as Modernity or the Modern State – considered to be constitutive – are products of complex, connected histories that include colonial pasts, which are, however, often subject to elision in sociology. But to understand more comprehensively what was done and from that point forward elaborate better what is to be done, sociology needs to take these histories "into serious account" and, of course, understand that it is implicated in the colonial and imperial violence on which Modernity and Modern Nation-States are founded, aka Modernity is a "Colonial Modern": this entails the issue of how the relation between societies and the nation-state constitutes and is co-constituted by the conceptualization of the "Modern subject", as Bhambra reiterated for her audience. This Modern (aka Euromodern) subject – in the discourse from Locke to Hegel – becomes the "subject capable of property" (see also Bhambra 2015; and introduction to Bhambra/Narayan 2017). Subsequently, three relational positions emerge: (1) the position of self-ownership as the realization of the Euromodern subject, wherein the possibility of owning one's self is predicated by the possibility of taking ownership of Others, namely (2) those "incapable of property", while also, however, (3) silencing, othering, and, thus, dispossessing those whose "property" is (radically) different and who, therefore, also do not or would not want to become Euromodern subjects. Bhambra concluded, in her talk, that, in consequence, "dispossession becomes central to the emergence of Modernity"; and with Saskia Sassen (2014) we can speculate on its centrality (in the form of "expulsion" – which now includes all "land factors") for the geopolitical maintenance of Euromodernity.

The work of Tocqueville and Beaumont (see here: Kohn 2002; Janara 2004; Welch 2007; Turner 2008; Kemple 2012; Noll 2014; Sunar 2014; Benson 2017; Krause 2017) is particularly impressive. Taken seriously, their work does implicate sociology and empire, and even predates, in one manner or other, ideas that I have attempted to give articulation to here – such as the importance of thinking through a differentiation of forms of slavery in order to understand how cruel and violent the white European Americans form of slavery was and that it held an underlying dynamic which (as they could speculate it would) persisted even after abolition and which still shapes much of today's US-American society. Schaub (1998: 622) summarizes as follows:

> As Tocqueville points out in discussing the differences between ancient and modern slavery,[12] the racial basis of modern slavery means that even after emancipation the former slave and his descendants bear an enduring mark of their past degradation. The freedmen still belong identifiably to the race of slaves, and that allows for mistreatment to continue under the guises of second-class citizenship, segregation, and exclusion from professions. After saying that "The black race is despised in America because it is a race of slaves," Daniel Nelson goes on to add:
>
> "it is hated because it aspires to liberty." That aspiration is expressed more among the freedmen in the North than among the slaves and, as a result, Northern racism is more virulent. Because there was no barrier in law to the social and political equality of blacks, public opinion alone upheld the separation of the races and the disfranchisement of the freedmen. Accordingly, public opinion was all the more assiduous in discerning traces of "hereditary ignominy."

Subsequently, Janara (2004) is among those who draw the conclusion that, along with Tocqueville and Beaumont, the link between US democracy and racism becomes clearer as she finds

> U.S. racism entwined by deep psycho- logical affinity with dynamics central to both a democratic social condition in general, and the particular European American postaristocratic, post- colonial democratic condition. (775)

Janara's essay is particularly insightful as she – like Bhambra – is one of the few who also takes explicit notice of the epistemological and practical differences in the injustice found among European Americans when it comes to both Black and indigenous people: In a sense, both regimes of injustice, toward Black people and towards indidigeneous people are different tragedies, but they are also both tragic in some ways that are similar and also tragedies that exist in an interdependent state.

Krause (2017) does emphasize that Tocqueville (Beaumont is not made into a central figure by her) manages to predate sociological tools of status and habitus analysis for explaining distinction and social reproduction. In Krause's account, Tocqueville appears as a precursor to Bourdieu, excavating the (racist) mechanisms that intensify existing social differences.

It was Cheryl Harris (1993) in her seminal study on "Whiteness as Property" who produced a thorough conceptual account in terms of legal history that not only linked the Critical Race Theory with Legal Realism (see on this Tillery 2008) but also genuinely connected the US-American (and thus Euro-American) racism with the legal conception of "property" and the narrative dialectics of the legal practice – similar to but different from the narrative dialectics of Settler Colonialism and the *property legal practices* regarding indigenous people and "their" land (qua *terra nullius*). Thinking from here on, along with Gurminder Bhambra, Cheryl Harris, and Lewis R. Gordon, I want to put my main focus on the issue of political participation in the *polis/oikos* conjunction. Lewis Gordon reminds us that

> For politics to exist, there must be discursive opposition. Such activity involves communicative possibilities that rely on the suspension of violent or repressive forces. In effect, that makes politics also a condition of appearance. To be political is to emerge, to appear, to exist. Colonization involves the elimination of discursive opposition between the dominant group and the subordinated group. A consequence of this is the elimination of speech (a fundamental activity of political life) with a trail of concomitant conditions of its possibility. It is not that colonized groups fail to speak. It is that their speaking lacks appearance; it is not transformed into speech
>
> (2011: 99/100)

There is, then, a connection between property, political participation (qua speaking), and appearing/existing in political history. In *Human Condition* (198), Hannah Arendt states that

> "Wherever you go, you will be a polis": these famous words became not merely the watchword of Greek colonization," they expressed the conviction that action and speech create a space between the participants which can find its proper location almost anytime and anywhere. It is the space of appearance in the widest sense of the word,

> namely, the space where I appear to others as others appear to me, where men exist not merely like other living or inanimate things but make their appearance explicitly.

Thus owning and, also, inheriting property (oikos) means that one is considered as being in a polis, one is already a citizen (with rights) and can exert political power over others – whom one colonizes in terms of recognizing them as citizens or disavowing them, i.e. constructing the other according to this episteme with real life effects through practices. But who can own property (*oikos*)? Not "every body", and only those who can own themselves (Euromodern subject). The 19th century complicates this marginally by the scientific theory of *energy* – of which the white/male-centric coloniality of power claimed neither Black people nor women[13] had enough to own and manage (*oikein*) property (*oikos*) or participate in public (*polis*) (see M. Norton Wise in Riskin 2007; also conclusion to my book *The Digital Coloniality of Power*) – and by the question of personhood of corporations, which is part of the story of dispossession of indigenous people in the United States.

There is, in other words, a knotting together in a vicious circle of "political participation" of (a) the modern subject qua self-ownership having the property of being capable of owning property, (b) owning property (*oikos*), (c) being able to speak freely in public (*polis*), and (d) thus appearing and existing.

On a side note, this extends to non-human and extra-human agents in the law as well, when we are dealing with the very ecosystems from the microbiome to the entire planet (or *Gaia*, if you prefer), as Yoriko Otomo notes[14]:

> The human and nature separate conceptually in order to interact dialectically, resulting in a split between economy (*oikos nomos*) and ecology (*oikos logos*): the law that applies to our dwelling seems to exist in opposition to the reason we dwell there at all.
>
> (Otomo/Humphreys 2014: 8)

This *knot* of "political participation" – in which self-ownership is so central – can be otherwise known as *Liberalism*, whereas Neoliberalism transforms the *Euromodern subject* into a *subjectivity* of self-entrepreneurship (within a state capitalism).

Now, *Liberalism* is a strange proposition, in light of this genealogy, because it evokes a number of associations that include the idea of human rights as much as they do neoliberal capitalism. But all these ideas that they evoke today (and which they have evoked some 50 or 60 years ago) are far-flung from the 18th- and 19th-century origins of *Liberalism* as a programmatic. Jürgen von Kempski in 1953 (in Kempski 1992, Vol. 2: 300–320) gave an excellent summary of the promises and discontents of *Liberalism* in light of its "modern" genealogy. Even with certain references to the political *Zeitgeist* in Germany at that time, some of the works of von Kempski are insightful and fresh and allow us to forge a few conclusions for the situation that, in particular, sociology finds itself in, with regard to the tradition of *Liberalism*.

> *Not only*, he says (300; my translation), *Christian-conservative and socialist democrats declare Liberalism obsolete, even the Neo-Liberals themselves declare the form that Liberalism has donned in the 19th century for hopelessly discredited*.

The double-problem, all of these political-ideological flavors deal with, or, rather, the double-solution that they witness for all other problems, is the idea of "Freedom" and the notion of *belonging* (*Bindung*[15]).

But "freedom" in this discourse is so clearly bound to "property" and reducing belonging, participation, and, by the way, provisioning to relations of ownership (which are inherently exclusionary, including their conceptualizations of goods [see *The Digital Coloniality of Power*]) What I argue for, however, with Sylvia Wynter in mind, is that the interrelated cognitive culture and organization of belonging, participation, and provisioning that currently and for the longest time – and especially within the (White Collared) social sciences – has existed in the form of *Cognitive Empire* (Santos 2018) must come to an end. Social sciences in particular must understand that there is (and has been) a Pluriverse (Escobar 2018) of a variety of cognitive cultures and organizations of belonging, participation, and provisioning.

Digital Culture's Filligree: It Ain't Finished Until They Can Monetize Poverty, Too

The cave, says Blumenberg (1989: 200; my translation), *that Aristotle took over is a lushly accessorized vault, a gallery of works of art and of refined taste*. This Aristotelian cave, as realized in Digital Culture, now demarcates a layer in the mytho-historiography of the *cave myth* that emasculates its own logic. It emasculates the creative power of what I have elsewhere called the *filigree* (Stingl 2015) that, rather than operating on fact and (right to) justification, favors promoting *ekphratic forces* of matters of *entitlement* and *truthiness*. It is an interesting irony that this is the era proclaimed as that of a global – or, actually, to speak with Ulrich Beck: globalistic – *Digital Culture*, wherein many had thought of the Internet (first of all in terms of a collective actor, and second) as a conveyor belt for "more democracy" and, as a consequence, of "more equality". And yet, looking at the numbers, the opposite seems to be true: With more *digitality/globalism* we did get more (of the "same old same old" kinds of) inequality; we got, for example, expansion of property rights, as well as more and very expensive extra-legal courts such as the International Centre for Settlement of Investment Disputes (ICSID, [Trakman 2012; Lhuilier 2016]). A number of these developments derived from the neo-liberal, capitalistic engine of the middle of last century, the military-industrial complex, which was instrumental in the creation of digital technologies themselves. There is an uncanny resemblance to the diagnosis of the effects of democracy in ancient Greece, if we follow Weber, as McCarthy puts forth quite nicely (even if I do not agree with his account of connecting Weber to Nietzsche):

> With the Peloponnesian War between Athens and Sparta (431–404 B.C.), there was further development toward democracy as military tactics and naval technology required more participation on the part of the general population. Payment for jury courts was instituted. The restrictions placed on the sale of land, properties, mortgages, and slaves were abolished. Weber is well aware of the historical irony that, with the implementation of democratic rule, there was a corresponding social transformation producing greater social and wealth differentiation and unlimited land accumulation. Democracy pursued its own demise by unleashing the power of capitalism through free trade. But it was also undermined by the growth of slavery during the classical period. Weber maintains that slavery and property concentration were the major factors in the fall in the urban deme population, drop in the hoplite army population, limited opportunities for free labor employment, decline in living standards and consumer demand, and restrictions on the development of a market economy. There was also no improvement in the technology and social organization of production in the "industrial capitalism" of antiquity.
>
> [...]
>
> The economy of land rents, mortgages, and slavery undermined the viability of ancient capitalism to develop its own industrial base and consumer market and to move beyond a rentier and state capitalism.
>
> (McCarthy 2002: 74, 75)

What has effectively changed between Ancient Athens and today is that slavery is not a "legal option" in the countries of the Global North, although *racism* and *disposability* remain. Other forms of *binding* forces to be held over and transform people's lives certainly exist alongside them with a variety of modes of "Othering" and "silencing" that contribute in terms of the *coloniality of power and Being*. Many lives have been rendered disposable through coloniality: the lives rendered disposable in the Settler colonial pursuit of native lands, the lives of Black people transported into slavery in America, and, recently with refeudalization, worker and social welfare lives (of often already "disadvantaged" people) in Late Liberalism. Furthermore, the category of *land factors* includes today more types of property than merely land (this expansion was already noted by Talcott

Parsons Marshall Lectures 1953 [1991; also Parsons/Smelser 1984/1956]), an expansion which we can understand archetypically as a pure form of *digital colonization*, turning tangible but marginally material or non-material factors into equivalents of *land factors*. Kean Birch, for example, has studied the new forms of rentiership that emerged (e.g. 2019)[16] in the form of *technoscientific capitalism* – a term coined by Jean-François Lyotard. In this respect, we can make a few adjustments to McCarthy's statement:

> The economy of (extendend) land factor rents, mortgages, precarization/disposability undermines the viability of contemporary capitalism to sustain its own industrial and resource base as well as a solvent consumer market and this has further reinstalled the *oikos*-system of rentier and state capitalism – otherwise called "neoliberalism".

In this line of a "neoliberal project" of a transformation of society, according to the rules of *oikos*, falls the idea that one of the finishing touches to the project is contained in the sentence: "It ain't finished until they can monetize poverty, too."

This *monetization of poverty* proceeds on the global/international level and on the "individual member of society" level. Konkel in a study of the World Bank (2014) has informed us about an important chapter in international institutional history. But especially in the United States and the United Kingdom, where the neoliberal mentality has reached its heights, there are policies in place where the plight of those suffering from inequality and injustice, those living in a precarious condition and under increasing precarization, is *rendered* profitable or re-rendered into rentiership. This, too, makes precarization itself a politics of surveillance, control, and discipline. The *assemblage* processes of monetization of poverty and precarization can be found in the ways and technologies that allow, above all, corporations to monetize poverty; see the use of ATMs and cash-checking and other dependencies that poor people and social welfare recipients experience when trying to gain access to their money[17], see the turning of schools and universities into profit centers, or see the ways that US prisons and Immigration and Customs Enforcement (ICE) detention have been privatized.

In this regard, I think, the Feminist science studies through the work of Donna Haraway allow us to view these current developments and others in terms of a spectrum between two poles – poles that each demarcate a dynamic process and relationality of learning and becoming, of generating an attitude towards the world and others, and of being *enabled/constrained*: This is the distinction between *rent-ability* and *response-ability*. A relation is *rentable* if one entity is rendered so that rent can be generated and drawn from it, whereas another entity is able to extract that rent. *Response-ability* (see e.g. Haraway 2016) is about a relational ethics of care in mutually capacitating one another to be able to respond and be responsive. Precarization and rent-abilization, however, are strategies of muting and silencing responses and the ability to respond. Rent-ability as a relation, however, not only renders the agency of those rendered into land factors mute but also constrains the agency of those who draw rent. It is the parasitic logic that the system and its direct agents (those who work for the logic of the system) that profit more than the rentiers themselves – they are more like addicts or subject to a parasite: Rentiers and investors are not themselves *locusts*, as a German politician once mused. They are, in a sense, victims of a (social) pathology. They, too, cannot give response. The logic of the system, neoliberal governance aka bureaucracy, has taken over their action and they are suffering a weakness of will (*akrasia*): *Bureaucrasia* (see Stingl/Weiss in Weiss et al. 2014), a *stählernes Gehäuse*,[18] indeed.

In contemporary capitalist neoliberal/neocolonial economy and its social scientific warrant qua economics, we find that there exists, what I would call the analogy to physics, the *Dark Matter of the Economy, which consists of unpaid work of Prosumers, of the so called grey and black labor, of blood resources (e.g. diamonds), of profits from (Sex) Trafficking, and of producing and dealing in cocaine, etc.* Just think about it: While financial capital makes up most of the global capital, all capital is still, at least theoretically, exchangeable at any moment for "substance", including cash, retaining if not its values, then its worth. The amount of "substance" is not equivalent, however, to the amount of financial money. This is, however, a problem with the capitalist universe, which is, yet, partially solved by capitalism's *dark matter*. This *dark matter* is an infrastructural problem.

Towards the end of this book I will introduce the infrastructure for BluesCollarship, a somewhat different approach to think with and about *infrastructure* than what most people associate with the word "infrastructure": an understanding that – while not forged in ephemeral idealisms – cannot be reduced to *ownable* "materialities" nor "patentable idea" but follows my understanding of tangibility as a genuine *Thirdness* complementing the material and the ephemeral, being neither reducible to either one yet also not independent from either. Infrastructure, in my approach, is compatible with Michael Mann's concept of infrastructural power, wherein Mann's concept of power has been deliberately infected or inoculated by a pharmakon-noetic Foucauldian parasite or symbiote.

But that does not, I think, free us of the need to consider different notions of infrastructure first; for even though my approach to what infrastructure is and does and how to *think with* it may differ from them, both the generally accepted common intuition of infrastructure as well as the important and influential work by sociologists and STS scholars – especially by Geoffrey Bowker and the late Susan Leigh Star, along with their collaborators – "do, have done, and will still do work", i.e. the work that concepts do. Many of the concepts discussed here "do work" – as Donna Haraway says, for example, about the concept of the Anthropocene, a concept which she seems not in favor of, but acknowledges its *mattering* existence. For the colonization of the mind in the "energetic" history of Liberalism and capitalism that establish Digital Culture, mired in *digital coloniality of power*, the concept of attention and its history are of exemplary importance, establishing an economy of attention, wherein attention is a kind of currency, as we *pay attention*.

Techno-Somatic Body Interrupted? The Political Economy of *Attention@Fatigue. Pathology*, the Embodied Mind, and the Cyborg Gaze

Think of this book and its chapters as conceptual-empirical laboratories that function as theoretical-particle accelerators, and which create collisions/collusions of new connections that enable windows of opportunity. Herein, I am motivated by a line of flight, which was proposed by Alfred North Whitehead, and which recent thinkers as diverse as Isabelle Stengers or Alberto Toscano have promoted, namely that "a society may be evaluated by the uses to which it puts abstraction" (Toscano 2008).

While taking Toscano's Whitehead seriously – specifically the importance of abstractive modes of communication across disciplines and fields – I will try to go "before and beyond" both the pejorative uses of abstraction and the Marxist-oriented notions of "abstraction". The notion of abstraction, I am interested, cannot be seen apart from what Michel Foucault has called *psychagogy* – which is precisely practical and political in the way Toscano demands – nor, simultaneously, from (the body in) embodiment.

Body, mind, and body-mind are three related and yet astonishingly disparate points of access for humanity's current techno-scientific condition – or at least the Western/Northern idea of humanity: Children develop a techno-somatic environment that affords them with aesthetics, economies, and politics, in short, techno-normativities that define normal/pathological performance in these ecologies. The emergence and transformation of the idea of "attention" and its pathologies are a testament to this historical process: From (post-)Enlightenment philosophy and experiment (such as in Kant, Schelling, Reil, Hoffmann, and Cuvier) to industrial fatigue research (Mosso, Kraepelin, Henderson), a techno-normative and biopolitical shift transformed the disciplinary techniques and milieus that immunized the body and the mind against fatigue: Attention was turned into a psychological and physiological technology of the self, which characterized the 20th century in terms of a mind-body split. In recent years, not only has the notion of the neuro-chemical self (Nikolas Rose) transformed the relations between body, mind, and technology, but more importantly, we are now forced to become "experts of our self", review, supervise, and deploy surveillance to our body-mind through a lens of cyborg gazing. This represents, simultaneously, a return and making-explicit of the psychosomatic experience and a radical transformation. We encounter the intertwined development of attention/fatigue and of techno-somatic ecology in modern history as a narrative dialectics shaping people in a psychosomatic process and shaping institutional practices in the development from medicalization to biomedicalization and the cyborg gaze (see *The Digital Coloniality of Power*). Contemporary society deploys a cyborg gaze – a novel techno-normative mode – that evinces the psychosomatic dimension of normalcy/pathologies while also reproducing the structures of fatigue research socially.

Part of my critique is that I do not think that Digital Culture and its media represent a (recent) rupture of any kind, but are merely an extension of much older forms of coloniality. I think Jonathan Sterne (2003: 368) was really on to something over a decade ago, when he opined that to

> refer to digital media as 'new' technologies is to import the value system of advertisement into scholarship, where 'newness' is itself an index of sociocultural significance and transformative power.

Of course, this is not my main concern in this chapter as I have elaborated on this in more detail in *The Digital Coloniality of Power* (2015). In this chapter, instead, I am more interested in what can be called the *contact zone* (Linke 2006) or *ethical plateau* (Fischer 2003) between the political economy and affective ecology of children experiencing the violence of being pathologized in techno-somatic worlds (forged in economies of relevance and ecologies of attention). Accordingly, Sterne also points to the importance of the study of technologies,[19] which is one of my main fields of inquiry, with regard to both questions of the body and the kind of reflexive humility and caution that, for example, Pierre Bourdieu proposed (Sterne 369), wherein the question of body/mind makes sense to be posed in relation to attention deficit hyperactivity disorder (ADHD) and

> the study of technology: it too is a strategic research site for thinking about the relationships between embodied experience, organized movement and the organization of society.

Kant

I begin our *adventure* (I channel here *adventure's* etymology as "that which happens by chance", "perilous undertaking", "wonder", "the risk of a loss", "the thing about to happen", "the chance we take"), perhaps somewhat surprisingly, i.e. in taking a chance and a risk, but perhaps in speaking perilously (*parrhesia*) to Kant.

While there is good evidence and reasonable argument that Kant must be understood as highly colonial thinker (Maldonado-Torres 2007; Mignolo 2011) – for example, through his pivotal role in the emergence, the pervasiveness, and the continuity of race/racism in biology and modern geography, two disciplines that originate as academic disciplines historically during Kant's later years – there are also windows of opportunity in Kant's writings that (would have) allow(ed) for strategies of epistemic disobedience, i.e. of thinking *and* living otherwise. Kant, I argue, must be provincialized (Chakrabarty 2008), though in becoming provincialized, he is not entirely lost to us as a positive influence. Moreover, in provincializing Kant, we can learn a lot more about Kant, (his contemporary and later) Kantianism, and Kant's lifetime and geopolitical situatedness. And even though I am not making the move to unfold this in this chapter, it is an important aspect to make explicit, responsibly and transparently, the geopolitical crossroads – and it is indeed, presently, crossroads – from where I am embarking on this and other ventures, and by what means.[20]

Here, I am conducting my inquiry, by a move that claims an appreciation for both (a) Kant as a psychosomatic philosopher and (b) the comprehensive nature of his concept of mind as *Gemüt*, and, subsequently, one will arrive (c) at an interesting notion of a (non-spiritual) practice of mindfulness.

Ad (a) Already during his lifetime, but more (d)effectively in the various (Post-, Neo-) Kantianism, the life, teachings, ideas, and discourse of Kant became refracted as if Kant himself was a kaleidoscopic lens. There is not one strong account of Kant, but there are many strong and many weaker ones, many likely and some highly improbable ones. The account I am favoring is, I claim, highly probable but has enjoyed for the past 150 years only weak ties and support. Founded in a *semantic agency theoretical* (Stingl 2010, 2011) account, deploying close readings of Kant's writings and biography, as well as different perspectives from within the history of science regarding both Kant's lifetime and the preceding era, I understand Kant's disposition (etymologically: "way of arranging knowledge"/"state of mind") as entrenched in and saturated by what later writers would call "*psychosomatic(s)*" on the one hand, and as a politico/juridico-medical on the other hand.

In other words, I think it is a fertile and productive move to understand Kant's oeuvre as a brush-stroke that aimed at viewing, first, the human being as relating, in analytical terms, between mind and body dialectically towards an understanding what practices of mind and practices of body (because both are either analytically aspects of the same or at least interrelated [German: *Wechselwirkung*]) *enable* a human being to live a long and healthy life; second, Kant, I think, evoked the imbrication (*Wechselwirkung*) between the body organic and the body politic – both for individuals and collectives.

To begin with, the differentiation of academic labor and knowledges was negligible compared to the period beginning in the second half of the 19th century and climaxing in the Babylonian status of academia today. Philosophers in Kant's time were expected to have intense knowledge about all aspects of natural philosophy, including what would later, around 1804, be coined as "biology" and physics. Furthermore, we find various instances in Kant's biography that reaffirm his interest in what we would nowadays call psychosomatic medicine, beginning with his account of his mother's death – from a more or less "imagined" contagious disease –, his early desire to study medicine, and the dedication of his first publication to a professor of medicine at Königsberg (see Kuehn 2001). Over the course of his life, Kant would correspond with experts on medicine and dietetics, seek to learn about food, and experiment with various health and diet regimens. His complete writings feature life-long interests in "vital powers" (which some commentators assume pertains only to physics, while others – myself included – do not), in anthropology, in dietetics, in "maladies of the mind", which are connected to his work on aesthetics, *sensus communis*, judgment, etc. In addition, I think that while an increasing number of philosophers and scholars (in particular the extremely and lastingly influential Southwestern and Marburg schools of Neo-Kantianism) have deviated from such an account, I have proposed (2008 [2010]) that some important contributions – such as in 19th-century philosophy, biology, psychology, and medicine (for example, the works of Rudolf Hermann Lotze, F.H. Bradley, and Alfred North Whitehead), the emergence of early American Pragmatism and early American sociology, as well as a number of 20th-century notable scholars who did not, however, form an actual school (such as Karl Jaspers) – cannot be accounted for without (or, rather, inconsistencies in existing accounts can be remedied because of) an account of Kant's philosophy in terms of psychosomatics and a politico-medical outlook. I should also point out that my account is not as outlandish as it might seem to some readers; it is, actually, quite close to Michel Foucault's (2008) understanding of Kant and the role that Kant's anthropology played in relation to his oeuvre.

Ad (b) Subsequently, given both the historical importance of Kant and Kantianism for 19th-century history of science and scientific ideas (for example documented in: Lenoir 1989 or Richards 2002) *and* the deposition just given above under ad (a), I have tried to account for the concept of "attention" in Kant's work in a historically significant way that not only can but also in all likelihood must include a psychosomatic dimension. What I find fascinating is that in Kant's works, such as in the *Anthropologie in pragmatischer Hinsicht* (1983, engl. 2006) "attention" has a specific, interrelated function, which scholarship and public discourse in the 19th century will change irrevocably – including the values of its related concepts. In short, in Kant's anthropology *attentio* appears in conjunction with *abstractio* and *distractio*. All three are functional terms in the *Gemüth*, which is one concept of the *mind* that is used in Kant's works and at times.[21] Furthermore, in terms of the *Gemüth*, they are also mutually dependent on one another. While I merely have the space here to summarize, I think the salient aspects shine through: To be able to pay attention to an idea or association is one thing, but in order to exert power of will over one's idea, to form new relations, etc. one must be capable of abstraction. But to be capable to either attend to knowledge or abstract from knowledge, one should be able to relax one's mental faculties and be able to be playful with them, which is covered by distraction. One cannot have, therefore, one without the other two, nor are either in and of themselves viewed as particularly "good" or "bad", i.e. *distractio* is a necessity that makes the other two possible; abstraction is not viewed in negative terms like deployed today when people say "this is too abstract", referring to something they deem theoretical/impractical. If anything, Kant emphasizes, that while all three are necessary, interrelated, and there is no externally discernible hierarchy between the three (functional) concepts, there is, yet, the thinking that *abstractio* has a somewhat higher for generative value compared to the other two concepts. In sum, however, the *Gemüth* as *mind* should also be seen in contexts that contemporary philosophers would associate with concepts such as "enactment" and "embodiment", and the Kantian psychosomatic being as embedded in a complex ecology of affordances, technologies, food, etc.

Ad (c) My conclusion following the consideration of ad (a) and ad (b) as premises is that Kant's psychosomatic deployment of *mind* as *Gemüt*, with the three interrelated functional concepts, points toward a being that uses its "powers of mind" in its present place and time most efficiently, playfully, and openly. A good, old-fashioned German word for this is the word *Geistesgegenwart*, which can be translated as presence of mind, but which would best be understood as *mindfulness* (Ellen Langer) – herein, it should be emphasized, I mean a non-mystic, non-meditative concept, which justifies my attraction to the Ellen Langer's non-esoteric/empirical concept of mindfulness rather than to other concepts.[22]

19th century rise of 'attention', the emergence of fatigue research and the body-mind split of attention

If we think of Kant's concept of "attention" as a conceptual object in a discourse, this object collided with other discourses and was sufficiently changed – it became a remarkably different animal, a distinctly transformed assemblage, so to speak.

If we think of attention today, we think, for example, in terms of "focus versus the wandering mind". These are elements that were present, for instance, in notions of "wandering wit" or notions of rigor in applying mathematics.

The conceptual and related societal transformation, mediated in terms of the emergence of the "modern" techno-social imaginary, can be genealogically reconstructed in the discourse that saw the fragmentation of (natural) philosophy as an academic institution, the emergence of the (Western/bourgeois) public sphere (Habermas 1991 [orig. dt. 1962]), the normalization of education, the creation of psychiatry, psychology, sociology, and (bio)medicine and as academic disciplines and scientific practice regimes, including the *implification* (on this concept, see: Stingl forthcoming, 2015) of fatigue research, industrial relations, scientific management, neoliberal capitalism, and the human body. While it is not possible to give a substantial and ultimately satisfying historic account in this chapter, I will at least attempt to provide a sketch for underlying the genealogical logic, which results in the aporetic constellation of the emergence of the figuration of the psychosomatic/techno-somatic of childhood on the one hand, and the introduction of a mind-body split in and through the conceptual transformation of "attention" on the other hand.

In the final years of Kant's life,[23] "biology" as a scientific discipline and, therefore, distinct academic point of view and *dispositif* became a possibility. The term, which was juggled around during the 18th century, was eventually coined in a multivolume work (1802–1822) as an academic and, thus, institutional denominator by Gottfried Reinhold Treviranus (1776–1837). In the same biological context that was congregated by discourses of physiology, embryology, medicine, race/racism, psychology, and, of course, philosophy, we can find significant transformations of medial semiotics – understanding how physical signs are read and understood – and psychosomatics, emerging from Kant, or rather Kantianism, as a nexus that had to be simultaneously followed, interpreted, adapted, disagreed with, fought, all while remaining a major reference point. For good or ill, any attempt at provincializing, including the act of critiquing the colonial impetus of the "modern" life science, must take this nexus into consideration as a geopolitical fixture (see also the case made by Walter Mignolo [2011]). It is important to stress here that this is not a particular novel insight, even though few philosophers and science historians have entered into an entanglement of this genealogical mess. It was, however, precisely Michel Foucault who did so beginning in his earliest years of scholarly production, and who was quite aware of the connections. My engagement with this subject, such as in this chapter and elsewhere, is to be seen largely as a genealogical continuation of the work Foucault had begun in his archeology of anthropology in Kant (2008): The point made in this chapter is that while Foucault points out that the conditions of possibility of experience date back to the empirical existence of a subject in Kant's anthropology, the *empirical* subjectivity in question is psychosomatic.

To begin with, during his final active years, Kant corresponded intensively with a number of scholars that we would nowadays qualify as interdisciplinary in the most integrative interpretation of the term. Among them was the author of the most influential scholarly work on dietary regimes and well-being of the late 18th and a good portion of the 19th century, which still has some influence today: Christoph Wilhelm Hufeland (1762–1836) and the teachings of *Macrobiotics*. His life and scholarship are deeply intertwined with the disciplinary histories of nutrition, psychiatry,[24] and medicine, as well as with the hygiene movement, the professionalization of medicine, the institutionalization of social insurance, the introduction of vaccination, and the emergence of social medicine in terms of the correlation between poverty and physical illness.

Johann Christian Reil (1759–1813) is often described as the person who coined the term *psychiatry* institutionally. Just as with Hufeland and educated in an intellectual climate that was saturated with Kantianism – such as the home of Kant student Markus Herz and his wife Henriette – Reil's scholarship congregated psychological, physiological, and anatomical aspects (Richards 2002; Binder et al. 2007) around a conceptualization of *vital force* (which was part and parcel of the discussion of what comprises biological life, a discussion that had Immanuel and his correspondent Blumenbach at the center (Lenoir 1989; Richards 2002) which would culminate in the industrial energy theories of the late 19th and early 20th century (cf. Max Weber; Rabinbach 1992). However, with Reil, "attention" experiences a deterritorialization into the sphere of pathologicalization:

> Based on his reading of the Romantic philosophy of Kant and Schelling, Reil formed new concepts of the mechanics of mental illness. He regarded insanity as a disruption or fragmentation of the self. He distinguished three chief forces, the disruption of which could produce mental illness: self-consciousness (*Selbstbewußtsein*), awareness (*Besonnenheit*), and attention (*Aufmerksamkeit*).
>
> (Binder et al. 2007: 1092[25])

Reil's concept is, in terms of discourse analysis, to be interpreted with proper reference to Johann Christian August Heinroth (1773–1843), the first professor of psychiatry, who is considered the person who coined the term "psychosomatics" in a scientifically and institutionally significant way. Heinroth was concerned with the "whole human being" (quoted in Steinberg et al. 2013), including, but also superseding, the idea of the unity of body and soul in terms of a rich concept of the "person":

> According to Heinroth's concept, all mental illnesses are 'disorders of the soul'. A mental illness emerges if two different factors come together simultaneously. He differentiated an internal, predisposing factor, which he called the 'soul's mood' ('*Seelenstimmung*') from external ones that affected it in the form of 'stimuli' ('*Reiz*').
>
> [...]
>
> Thus, Heinroth found it essential to analyse his patients' physical wellbeing, their social situation and life style, their sanitary conditions, nutrition, sleep – briefly their whole way of life. For him all these factors had a major impact on the emergence of a mental illness.
>
> (Steinberg et al. 2013: 12)

In deploying his views on various disorders, Heinroth also attempted to tackle the problems of sleeplessness, the causes of which he attempted to understand as being "psychosomatic", in a way that makes him a legitimate precursor to George E. Engel's biopsychosocial model (1977). Of course, it should be stated here that already Melchior Adam Weikard in his 1775 treatise on the "philosophical doctor" (see Barkley/Peters 2012: 625)

> discusses the influence of the body on the psyche that is quite reminiscent of the psychoanalytical theory of Freud and modern psychotherapy. He did not view diseases as being distinct (as he did in his later medical works), instead viewing the whole personality, whether healthy or ill, in the sense of a more general psychology or that of a philosophical doctor trying to understand, as Schmitt put it, how the animal machine was set into action.

These discourses should not be seen as merely academic, as they had great effect in the emerging public sphere. Mind you, it is precisely in this era that the very ideas of public sphere, civil society, modern education, and bourgeoisie usurp their (Euromodern) shape (Habermas 2015 [orig. 1962]; Richards 2002), as it is characterized for Western society and, subsequently, the Global Northern hegemony of what "political (agency)" means.

The emerging public sphere, the establishment and normalization of education in the form of (urban) schools, industrialization, rural exodus (*Landflucht*), the emergence of modern psychology/psychiatry and the social sciences, the idea of upward mobility, and social/moral hygiene form a complex in the 19th century that will bring about a significant shift in the meaning of the conceptual triad distraction-attention-abstraction, transform the techno-cultural and political economic status of attention, and induce a regime of normalization and pathologicalization around it, a regime that rests fundamentally on the Cartesian mind-body split, which is still characteristic for (bio)medical practice of the 19th and 20th centuries (see for this purpose the seminal works of Thure von Uexküll).

In a nutshell, and I am aware that this is somewhat oversimplifying, with an increasing number of people experiencing what we would call today social upward mobility, the pressure to maintain the status achieved and to possibly continue the upward momentum across future generations, education and, specifically, schools became an important apparatus of social transformation and reproduction geared around (an imaginary of) upward mobility. Michel Foucault has famously undertaken the study *of* and made available a frame *for* the analysis of this mechanism appearing in the military, prison, and the school in terms of "discipline". Of course, not all the educational/disciplinary labor was restricted to the school, but instead created boundaries for a wider social space which can be variously called childhood, child education, socialization.

In the 19th century, a particular nexus evolved around the use of stories to teach children morally acceptable behavior, which was often framed in a somatic imaginary – not adhering to a rule of accepted behavior would naturally lead to physically unpleasant consequences, which decidedly did not involve punishment. The most famous and one may almost say genre-defining tales were written by Heinrich Hoffmann (1809–1894), anatomist, physician, psychiatrist, who was from 1851 to 1888 the director of the lunatic asylum of the city of Frankfurt am Main (Germany). Hoffman is also often credited as being perhaps the founder of the field of child and adolescent psychiatry. Due to the lack of proper moral tales for his children, he wrote and illustrated a collection of such tales for his son, which he subsequently published in 1858 under the now world-famous title *Struwwelpeter*. This collection includes what is considered today the archetype of ADHD, the story of the *Zappel-Philipp* ("Fidgety Phil").

The story of *Zappel-Philipp* – with text and illustrations – shows a boy at the dinner-table with his parents. The boy, unable to sit still, keeps whipping on his chair until it finally topples over backward. The boy, now helpless to prevent his fall, grabs the table-cloth in a final attempt to stop the motion, but only ends up cleaning off the table of all food and drink, destroying the dinner and the table-ware alongside (when I was told the story as a child, grown-ups would often add the element that the child was burned by the hot soup). Another story of the *Struwwelpeter*, added by Hoffman to a later edition, is the story of *Hans Guck in die Luft* ("Johnny Look in the Air"), which should be mentioned here, because it mimics another famous story, and plays a main part in the prevailing anti-intellectualism of industrial/neoliberal[26] society. Hans will not look at where he is going, but look upward into the sky instead. As a consequence of not paying attention to where he is going, he trips and falls over a dog, and later falls into a river, having to be fished out of the water to much ridicule.

This story has a famous precursor, of course, namely the laughter of the Thracian maidservant (see here the important metaphorological study by Blumenberg 1987 [engl. 2015]), ridiculing – according to a story told by Socrates in Plato's account – Thales of Miletus, who fell into a well because he was so busy stargazing while walking. This is, of course, one of the original stories setting up the practice-theory antagonisms, anti-intellectualism, and anti-science, also Marx's 11th Feuerbach thesis, philosophy and humanities bashing, and so on. *Hans Guck in die Luft* is pretty much a story that falls right in the middle of a discourse that runs from the Thracian maidservant to the Neo- and Late Liberal war against education and critical thinking.

Tales such as Hoffmann's and related medical, moral hygiene, health/medical, physical exercise, and education advise became subsequently distributed in various popular and medical publications for the emerging, newly literate, upward-migrating public, publications be it *Die Gartenlaube*

(Germany) or *The Lancet* (United States). As such, they have become integral as images, myths, vignettes in the *social imaginaries* (Fujimura 2003; Taylor 2004) of the Global North, regarding the idea of normal and pathological mental and somatic functioning. Today as well as in the 19th century, the difference, whether attention (and/or hyperactivity) is pathological – i.e. as attention deficit – in a purely psychological, purely physiological/somatic, or psychosomatic dimension, is simultaneously an eclectic and often arbitrary choice, precisely because of an inherent Cartesianism that focuses on the question of whether or not a child conforms to the very practical norms and demands of its socio-economic environment and the respective (disciplinary) discourse on health, morality, and hygiene that was (and is) precisely afforded by this environment.

In the 19th century, this environment is strongly influenced by industrial relations and the emerging field of fatigue research, which differs geographically. The Swiss historian Philipp Sarasin (2001, 2014),[27] who I follow here in his ground-breaking and erudite depiction, follows this development from *scientific management* to *biological rationalization*, from Frederick Taylor to Max Weber, from the United States to France and Germany. Taylorism, the idea of scientific management and the industrial-physiological (*arbeits-physiologische*) *analytics of the body*, accordingly (Sarasin 2014: 67), influenced Western culture in terms of its understanding of how to deploy *time, labor* (*Arbeitskraft*), *energy, and the body*. Sarasin qualifies Taylor's influence in the face that he managed to extend the reach of his (modern) point of view as an engineer into and onto a different field of knowledge, i.e. the human being. Taylor's point of reference in terms of this discourse (ibid.: 68/69) is not

> *the optimization of the machine or in improving the human being, but it is in the 'system', the 'organization', that conjoins man and machine into a unit of maximum performance and efficiency* [my translation].[28]

Taylor's approach, as characterized by Sarasin's account (ibid.: 69f.) of Taylor's writings, is political economical: Taylor sees the problem in workers intentional (explicit or implicit) consensus to execute their power over the labor-process through deploying tactics of delay, sloth, and waste.[29] Taylor's "enemy" is the waste of time and energy, which fits an intellectual climate in the science that is obsessed with "energy" (and horrified by the idea of "entropy").

What Sarasin (72ff.), however, makes clear is that with regard to Western imaginaries of science, culture, and the political economy that connects them in order to build the (imperial-industrial) Global North is that the connection with Fordism – the "Henry Ford"-inspired nexus of factory floor, serialized industrial production, and factory-internal differentiation of labor process – is ambiguous. In other words, while they are inherently connected, they are also internally opposed not only in terms of Taylors dissecting and reassembling the physiology of time of the labor process within the single worker's body, where Ford distributes it across the bodies of several workers, therein shifting the center of intervention from the body (Taylor) to the machine (Ford). More importantly, as Sarasin (74) notes,

> *the means of rationalization and control for Ford is the assembly line and the rhythm of the machine, whereas for Taylor it is the 'system', it is 'science'* [my translation].[30]

To this day, there is no adequate comprehensive view on the transatlantic capillaries of transdisciplinary (in the sciences), transcultural, and cultural technical[31] knowledge transfer between Europe and the United States for the 19th century or even for what I have elsewhere called the super-sized[32] 19th century (ranging from Kant's first lectures [1772] on anthropology to Elton Mayo [1880–1949] joining fatigue researcher L.J. Henderson [1878–1942] at Harvard Industrial Relations [1926], which can be considered as a foundational event for the *Human Relations Movement*). But in terms of fatigue research, Sarasin's account suggests a number of entry- and connector-points: France (and, I would add, to some degree Italy[33]) was differently receptive to Taylorism than Germany. As Sarasin points out (77), France's science-industry nexus should be seen in terms of its connection to Cartesian, materialistic ideas focusing on the physiological aspects of labor and body. The German discourse, which culminated in a development from psychophysics to psychotechnics (from Wilhelm Wundt [1832–1920] to Emil Kraepelin [1856–1926] and Hugo Münsterberg [1863–1916]), was characterized by a focus on the effect of fatigue on mental labor. Besides Wilhelm Wundt's interesting role from being what we would call a transdisciplinary researcher to becoming the man who laid the foundations for what would be an interdisciplinary research differentiation within psychology in basically founding different branches of psychology and more or less founding ethnology/cultural anthropology (see on this complicated history Valsiner/van der Veer 2000), Emil Kraepelin deserves particular attention. Kraepelin, besides being an experimental psychological researcher, was also engulfed in field-ethnological research in Indonesia and, respectively, an important and influential interlocutor of Max Weber – who himself was exceptionally well versed in psychiatry, industrial physiology, and psychophysics as academic fields and professions, while also writing expertly about (and against) prevalent theories of "energy" in industrial relations and what we would now call cultural anthropology/sociology. Furthermore, Kraepelin had, (in)famously, a major and lasting role in medical and psychological illness classification to the point that prevalent diagnosis practices and models are summarized under the label *Neo-Kraepelinism*. Hugo Münsterberg – a student cum lab assistant of Wundt's and friend of William James and James's student

Edmund Burke Delabarre (1863–1945)[34] – is not only an important figure in this account for his innovative ideas but also an international bridging figure with his migration from German-European universities (Leipzig, Heidelberg, Strasbourg, Freiburg) to Harvard University.[35] Münsterberg stands for the bridge between the American and European worlds of 19th-century ideas surrounding fatigue and energy, and what I consider the parallel research directions of distinct mental and distinct somatic approaches, rather than any holistic, comprehensive, or complementary psychosomatic ones. Public discourses taking notice of scientific progress on the issue of fatigue constituted as largely "legitimate" knowledge that was either about mental/psychiatric or somatic/physiological (and today we would include a distinct neuro-physiological) knowledge about fatigue, normalcy/pathology, and well-being, whereas psychosomatic views were rendered continuously illegitimate, obscure, and esoteric. Not that these latter kinds of knowledges didn't have followers (Harrington 2000, 2008; Bischof 2010), but the main discourse allowed that legitimacy procedures were deployed largely with regard to the mind-body distinction.[36] What is, however, a crucial and distinctive feature in all of these approaches, and – as Sarasin notes in passing – particularly in the work of Münsterberg, is that these developments have emerged with a focus on the urban/city imaginary, indicative of the "urban normativity" that governs Western/Global Northern science and social science as Hans Bakker and others have shown.

In the 1920s, fatigue was an established (physiological) research arena as was its in-depth connection with industrial relations (for example in terms of what was called "industrial hygiene"; and turned into "public health"), which could be seen in the integration of researchers such as Elton Mayo, Lawrence J. Henderson, Chester Barnard, and Walter B. Cannon (who pioneered research on the question of *homeostasis*) into the ecology of Harvard University's Business School in the name of "human relations". In terms of the techno-social imaginary, roughly the resource pool of cultural, metaphorical, and phantasm imagery that people "lived by" (Lakoff/Johnson 2003 [1980]), this *constellation* of physiologists (Henderson), biologists (Cannon), industrial relations professors (T. North Whitehead), business leaders (Barnard), as well as philosophers (Alfred North Whitehead), who proved fluent in a shared vernacular that was characterized by biology, has been extremely influential within and beyond the academy. Consequently, it is no surprise that after Kant's Enlightenment dictum that enemy of reason was complacency (i.e. being too complacent to use one's faculty of reason) for Alfred North Whitehead, who was schooled in modern biology and physiology (see for example the role that Schwann's cell-theory plays for *Process and Reality* [1979 {1927/1928}] and *Science and the modern world* [1997 {1925}]), the anti-thesis of Reason "is Fatigue" (1971 [1929]).

On a more "cultural" level of *oikos*, and one should here point specifically to the context of bourgeois culture, the question of countering fatigue (degeneration and distraction) through hygiene became related to the question of sexuality on the one hand (see, following Foucault Sarasin 2001), and physical exercise/sports (or following the ancient Greek terminology: *gymnastics*) on the other hand. The body and the body-mind question appear prominently in both these discourses and have been the subject of multi- and transdisciplinary discourses throughout the century, including ideological and political debates regarding the body, sex, and gender. Because of these it is impossible to write any history of the body that isn't qualified as political history (Sarasin 2003: 102f.). I will not explore this aspect here any further both for constraints of space, but also because there is a larger amount of research and literature available. But overall, a more comprehensive view on the history of the body, that takes "*the* body" in terms of its biological *and* historical entanglements *politically* seriously, means to emphasize some of the less explored adventures in terms of the energy-paradigm of the 19th century: In the Euromodern *oikos*, both abstraction and distraction must appear as wasteful in comparison with the – now privileged – economy of attention.

ADHD from Bradley/Still to the recent editions of the DSM and the recent attention economy.

As I have attempted to illustrate, "attention" as a concept has been subject to some conceptual and contextual displacements, which figure prominently within the biomedical and mental health sciences, but also for parents and teachers, and, above all, for children and adolescents who are subjected to these concepts and practices as a normative context. In the biomedical/child psychiatry discourse, attention as a benchmark of what is normal, and the idea of a "deficit" in attention as a pathological state, has an discursive institutional history, which we can only, but also must, briefly sketch (see also Baumeister et al. 2012):

In 1902, a lecturer named George Still (1868–1941) in London presented a set of ideas, which were about to become extremely influential. He was studying defects in *inhibitory volition* displayed by children, specifically children who were already showing a variety of symptoms that could be labeled as disruptive, aberrant, or deviant behavior. Such behavior could be found in the archetypical *Fidgety Phil*, i.e. the inability to sit still, but go as far as stealing. In line with the tendencies of Victorian medicine, this was a medical and moral disorder. This rather heterogeneous bunch of symptoms and phenomena was of course part of an emerging medical consensus rather than the spontaneous invention of George Still (see Lange et al. 2010[37]). But following a research tradition spanning William James and Theodule Ribot (1839–1916) physicians and researchers had recorded and increasingly explained aberrant behavior for different types of "brain lesions". While attention pathologies would eventually be institutionally registered in the medical/psychiatric canon in one of the later editions of

Leo Kanner's (1896–1981) influential book on *Child Psychiatry* (1972) or in the second edition (1968) of the *Diagnostic and Statistical Manual for Mental Disorders* (DSM II), the disorder regime was originally coined according to its most visible symptoms, such as the inability to stay still, as *hyperactivity* and *hyperkinesis*. The latter term was taken from Charles Bradley (1902–1979), a Providence, RI-based researcher who had begun treating phenomena he considered *hyperkinetic* with drugs in 1937 (see Strohl 2011; Connolly et al. 2015). Bradley used a stimulant drug, Benzedrine, which set the stage for the types of drug which would become and still are the supposed the first line of defense in the "war against ADHD", stimulants such as methylphenidates and their most famous brand name Ritalin.

In the history of treatment and diagnosis of mental health problems, the question of standardization plays an important role on the one hand, and the pendulum swings between strongly positivistic classificatory and phenomenological approaches on the other hand. Two extremely classification regimes became internationally recognized and influential over the 20th century: the World Health Organization's (WHO) International Statistical Classification of Diseases and Related Health Problems (ICD), currently in a phase of transition from its tenth iteration (ICD-10) to its eleventh, and the American *Diagnostic and Statistical Manual for Mental Disorders* which has been recently upgraded – with much controversy – to its fifth version, DSM-V. Globally, the DSM, and specifically its third and fourth editions, has been the most influential in disseminating and popularizing diagnostic classifications. It is important to notice the specific history, namely, that researchers of the so-called Neo-Kraepelin school took over responsibility for the production of the third edition of the DSM (see Decker 2013;). With that institutional rupture, a group obtained defining/classificatory power, which culminated in a widespread understanding of psychological/psychiatric practice as a categorical approach of operational diagnostics. In this context, *Attention Deficit Disorder* (ADD) obtained both its name and diagnostic practice. This move also indicated the creation of research programs and the fact that case studies had now turned into a phenomenon that was now finally part of public health and social medicine. Between the 1960s and late 1970s the question for the causes of ADD shifted away from the role of the family and the social environment towards the question of genetics. Now, ADD was not only a public (and also popular) phenomenon, it was also part of the medicalization-geneticization complex that constructed this entire group of child and adolescent patient as faulty neuro-chemical machines open for physico-chemical intervention to create social fitness. For most lay-people and many practitioners the DSM-IV cemented this line of thought in the 1990s in correlation with a surge in the number of children diagnosed with what now became ADHD and the realization that ADHD was not restricted to children but also present in adults (Stingl 2010; Stingl/Weiss 2013).

Today, we can say that the genealogy of the biomedicalized imaginary and the genealogy of the digitalization of lifeworlds – begun with the emergence of mass-media – which have led to attention as its currency, have, in a sense, made the scarcity and poverty in this currency into a (both social and individualized) pathology, the historicity of which has and continues to structure the socio-economic condition of the "global world", by which I mean the Global North and the geopolitical epistemological *influence* (Westernization) that it exerts. Tim Wu (2018) in a legal studies appraisal of the emerging world of "attention industries" says,

> [h]uman attention, valuable and limited in supply, is a resource. It has become commonplace, especially in the media and technology industries, to speak of an "attention economy" and of competition in "attention markets." There is even an attentional currency, the "basic attention token" (BAT) which purports to serve as a medium of exchange for user attention. Firms like Facebook and Google, which have emerged as two of the most important firms in the global economy, depend near exclusively on attention markets as a business mode. (1)

This is an interwoven strand in the filigree of *Digital Culture*, extending the *coloniality of power* through the setting up of ecologies and economies of attention.

Notes

1 In Stingl (2015), I introduced the concept of *implification* as a theoretical assemblage that stands for process, wherein implication and intensification are interdependent, co-productive, and effectively intertwined.

2 This "definition" is tentative. It serves as a bridge between the classic views on *techne*, which I am about to discuss a little in bit on the following pages, my own conceptualization of the status of the technological object as *tangible* (see Stingl 2014a), and Bernard Stiegler's discourse. Stiegler in particular attempts to design a concept which focuses on exteriorization processes, i.e. that humans "extend" via prosthesis and that the relation to the exteriorization as technics *in*, *through*, and *as time* becomes interesting. For me, I am less interested in the humanly exteriorization than I am in exteriorization of agency and the question of temporalities (so *implified*). See here, also, the 2009 special issue on Stiegler of the online journal *Transformations* (www.transformationsjournal.org/2009-issue-no-17-bernard-stiegler-and-the-question-of-technics/).

3 A related issue to this question of representation is the problems of internalism/externalism that emerged in a sort of historic succession in philosophy, history of science, in sociology, and in philosophy again. Herein, too, we find the discourse both distinct and yet interconnect with regard to each discipline, as well as in relation to the discourse on representation. Simultaneously, it cuts across the repeated and persistent *Werturteils-/Positivism Debates* which re-occur in the social science since the late 1800s (for the latest rounds see the interesting reader collected by George Steinmetz 2005).

4 A different take on Nietzsche in context of Schmitt (Siemens 2013); Thomas (1986) already argues, to the contrary of McCormick's claim that Nietzsche wasn't directly present in Schmitt but mediated through Weber, that "we know" (66f.) that Schmitt had studied Nietzsche and was "deeply attached to his philosophy".

5 These tensions between inclusion, exclusion, and silent Othering/silencing (radical) Others can be found, for example, throughout the papers collected by Patricia T. Clough and Craig Willse (2011) in the book *Beyond Biopolitics*.

6 Northern theories, sociological theories in particular, have been extremely "afraid of equivocations", leading me to ask the question: Why?... The question is important, because sociologists have so far only responded to the question of *how* they are afraid of equivocation, for example by *collaring-up White*.

7 https://aeon.co/essays/how-work-changed-to-make-us-all-passionate-quitters.

8 While, of course and in implicitly or explicitly upholding a Habermasian utopia of discourse community, they *implicate* their "others" as not engaging in serious argument, they themselves intensify the problem, and the whole process of *implification* enters into a perpetually aggrandizing cycle.

9 Kemple (2012) and Krause (2017) are among the few notable exceptions with regard to Tocqueville, although Beaumont seems still underappreciated even with them.

10 In the *Séminaire Pour une sociologie mondialisée non hégémonique* – Fondation Maison des sciences de l'homme, Jeudi 23 mai, 2019.

11 However, I should remind my readers (and myself) here that I am intentionally producing a writerly text that is meant to provoke (somewhat permanently) thinking and discussing together with others precisely by "being unfinished", "raw", and, at times, a bit on the *daft punk*-ish side.

12 However, based on my review of respective literatures, my impression is that among "ancient" slavery this is mostly true for Rome but in Ancient Greece and Athenian law, there seems to existed a racism, which then plays into the constitution of *oikos/polis* – and from thereon reintroduces repeatedly into history the *notional possibility* of "disposability".

13 As Marianne Weber has elucidated in her history of laws concerning wives and mothers, the British Common Law didn't consider women as a person who can say of herself "I" (Radkau 2011), same as in Athenian law.

14 Exemplified:

> Payment for Environmental Services (PES) is currently one of the most popular market-based policy instruments designed to conserve natural resources. PES projects are reward schemes in which landowners are compensated with money or in-kind direct payments for the land management practices they contribute By providing economic incentives, PES encourages landowners to build nature's economic value into their management plans, thereby improving the provision of specific environmental services.
> (Rodriguez de Francisco/Boelens 2015: 481)

And in recent discussions of including the ecological aspect and environmental crises into the concept of citizenship, Latta (2013) addresses the issue as what can be understood as its Global Northern geopolitical standpoint:

> A notable highlight of debates on environmental citizenship is the focus they place on the spatial mismatch between ecological challenges and nation-states. As Kuehls (...) asserts, the nature of ecopolitics forces us to re-evaluate the concept of state sovereignty as an organizing principle of political life. Biophysical systems, economic relations of resource

exploitation and the impacts of ecological change all cross over the national boundaries that [*geopolitically; A.S.*] fragment our planet. But if political subjects are constituted within states, how might citizens respond to ecological crises that overflow sovereign territorial jurisdictions? In the light of this conundrum, numerous theorists have argued that environmental citizenship must burst the containers of the nation-state and reach towards a global conception of citizen rights, responsibilities and action. (566)

15 The German word *Bindung* is difficult to translate here: In general, it refers to attachment or connection, but also relation and, of course, binding; yet in the sociological/political philosophical context, it also refers to social cohesion. However, I do think that the word *belonging* is a good translation here, precisely because it captures, in my view, the original meaning and context of 19th-century Liberalism *and* the historical context of von Kempski's paper, but also – and, due to the prior, not ending up a presentist imputation – working in line with the argument that we must inquire – as suggested by Walter Mignolo and others – after the political modes of belonging, and whether there is, as Western political epistemology and Global Northern geopolitics suggest (mono-culturally), truly one form possibly, namely (political) citizenship, or whether there are a variety of forms possible.

16 Importantly, both Parsons and Birch evoke Ricardo's theory of rent as the historical basis of their deliberation.

17 See also: www.newyorker.com/business/currency/the-cost-of-cash-for-the-rich-and-the-poor, www.huffingtonpost.com/2014/04/03/bank-fees-welfare_n_5072087.html, and www.vox.com/2015/5/22/8643921/kansas-poor-welfare-atm.

18 This turn of phrase of Max Weber – appropriated by Heidegger, as he misappropriated ideas from so many scholars of his time, including Weber and, above all, Simmel – is often translated as "iron cage". The translation somewhat misconstrues one or the other dimension of the German term, but captures others quite well. The main issue is that in a cage, rather than in a "(livable/dwellable) shell/casing", the one housed in it is aware that s/he is captured and, thus, caged. In the *stahlhartes Gehäuse* one may or may not be so. Second, *Stahl/steel* is a product of industry whereas iron is an element. Weber's choice of words manages to implicitly emphasize the artificial nature in this way.

19 However I will say, equally cautioning, that there are, indeed, some details wherein I may disagree with Sterne's (2003) account of Bourdieu's concept of *habitus*. I found his ideas overall stimulating and still noteworthy over a decade after its publication. His diagnosis regarding certain shortcomings in the various fields of study of science and technology is still pretty much appropriate, and the adoption of Bourdieu in the core field of STS has been as bad if not worse than it has been in sociology. An exception in American sociology can be found around the intersecting groups of socio-historically inclined sociologists around Phil Gorski (2013) and postcolonial sociologists around Claire Decoteau and Julian Go (see both in Go 2013a).

20 Kant, I propose, allows for an account of psychosomatics that counters the prevalence of the Cartesian model of the body in medicine and shows that contingencies were at least theoretically possible. Of course, in medical practice, the Cartesian model of natural science and the body have dominated, as Thure von Uexküll and others have repeatedly argued.

21 *Gemüt* is just as notoriously difficult to translate into English as *mind* is to translate into German. There is a lot of historical and disciplinary baggage.

There is, however, substantial extensional overlap between *mind* and *Gemüt* to justify the translation of one into the other in the Kantian context.

22 This follows the chronology of the development of my position, which migrated in a genealogical move from Kant to Langer (and not, as some critics have implied, accusing me of presentism and imputation, the other way round).

23 Of course, any work has shortcomings, and any chapter-length work more so than a book-length work. Given my line of inquiry here, there are historic figures I cannot connect here to the discourses I bring before you. Georg Ernst Stahl (1659–1734) would deserve mentioning because of his involvement with the notion energy and psychosomatics. So would Alexander Crichton (1763–1856) who in 1798 is said to have delivered the first more widely distributed description of "attention and its diseases" in a book – although Melchior Adam Weikard (1742–1803) is credited with the first description of the symptoms in 1775 (see Barkley/Peters 2012) in his treatise *Der philosophische Arzt* ("The Philosophical Physician"), which would have been forgotten had Michel Foucault not mentioned it in his early work on madness. Despite their importance for the historical account, I am uncertain whether they can be asserted enough gravity in the production of *attention (deficit)* as an assemblage in the discourse that I am looking at. I may, however, have to revisit this point in the future. While Barkley/Peters must be thanked for introducing Weikard's contribution to scholars interested in the history of ADHD, their conclusions – for example that Crichton must "likely have met" Weikard, are far from the likelihood they believe in.

24 He also had an instrumental role to play in the rise and fall of the homeopathic movement and its founder Hahnemann.

25 Binder et al.'s account, however, is misleading if not mistaken on the account of Reil's "construction" of a notion of *Gemein-Gefühl* as "unity"/"common sense". This issue must actually be seen in relation to Kant's re-interpretation of the *sensus communis*, departing from the "organ of the soul" discourse but within the question whether the soul and/or the mind can be distributed (Stingl 2010 [2008]), which was a discourse still ravaging in Rudolf Hermann Lotze's (1817–1881) disputes with medical and philosophical interlocutors.

26 Late Liberal (Povinelli), Near-Liberal (Ecks).

27 Sarasin's works cited here are currently only available in German to the best of my knowledge. Sarasin has come to the attention of American readers with a study on Anthrax and the phantasms contained in bioterror discourses (2006). However, the research and publications by Sarasin – who is also one of the leading Foucauldian historians in the world – in this research area are seminal works, which should be accessible at this point to a larger international academic readers through translations.

28 Sarasin (2014: 68/69):

> [D]ie Gründe für den nachhaltigen Einfluss lagen nicht in einer spezifischen ingeneurswissenschaftlichen Leistung, sondern darin, dass er als Maschinentechniker einen einen ihm fremden Wissensbereich okkupierte: Den Menschen.
>
> [...]
>
> Der Fluchtpunkt von Taylors Diskurs liegt daher weder in der optimierten Maschine noch im verbesserten Menschen, sondern im "System", in der "Organisation", in der Mensch und Maschine zu einer Einheit mit maximaler Leistung und Effizienz verknüpft werden".

29 In another context, we could say – with a nod to Walter Benjamin and Georg Simmel – that the enemy of or pathological antagonist of Industrial Physiology (and Taylor) is the *flaneur.*

30 "Das Mittel der Rationalsierung ist nicht wie bei Ford das Fließband, der Maschinenrythmus, sondern das 'System', die 'Wissenschaft'" (Sarasin 2014: 74).
31 In the sense of *Kulturtechnik* (Winthrop-Young 2014).
32 This is clearly my play on Eric Hobsbawm's famous notion of "the long nineteenth century", which stretched from 1789 (French Revolution) to 1914 (World War I). My point is that these "events" are framed on both ends (a prologue and an epilogue, if you wish) by open boundaries in the form of knowledge regimes that constitute a discourse on the political economy of the human, Nature, and practices/matters of care, which make the long 19th century possible first and reset the stage at the end. (The years 1772 and 1926 are, thus, at the same time random while there is also good justification for choosing them as the point for analytical cuttings.)
33 In particular, I am alluding here to the – still not fully reconstructed and understood – role of researchers such as Angelo Mosso (1846–1910), who studied and worked across Europe and undertook an influential lecture series in the United States, while developing influential research ideas and devices from nutrition, concepts of "degeneration" to neuroimaging.
34 Another extremely influential and prolific researcher and teacher in his time, whose influence on generations of students is largely obscured, as is the fact that he authored almost an entire chapter of the second volume of William James's seminal *Principles of Psychology* (1890).
35 Münsterberg among his many accomplishments is also one of the first psychologists to be concerned with mass entertainment media (movies) through his study the *Photoplay.*
36 Of course, this is subject to analysis with Foucault and Lyotard in mind.
37 Lange et al. create a good Whiggish depiction of the chronology of ADHD diagnosis. It is, however, as I said, Whiggish. It also is entirely presentist in the very notion that Foucault has criticized. Earlier attention-related/hyperactivity or similar concepts are compared to the present ones, *as if* the present positivistic notion of AHDH were *the* ontologically existing entity that earlier ones were approximating towards. This view neglects the genealogical input in the construction of disease categories.

5 Care Power Information

A B'n'B for *Oikos?*

Value-Mattered Narratives

In his contribution to the Habermas Festschrift from 1989 *Zwischenbetrachtungen im Prozess der Aufklärung* (Honneth et al. 1989), the theologian Johann Baptist Metz contributed a short but sharp essay. One of Metz's core issues was to argue that the crisis of the humanities and arts (*Geisteswissenschaften*) was what I would call "an archival problem": Its *reason* (*Vernunft*) had been halved due to Christianity's focus on Greek thought and obscuring its Other for the history of intellectual tradition – for Metz that Other half was the intellectual tradition of Judaism. Of course, I think it's worse: The (Christian) West/North has rejected and abjected all others that do not fit within its Greek origin story, which is told and progresses in the narrative of *oikos*.

The logic of *oikos* is not a logic without *care*,[1] but as *matters of care* go, to speak with Maria Puig de la Bellacasa and evoking Donna Haraway, it matters what *matters matter the matters of care*. The managed and governed (*oikein*) care that we encounter is for those and by those who belong to an *oikos/polis*, where the demarcations whom one does not (need to) care *for* and *about* are clearly drawn and conjoin dispossession and disposability, which are – even if not maintained throughout history they are rediscovered – retained qua colonialism and forward until today.

Nonetheless, other forms of care and, thus, of belonging, participation, and provisioning are possible, of course. Donna Haraway has shown us one way among many:

> String figures led me there. Well, the sheep are really what led me there. Well, no, actually Cayenne brought me there because Cayenne was an Australian Shepherd, a dog of white settlers in the U.S. West. So, in a non-trivial way it's Cayenne who linked me to all that. That's what I mean when I say things like "dogs make us more worldly, not less." If we track them and their histories, they lead to other tangled histories, and each of these histories leads us to caring. Every time we track a thread, our universe of caring grows. And then accountability. You can't forget that you know. I may not have known once, but I do

now. And so once I know, then what? Then you owe something back. Your caring has consequences.

(https://brooklynrail.org/2017/12/art/DONNA-HARAWAY-with-Thyrza-Nichols-Goodeve)

As for using the knowledge of alternatives to go before and beyond *oikos*, I propose that care exists in one breath with information – for example the "information order" problem (i.e. how information is ordered by us before it becomes explicit knowledge, already in terms of perception; I am including here, for example, the effects of [geo]onto-power [Povinelli 2016]) – and power (which includes the fact that power is not a simple concept that can be reduced to the Marxian-Weberian equation of power with coercion, but a complementary, equivocative concept).

Geopolitically, it is difficult to develop and institutionalize alternative ways of belonging, participating, and provisioning. But examples from art, fiction, and scholarly thought across time and space can help us here. There are many examples from Africana philosophy (Gordon 2008) to Japanese anime. I want to take a brief look at the ideas from the anime *Eden of the East* (東のエデン *Higashi no Eden*). I am not saying here that it is about mimicking and realizing the anime's story that would not be generative at all. On the contrary, the point is using the anime as a *writerly* form that enables to generate alternatives and, perhaps, to do so *together with others*.

In *Eden of the East* (Kamiyama 2009b), an anime series of 11 episodes and two films that conclude the story, the two main characters are Takizawa Akira (滝沢 朗) and Morimi Saki (森美 咲). Saki is a tourist, who has traveled to Washington, DC. There, she encounters Akira in front of the White House. Akira is, at this point, naked and holding both a gun and a cell-phone, but he suffers from amnesia. Saki lends the confused young man her coat and the somewhat unusual cell-phone connects him to a "concierge" by the name Juiz, who instructs him to a place where he finds among other things some fake passports, one with the Takizawa Akira (滝沢 朗) identity, which he chooses. Meanwhile, a series of missiles hit Japan without, however, causing any deaths, leading to this mysterious occurrence being called "Careless Monday". As the story progresses, Akira finds out that he was somehow involved in this attack, and that he is part of a game, in which a group of 12 people called the *Seleçao* are each given 10 Million Japanese Yen (stored via the cell-phone, and Akira has still 8.2 million left of the original amount) to spend on a vision to "save Japan", which Juiz helps them realize as they can order him how to spend the money. If, however, they have eventually spent all their money for selfish reasons, they will be eliminated by the "Supporter", the twelfth person among the *Seleçao*. The motto of the game, brandished repeatedly over the cell-phone, is *noblesse oblige*. Akira's involvement in Careless Monday is itself quite strange in that he helped transport 20,000 people "Not in Education, Employment, or

Training" (NEETs) to evacuate in advance the areas where the missiles would hit. Saki becomes increasingly involved in Akira's role in this game, and drags her friends in it, with whom she co-owns a small company (formerly a group of people dedicated to recycling) named *Eden of the East* – a social network app that recognizes people and things and allows the user access to information via cell-phone. The eventual resolution, which after numerous clever ploys, resolves at least in the hope for a better society. A lot of the story hinges on Saki's uncanny ability to improve the value of an item, even if others view it as junk, which becomes a basis for the company (and app) *Eden of the East*. The story enables thinking about the *noblesse oblige*, the use and importance of money, as well as "value" in general as well as in Digital Culture that centers around apps. Eventually it becomes quite clear that value generation comes before the structures that circulate it (e.g. markets), and that "value" is not a fixed concept. In making the NEET a central component of the story, the members of *Eden* as a community manage to help the NEET and others, and explicitly evoke the question of who is valuable to a society and why. In citing *Eden of the East* as an example, my goal is not to embark on a lengthy, academic analysis of a cultural product. I am merely pointing to the power of stories and narration to help us rethink the basic categories and even the ideal-types we take for granted. To maintain focus and have belief in a certain kind of "value", we have today created an economy of relevance and set aside infrastructure depending on what is considered relevant, and not the other way round. This is also a story, the "story of capitalism". But it is only one story; we should (and could) have others to tell our own stories with – to stay with the trouble made by Ursula Le Guin and Donna Haraway.

If we look through policy documents that describe, for example, the current transition to what is called *the* BioEconomy, we find an abundance of rhetoric of and metaphors for "relevance": The BioEconomy as a political agenda, epistemically situated in Western thought and geopolitically exercised from within the Global North, is above all mired in the *economy of relevance*. In this context, Isabelle Stengers (in Latour et al. 2018) describes a double narrowing of "'what kind of people are important?' and 'what kind of knowledge is important?'" in

> this sense, the story of capitalism is also a story where some people become very – I would not say 'clever' but 'productive' – inventive in a very specialised way on the one hand, while more and more people, on the other hand, become defined as 'beneficiaries' – now even workers are said to enjoy the 'benefit' of a job, which they have to 'deserve' it – of something made possible 'elsewhere', out of their reach.

Sociology, I thus propose, must become a sociology of possibilities, by which I mean of possible and impossible stories. However, traditionally and in a widespread White Collaredness, sociology in particular and the social sciences in general are *Wissenschaften* and, thus, full of contradictions; moreover, sociology is representative and co-productive of the imperialism and hegemony established and excreted by the Western epistemological episteme, throughout and from the Global North. Even its most profound, creative and insightful figures, such as Max Weber, in part because they have become *canonized* – canonization being one of the pillars of how intellectual hegemony is produced – are not innocent but problematic and more often than not complicit: Weber, who offers so many "ways out", at the same time has delivered so many of the social sciences' basic tenets that have against their potential for *provincialized generality* – for being used with caution by one group to understand the world from their geopolitical position –been rendered into universals *en mondialisation*, which are being reified, which turn into self-fulfilling prophecies, which colonize, which are absolutist explanations – however, mainly only of the very things they themselves produce. Their foundation is the Greek *oikos*, as above all Weber is witness and testimony for, when he schooled a young Arthur Salz that in the "domesticity" of one's home the contradictions, *discontexts*, and dissociativeness of the *Roman* home become equally visible. As Radkau (2009/2011: 7 [orig.2005:27]) states,

> For Weber, then, 'domestic' or 'family' communism had something primeval and homely that distinguished it from political communism. 'Society', in the sense of a specific gathering of people crystallized in the family (independent of state institutions), was a primary experience – although, to be sure, it was not only natural instincts but also capital that held the family together. 'A domestic community binds only if it is geared to indisputably common tasks', Weber taught the young Arthur Salz in 1912 (MWG II/7–1, 428). Yet for Weber the family never had the mere function of an economic system; it always preserved a vitality of its own. Which was not at all to say that it was a harmonious idyll, as it showed its strength precisely in the midst of a quarrel. With an eye to the slave barracks of ancient Rome, the 32-year-old Weber let slip a phrase that could have come from an instructor in Catholic social thinking: 'Man can develop only in the bosom of the family'.
>
> (K 57)

Weber is telling a story here, a story mired in *oikos* (family&property), anchored in coloniality and in a vision of *oikos*. We need to go before and beyond the story that these stories are told with.

Dussel and the Ethics of Liberation

One way to go before and beyond is to provincialize and decolonize. For philosophy and scholarship, the work of Enrique Dussel helped me much when it was necessary to understand from where Western/Global Northern/Euromodern ideas of Care Power Information took their "rationality" from and how they were deployed: urban normativity and metropolitanism mired in the legacy discourse of *oikos/polis*.

> The "rationalization" of political life (bureaucratization), of the capitalist enterprise (administration), of daily life (Calvinist asceticism or Puritanism), the decorporealization of subjectivity (with its alienating effects on living labor, criticized by Marx, as well as on its drives, as analyzed by Freud), the nonethicalness of every economic or political gestation (understood only as technical engineering, etc.), the suppression of practical- communicative reason, now replaced by instrumental reason, the solipsistic individuality that negates the community, are all examples of the diverse moments which are negated by simplification, apparently necessary for the management of the centrality of a world system that Europe found itself in the need of perpetually carrying out. Capitalism, liberalism, dualism (without valorizing corporeality), and so on are effects of Europe's management of the function given it by its role as center of the world system.
>
> (Dussel 2013: 36f)

Dussel (2013: xv) takes as his "point of departure [...] a world system of globalized exclusion". For the question of *the* BioEconomy as an escalation – yes, even as an actual *implification* (as it implicates us humans in the Anthropocene but then paradoxically intensifies the extraction-exploitation-"[inclusive] exclusion"-complex of the *precarization of everything*) –I propose and therein hope to suspend ontopowers that exclude non-humans and even colonize or hegemonially perform the practices of speaking "what is (even) life), speak of a world system of exclusion, extraction, and exploitation".

Many ethical debates, that are subject to the intervention by Dussel, are characterized by a denial that normative ethics can accomplish much or anything in and for this state of the "world", since normative ethics – in the general view – in this portrayal cannot be based on a "rationality with empirical validity given that its deployment would be grounded at the level of mere value judgments" (xvi); Dussel counters with a proposal that is grounded on "universal material principles", "principled" criteria for the "application" of that principle, and the principle of "feasibility". These three "principalizations" are drawn in the face of the problem of the "dialectic of the *goodness claim*" (vxii), which is itself the

> end result of a complex process of wherein the content of truth, intersubjective validity, and ethical feasibility have the effect of producing or enabling the fulfillment of 'the goodness claim' (*das Gute*), [...] the good person is a concrete ethical subject, but only once this subject has brought 'goodness' into action upon a normative basis

Yet – and hence the "dialectical inversion" of the goodness claim in (what, from a critical realist or Deleuzean point of view, we could call) *actuality* – the goodness claim "becomes inverted and is dialectically transformed into evil because it has produced a victim", "a concretion of the effects of the application [of norms, acts, microstructures, institutions, or ethical systems that assume to fulfill]" the goodness claim.

By grounding his approach in the daily life of people and attuned to the, however, non-intentional "negative dialectics of the goodness claim" in practice that produces (in "every kind of autonomously organized and regulated structure" [xviii]) victims aka "the oppressed", Dussel argues that the normative basis of an ethics must not only adhere to a *pure rationality of said ethics* but also attune to motivations,

> that include affective drives that are deeply rooted in the critical superego of the oppressed, and that are often nonintentional in character, grounded in social contexts and cultural values, and in historical and biographical causes and factors, and in the impetus of principles such as responsibility and solidarity.
>
> (xviii)

I have addressed a similar issue previously in my work on the empirical arena of the digitalization of medicine and the production of digital divides and digital health care capital (2014): There, I saw a need for, on the one hand, evoking how Talcott Parsons's emphasis of the notion of "motivations" in medical sociology (and in the much older tradition he was relating to, that of discourses in US sociology construing sociology as being the sociology of institutions, which I view as referring to what today we would call "cognitive cultures and cognitive organizations", wherein "cognitive" is not restricted to humans) and, on the other hand, addressing the role of "motivation" in the work of Jose van Dijk and Victoria Eubanks on Digital Divide and Digital Inequality, respectively. As one of many consequences, it is particularly *responsibility* and *solidarity* that occupy unimportant roles in my work: responsibility appears as an epistemic responsibility (e.g. using Lorraine Code) in what I call the "right to being justified to" as well as in my adoption of Donna Haraway's suggestion of "*response-ability*", whereas solidarity is tied to what in *Worlds of ScienceCraft* Weiss, Restivo, and I conceptualized as solidarity as *ScienceCraft* and integration.

Dussel creates an empirical argument – and, hence, his proposal is not only a purely theoretical endeavor but also a situated, empirical ethics/philosophy, although theory is, of course, important – against any (a) "absolutely postconventional" (Dussel 2013: 1) and (b) purely globalist morality that claims prescriptive universality over ethics.

Ad (a) Ethical systems are to some degree "conditioned" by the "historical sequence", or perhaps we could deploy here Deleuze's notion of *series*, which has two sides: interdependent yet also adversarial (see, on this, also the important difference marked as "the political" in the agonistic mode by Mouffe and following her the work by Marchart and Selg [see earlier chapters], which makes Dussel's proposal open to a conversation with post-foundationalism , even with Dussel's claim to a certain universality of his ethics). Before that, though, one should not fall for the fallacy of mistaking any situated content (of a "cultural ethical life" [ibid.] for its formalism), to some degree, which is precisely the problem that Carol Gilligan identified in Kohlberg's approach. He mistook a specific rationality for one he thought he could "naturalize" as a postconventional universal, but instead simply fell for the fallacy of reification and misplaced concreteness (see Morris Cohen, Alfred North Whitehead, or Talcott Parsons) that had plagued so many before him – again, this is not to say that there wouldn't be a usefulness to some neo-Kohlbergian thinking such as by Jim Rest, Mickey Bebeau, Steve Thoma, Darcia Narvaez (Rest et al. 2000), and others – an approach, however, that is wise to Gilligan's criticism and decolonial options.

The two sides of any ethics as an ethical system now are that they are made from the ethical lives lived, that wherein the latter's "ethical material and formal moral levels" attain an "abstract universality" on the one side, and enable and grow due do (ethical) "criticism". In short, I would conclude that an ethical system's "abstract universality" without the explicit possibility and reflexive integration of criticism is (totalitarian, antagonistic, imperialistic) ideology.

Ad (b) Globalism, other than cosmopolitan globalization (Beck) or *mondialisation* – although neither of these two concepts is entirely without problems (of a certain coloniality) – is strongly dependent on a center-periphery model that co-produces Europe as center against the "rest" as periphery, along with various other products, such as capitalism, that were – in repeating the fallacy of reification – portrayed as causes (rather than products) in the narrativization of "(Euro)Modernity", thereby all-too-often re-narrativizing the fact of historical developments and ethical lives lived into something beyond something covered by their actual facticity. Ideologically, this is shown in the Western obsessiveness with its own narrative: A case in point is not *that* one cannot critique and prod and question, for example Martin Bernal's (1987) account in *Black Athena* (Vols. 1–3), but that there is such a ferocity with which Bernal's and others' accounts – namely that Greek philosophy is itself a product of a longer history (one that, too, has no discernible, absolute foundational cause), in which Egyptian and other Northern African knowledges have an important, formative role to play – are met, a ferocity that does not seek argumentative reconciliation (not even consensus) and/or agonistic discourse, but that is largely annihilative, delegitimative, and destructive. This is not to say that, as a product of this ferocity and its underlying causes, the attitude of some of the participants in this debate who advocate a position similar to Bernal's has not also often become no less antagonistic and destructive themselves. The "center-periphery" moment, which Dussel describes in a process that actually produces two Modernities in Europe, of which the "Anglo-XXX" one is often given precedence by many philosophers, is equally well in its geopolitical epistemology and geographic imagination captured in Michel-Rolph Trouillot's seminal paper "North Atlantic Universals" wherein he interrogates and troubles precisely the two core concepts of "modernity" and "globalization"

> The world became global in the sixteenth century. Europe became Europe in part by severing Latin Christendom from what lay south of the Mediterranean, but also through a westward move that made the Atlantic the center of the first planetary empires. As such empires overlapped or succeeded one another within the modern world system, they brought populations from all continents closer in time and space. The rise of the West, the conquest of the Americas, plantation slavery, the Industrial Revolution, and the population flows of the nineteenth century can be summarized as "a first moment of globality," an Atlantic moment culminating in U.S. hegemony after World War II.
>
> (Trouillot 2002: 839)

Dussel, subsequently, identifies two moments of European Modernity:

> If Modernity is was the fruit of the 'management' of the centrality of the first world system, an this is our hypothesis, we now have to reflect on what this implies.
>
> One must be conscious that there are, at least, in origin two Modernities.
>
> a In the first place, Hispanic humanist Renaissance Modernity was still linked to the old interregional system of Mediterranean and Muslim Christianity. [...]
>
> b In the second place, there is the Modernity of Anglo-Germanic Europe, which began with Amsterdam, and which frequently passes as the *only* Modernity (this is the interpretation of Sombart, Weber, Habermas, or even the postmoderns, which produced a reductionist fallacy that occluded the meaning of Modernity, and, thus, the sense of its contemporary crisis). This second Modernity had to accomplish or increase its efficacy through simplification in order to 'manage' the immense world system.
>
> (Dussel 2013: 33)

Now how to reclaim ethics against the "management" culture embedded in reductionist, simplicist Modernity? Reading through the biological evolution of humankind, from the cellular to the cerebral with Humberto Maturana, Dussel defines what he calls *the material moment.* In this discussion and its relation to the history of Western philosophy, Dussel, I think, illustrates the simple beginning of the notion of disposability in the following quote:

> Dualism and Kantian formalism (as well as Cartesianism and so many other modern, purely formal ethical systems) have dealt a bad hand to ethics. The negation of the 'body' in favor of the decorporealized 'soul' (foot he Greeks up to modern ethics) speaks to us of a very specific tradition, [...].
>
> (2013: 65)

Disposability (from the Greeks to the colonization of the American and African continents to the slave trade and even to the precarization of everything but especially of black and brown bodies) hinges precisely on the notion of disposability historically rendered so, that one can deny the existence of the Other's soul (even if the soul was re-coded as "energy" [in the 19th century] or as lately, again and quasi-Greek recoded as agency, and only if such was said to having been diminished [see Stingl 2008/2010 on the question of the location/seat and the divisibility of the soul]), but one couldn't deny the existence of the Other's body: Indeed, the point is her that ethics that begin to obtain warrant from the body to begin with have a much harder time to justify (perhaps even a hard time to entitle) *disposability*; for example, to locate the ethical foundation in the soul (etc.) was the Western epistemological trick that allowed disposability and its effect of building and maintaining the geopolitical empire named Modernity on the backs of those it deemed disposable. This is, of course, today re-rendered through the geontopower of denying something to be qualified as "life" or by reducing certain forms of life to exploitable/extractable biomass in the so-called BioEconomy. [It should be mentioned here, though, that Dussel does seem to give a certain ethical primacy to the human throughout, and non-human agents do not play a significant role in his ethics aside from their (material) relations to humans. Humans are both subject and object of ethics, and non-humans for Dussel cannot, I think, be subject of ethics, and object only in a human-related way. This is where I disagree with Dussel, precisely because I think that there are non-human embodied, inactive, extended *cognate*(!) agencies in play, which must be viewed in terms of ethical systematicities; see for example the Environmental Humanities discourse.]

With the emergence of Anglo-Saxon political-economical-philosophical thinking, particularly in utilitarianism (69f.), we find the emergence of a regime that makes bodies (with and without "soul": from "soulless" populations that serve as disposable workforce to one's own body) *manageable* (*oikein*). I use this word here quite deliberately, because it has become so common for us; yet its etymology should not be discounted, since through its relation to "hand", i.e. something being "handled" "controlled by means of hands", it is a word used for training animals, especially horses, to perform certain *maneuvers*. Thus, it stems from the realm of animal husbandry, where husband is a verb that, as a noun, seems to exist in a different context – that of marriage. But the husband and the role of "his" woman – "woman as Other" is constructed in the history of Western mentality too often as a kind of property (Maldonado-Torres 2007) – were historically equally defined by metaphors and ideas about women not being fully in control of their passions nor about their leaky and unruly bodies; thus the husband was called on to husband, i.e. "to manage" his wife, in other words "to train" her if her parents did not already "properly" do so, like one would a horse. In this *oikos* story, the husband is narrated and made to matter as the master of the house, *the oikos*, which means he is the master of his wife – socially constructed as someone who has not full control over her own body – as well as the household's many "disposables" who have little to no soul to control their body: For instance, in fictional tropes of literary works as well as in the (pseudo) scientific discourses well into even the latter half of the 20th century, the "savage", the "black", the "native" are narrated by white (male) authors to commit perceived(!!!) transgressions beyond what the white, male Master/ Manager considers proper. These transgressions in the racist and misogynistic fictions were explained through notions of "they can't help themselves", and it was up to "the manager" to manage,[2] to husband, to train, by means of punishment (including making an example of someone for others to see): Kate Manne has described the mechanisms of the general "logic of misogyny" in her brilliant book *Down Girl* (2017). Is there really any doubt, that in 21st century welfare policies supposedly for (but rather, in reality against) Black Americans or sensationalist press reporting on refugees in Europe, these tropes do not come up? Why were, for example, work requirements deployed against largely Black urban neighborhoods and lifted for largely white rural areas?[3]

And to come back to women being considered as property: Is it not interesting, in which kinds of ways Western society portrays so-called "honor killings" among Muslim migrants, namely with the "these are primitive cultures" trope? But at the same time, on the various and many occasions that a white man kills an ex-wife or ex-girlfriend, this is not attributed to Western culture or "white man's" general nature. And we could speak here, in an expansion, about incidences of domestic violence in general. [Not to be misunderstood here, it is not my goal to simplify these complex issues. But the construction of the Other as property, as disposable, thus, the discourses in play, do play their role here.] The discourse subject of "woman as property of a master" can also be identified in the idea and trope of the "right to first night" (*droit du seigneur/jus primae noctis/ droit de jambage*) – even if that trope is largely an invention (Boureau 1998) for the sake of fiction, the fact that this was "imaginable", i.e. that this was an expression in a discourse that some could imagine to be true and even take it for a reason to rebel, should tell us something; similarly, we can identify the "woman as property" construct in today's internet threads on the "men's right to have sex" in what become the so-called *incel* discourse (*incel* stands for "involuntary celibate", and on related matters of colonial construction of Others online, see my book *The Digital Coloniality of Power*).

It should be said here, too, on the issue of "management", Dussel argues (2013: 26), that in setting up its "world system" version, Europe deployed the "center/margin"-form – with itself at the center – to "manage centrality" and create itself as "reflexive consciousness of world history" re-narrativizing many "displaced" ideas and inventions as if they were the sole product of European ingenuity on the one hand, and, on the other hand, re-narrativizing colonial effects (such as capitalism) as causes for development.

To return to the historic shift that utilitarianism presents, at its core lies an "empiricism" that does rewrite the relation to the body (Dussel proposed for the English translation of his book the use of the term *corpo-reality*); the body becomes simultaneously a new field of experiment (Kant would understand the use of dietetics to prolong one's life [Stingl 2010], while Mill would talk about the freedom to experiment with one's body [Forst 2002]) but also the locus of suffering, a suffering that could be deployed in a two-fold manner; it could be seen as something that needs to be vanquished but also as something that could be weaponized against populations – this use of suffering is, clearly, different from the suffering that Foucault describes in the famous beginning passages of *Discipline and Punish* (1995 [orig. 1975]) where it serves a purifying purpose. With the colonial-utilitarian move, suffering and to make suffer, we obtain a diversification: It can be used both as a blunt tool and a surgical instrument. It tells to the white European masses the fable that "the primitive"do suffer: they suffer their "lack of civilization" in the periphery, and they suffer if they do not obey; thus, the European population should be aware that they could equally suffer if they do not behave according to the centrist norms: In Auschwitz and other concentration camps, the Nazis would write above the gate: *Arbeit macht frei!* ("To labor makes you free!"), and one can ask, of course, how was and what made this enunciation possible to appear in this place, what did it evoke, what is the reason, the genealogy in play, but also why would the connection between labor and obtaining your freedom prominently continue to appear after Auschwitz? In *Modernity and the Holocaust* (1991), the late Zygmunt Bauman identified the emergence of a management style (*oikein*) to this effect.

Utilitarianism with its reductionist approach towards abstraction, such as reducing the "stability" and "happiness" of a society to mean "possession of goods as private property" within a scope of a methodological individualism, which is performed via a means-end control that follows a management style (*oikonomia*) of "an instrumental calculus that sites ethical experience with reference to passions as the ultimate point of reference [.....*allowing*] a certain disciplined empirical rationalization of decisions, actions, and their consequences" (Dussel 2013: 71). This, which Dussel calls a *consequentialist neostoicism*, allows, in a related footnote (FN 97: 487), a comparison of Roman and British Empire notions of learning "self-control" with a reference to the Greek *apatheia*, but I would point out, precisely, a reference to the Greek notion of *oikos* (see also on "management" in the form of *oikonomia* working as a theory [or possibly *technics*] of Empire: Nelsestuen 2017). For starters, it is precisely in tune with the idea that the (i.e. any) Other is to be rendered *disposable* in order to be used and abused and eventually disposed of, such as in slavery, because to do so requires a mentality of *apatheia*: It has been rightly pointed out by some, that an English translation of that term – *apatheia* – should not be "indifference" but "equanimity", for the disposability of Others (in their radical Otherness) is precisely not a general arrivial at the point where a colonial mind is unaware of the Other – even in their radical Otherness – being an Other being (even one with intelligence), which is precisely why some colonizers may truly be indifferent, whereas others are committing the violence and abuse that they are committing while fully understanding they are committing violence and abuse against some (one/thing that is a) Being, however Other that Being is. To be indifferent would mean, at its most charitable interpretation, that one acting against an Other doesn't truly see the Other as a being; one doesn't even note difference, however radical, as a difference of being. But in acting in equanimity, one is aware of the difference of being, but construes that very difference as one that makes the Other a killable/rapeable/exploitable/extractable, i.e. ultimately disposable: "The ethical order is *simplified* to the maximum in order to control rationally military strategy as well as economic and political *business*, both internal and *colonial*" (ibid.).

The fundamental Benthamite principle, then, that the goal of the determination between what is right and good and what is wrong was "the greatest happiness of the greatest number" – but who can be counted among that number and what is measured as happiness are, of course, only that which is relevant and those who are relevant.

Relevance in the Utilitarian scheme is, according to Dussel, then, appraised in a minimum of four dimensions (ibid.: 73):

- *a moment of pleasure*
- *a moment of utility*
- *a consequentialist moment*
- *a social effect* (i.e. fulfillment of happiness is meant for the majority)

Dussel points out the slew of issues that have been brought to bear onto the classic utilitarian: First and foremost, it is, of course, not as clear what "happiness" is supposed to mean, to measure, and how it transits from individual to a collective level. Finally, if utility means actual use or consumption, goods seem to be necessarily distributed in a capitalist horizon (74), which abstracts the commodity/buyer(-preference) relation to a point where the commodity producer's happiness is concealed and forgotten (ibid.) and, I would add to this, that on the one hand, the distinction between commodities and assets is also not properly reflected and, on the other hand, the emergence of prosumers and consumer-prosumer hybrids (Lamla 2013). The inherent paradox of utilitarianism identified by Dussel's account is that utilitarianism either does not grasp or ignores the "essence of capital(ism)" – which is its necessary condition – and, subsequently, its effect which is counter to the premised goal: Capitalism requires the existence of inequality precisely where utilitarianism seeks to establish equality, namely in the distribution of happiness-producing goods.

Beyond utilitarianism and formalism (from Kant to Habermas), Dussel argues (77–85) against a communitarianism that the MacIntyre-ean accounts themselves do not understand in their relativist notion of differing visions of the virtues of good/happy society; he argues as well as against Charles Taylor's notion of a common cultural horizon defined by the struggle for (authentic) identity and its recognition that fails to see that there is only one possible "universal material principle: the *mode of reality* that is human life itself" (83). And against Walzer, Dussel points out that while Walzer accepts the existence of localized domination in the respective "spheres" (of justice, wherein happiness/goods are necessarily unequally distributed), however: this leads Walzer, in Dussel's account, to affirm this local dominance (aka hegemony) and render it subject to management (*oikonomia*) in terms of compensatory regimes through other spheres's regimes of hegemony. Dussel's own *Ethics of Liberation* accepts the existence of dominance, but claims that once hegemony is revealed, work is to be done to overcome it; therein, I think, we can find a point of conversation between Dussel and Chantal Mouffe.

Going Further

Against the *oikos*-type theories – which often, such as in the economic doctrine of markets, falsely portray themselves as not being theory or ethics but claim to be practical or "realpolitik" – that are often at the heart of anti-intellectualism and of antagonism, respectively the *economy of relevance*, that fuels *White Collar sociology*, therefore, against the hypostasis and virtualization of hyper-positivistic, bald naturalism, self-fulfilling Leftist dystopianism, Neo- and Late Liberal economics, and against "STEM without steAm", we should set another kind of theory: a theory that is the *thinking with and learning together with radical Other(nes)s*.

But this kind of theory, which is precisely "not relevant" because it allows one to question relevance itself, as for its genealogy and criteria, is difficult to sell in "the marketization of thinking and of knowledges" – a phrase that should be itself oxymoronic, but sadly is performed as if this marketization was possible – when asked to "compete" in an *economy of relevance* against theories and ethics that style themselves as not being theories and ethics at all, which claim to be relevant because in having gained discursive/epistemic hegemony they can also determine what people are supposed to believe relevance is, before they produce that which is – fictionally – relevant. These relevant theories, which claim they are not theory (or, at least they claim to be parsimonious, or common sensical, when all they are is impoverished to the point of being propagandistic), make predictions that seem successful as long as they can keep the game rigged in their favor: As long as people believe in "*the* Market" and also believe that "*the* Economy" is the most import (sub)system of society and as long as people believe in the equation "money = power", for just as long will economics and Business Administration – the latter being nothing but glorified accounting – be considered the most important "theories/research areas" in the university along with those fields that claim to produce the most marketable and profitable goods, i.e. STEM. These fields claim to give us "predictability", but most people are not even aware how much this predictability rests on tautology and self-fulfilling prophecies that are grounded on people continuing to believe in this specific idea of *relevance.* And even in failure, even when the past 20 years rampant economic, social, or ecological crises could have removed any "veil of ignorance" from people's eyes, the opposite seems to happen in the Global North. Sure, we have seen movements such as *Occupy* or *Nuit Debout*, which are, indeed, important developments; yet, they also have not created *actual* political alternatives.[4] In part, many Global Northern protest movements fail to produce alternatives because – much like core Marxism did – they buy into the Western myths and logics of the *oikos* that prefigures their economy of relevance and, subsequently, the temporality for their actions: Much like any other group within Western society (as an epistemic superculture) and in the geopolitical sphere of influence of the Global North, these protest movements could not disentangle differing temporalities and allow for different processes with different temporal structures and durations: No matter how long a protest will last, protesters usually still assume a teleology that the successful end of the protest will result in an immediate econo-social (status) change (for their group, or, sometimes quite ambitiously, society or "the world"[5] at large). Rather than protesting for a "change", which is mostly a static kind of change, we must achieve a change in how we approach problems, find solutions, create opportunities from within the problems, thereby allowing us to understand what change itself is and should be in the first place. This is reconcilable with a post-foundational view to what I propose as an act of rethinking and delinking here: Most social protesters in the Global North start from assuming the existence of social and political foundations, for example that an economy of some kind is a crucial part of the infrastructure – not even, or especially not, the Marxists deny this, as they largely look for another distribution in the economy but do not contest the foundational status of economy. But quite frankly, it's not. What social protesters should be considering is that the change that must happen is that of a reopening of a discussion about *the foundations* that we can no longer afford to continually reify. Actual social change would mean to recognize that there are no inherent foundations of our societies and that the success of the *Late Liberal prophets of relevance* as well as of the neoliberal predator *profiteurs* and *rentiers* is genuinely helped by the general belief that the social and political foundations of societies are somehow self-evident or given (or natural) and untouchable; their success by intentionally abusing these foundations lies in widespread belief that these foundations are to be treated as if they were natural kinds. But thereby we do not uncover the true *vulnerabilities* of our societies – as uncomfortable as dealing with vulnerability may be, it is very necessary. Social protests are all too often about the *politics of sensibilities* rather than about the actual *vulnerabilities*. For example, many media outlets in 2015 and 2016 have been reporting that in the world of private liberal arts colleges in the United States, a fight over the cultural appropriation of food cultures in school cafeterias is being waged. While cultural appropriation is, indeed, in itself an important topic, what do these reports miss? They miss there is an important conversation required about cheap, unhealthy, and quite shitty food being produced and served in many of these cafeterias as well as a conversation about the precarious labor conditions that go with these regimes of cheap food production: We need to talk about the *oikonomic* sources for the *articulation* and *assembly* of certain food products, which are featured in these reports. The media, and probably also some student activists (although I am certain not the majority, because there are many smart students), focus on certain *sensibilities* rather than going toward the actual *vulnerabilities*. What should be called for, here, is an honest discussion, which may take time before it can change, over how the provision for livelihoods (including the production of food for the students as well as those who produce it) on college campuses should be created without exploiting vulnerabilities.[6] Unfortunately, what happens actually is that in most cases the outcome is just "more of the sane". To change, for example, the company that runs the cafeteria, replace the employees with new ones who earn 50 cents an hour more, and change the menu to include less "*in*appropriate" and more "appropriate (aka relevant)" food, is what is most likely to happen in such a case but it won't really change matters at their very foundation.

But that is precisely the issue that theorists who really understand theory as something *to think with*, and, thus, as going to think and fight about foundations – and do so in terms of *agonistics* and *epistemic delinking* – are just not as sexy (do not seem as relevant) as the STEMmy theories with their linear, immediate temporalities that promise to address our immediate sensibilities. It's easier to treat the itch with an ointment than to consider figuring out what in your diet may cause it and adjust your diet, or even listen to the person who is thinking about and trying out better diets, because the theory of *itch and ointment* seems more parsimonious, predictable, and relevant. That other theorist, the BluesCollar, who offers,

> *against the markmanship of this simplified kind of theory, a more fundamental kind of insight into matters* [of care; A.S.], *doesn't look too attractive at first. Through it [this kind of insight] he was rendered non-competitive. In contrast, the form of knowledge of the cave* [from Plato's cave allegory] *seems capable of progress: If left to its own devices, it should, in the long-term measure of its own overview, increase its own efficiency. However, those who are chained* [in Plato's cave] *do not recognize that the possibility of their competition rest on a* ***single*** *condition: the reliable of the machinations for the production of shadows from behind their backs.*
>
> (Blumenberg 1989: 155; my translation)

The crux for the BluesCollar, however, lies elsewhere: Unlike the *White Collar academician*, s/he cannot pretend to occupy a position of aloofness (whether as the god-trick of "objectivity" or "realpolitik"), as if they were not born or have ever lived in the cave (*The European tradition of the interpretation of the cave*, says Blumenberg [1989: 160; my translation], *has never had any comprehension and appreciation for the appeals of the cave*): The position of the BluesCollar, defying Plato's Socrates, is that of a genuine Third(ness); the BluesCollar knows what it is like to be to live in a cave and what it is like to be outside it; therefore, the BluesCollar does not resort to pretending to be either nor forges a false hybrid consensus between the two. Unlike the Socratic *White Collaredness*, the BluesCollar knowledge is not something "you" or "I" *have* (let alone can *have* in *exclusively)*; it is a knowledge that manifoldly and pluriversally *matters*, *worlds*, is variously *embodied*, *enacted*, and *extended*. It exists only in *togetherness*. And while it may be frustrating for some who are now looking for a precise definition what the kind of "member of the household of contemporary knowledge economies and future knowledge ecologies" the BluesCollar is supposed to be, that is just the very point – to define would mean to cut off and close [*clôture*] what is most crucial here: *becoming* – that the BluesCollar must be filled with life by living it to its fullest, by *becoming together with others*: BluesCollarship, a provincializing maneuver within Western knowledge, is a maneuver of *cui in vita* toward becoming *in plena vita*.

Foucault: Orientations for Thinking and Doing from Discipline, to Problem, to Opportunity

I feel it is important to say a word, in the end, about Michel Foucault. Why? Because in order to embark on the journey to provincialize the social sciences together with Others and to enter into constructive dialog, I need to talk about my positionality also, to show also, as well as I can and am aware, the implicit contexts where this book may be coming from. And I do see myself not as a disciple but as inspired by Foucault, even if I am aware of many of his faults and the limits of his work, as well as the overuse, misuse, abuse, and sloppy and/or malappropriation use of his works and ideas that have strayed from plausible accounts of his work. It is important to make these few remarks here, even if there may be some redundancy, to reflect on the geopolitical location and time-frame of what I have said in this book, and what I am still going to say in the final part of the book. Michel Foucault, who was both a strict Kantian and an inspired Nietzschean, was not a decolonial scholar (not even a postcolonial one) nor does his thinking lend itself without some active doing to recent and current discussions in de-/post-colonial and feminist theorizing, science studies, political studies, and sociology. That does not mean that Foucault's writings have not and do not still play an important role for scholars socialized into or operating within the Western academy, to make a beginning toward lines of inquiry that expose the coloniality of power and Being, the privileges of positions, hegemonic mechanics of discourses, and epistemic paternalisms within the Western academy, extruding from, or being amplified by it. An important and influential decolonial voice such as Walter Mignolo readily acknowledges and credits Foucault and his scholarship for various critical ideas he brought to the table that allowed to ask important questions that did and continue to help those who want(ed) to embark on uncovering and transforming the mechanisms of the colonial matrix of power. And yet, Foucault himself not only did not address, even seemed to have actively avoided (not only issues of coloniality but already) issues of colonialism – Peter Redfield and Ann Stoler, among others have expertly and studiously critiqued the Foucauldean enterprise on this ground (see my appraisal of their work: Chapter 1 of Stingl 2015).

Second, the original Foucault – despite a few attempts toward the opposite – does not lend himself well to being absorbed into either recent ecological paradigms or new materialism. While Brian Massumi (2009, 2014) is, of course, correct in alerting us to the brief mention Foucault made in his lectures to an idea of *environmentality* in his discussion of *gouvernmentalité*, some scholars – although I do not see Massumi himself really wanting to go that far – have attempted to use this as a point of departure for arguing that we should see in Foucault an influential scholar in the discussion of ecological/environmental problems as well as new materialist discourses who was already imbricated in these discourses at their, or even as their, very origin. This is, I am afraid, overstretching matters a bit. Sure, Foucault became aware of the growing concern with the environment among his students, among political activists, as well as among intellectuals such as Guattari or for other proponents of *deep ecology* and *ecosophy* ideas that date back as far as early 1970s (Fox 1995).[7] But it seems more likely, given the brief moment that Foucault toyed around with the idea, that he never did anything more than actually only "toy" with it and then discard it. There are several reasons why this is likely, but the most important one, from my point of view, is that it would not have lend itself to the kinds of methodological approaches and materials Foucault studied. At the time, there were no archives, as Foucault would have understood them, to study – it wasn't really, until scholars such as Dorceta Taylor (2009, 2014, 2016) did show us that there are. And while, indeed, the term *environmentality* itself might be useful and we can "happily"[8] use it, as a conceptual tool itself, it owes very little to Foucault himself but is fully intelligible, communicable, and deployable only today after a lot of transmutative doings. Unlike Simondon, who can be – as Erich Hörl suggests – "our contemporary" on issues of digital society, Foucault cannot be our contemporary when we are talking about ecology – he may be our contemporary on other issues, still, or, at least, be a useful classic to help make intelligible certain aspects in the same way as Max Weber's *Wissenschaftslehre* is. But, most importantly, Foucault also cannot be our contemporary with regard to the various *new materialisms* and associated attitudes and movements. Several scholars, among them Thomas Lemke (2015), have attempted to show how Foucault's works lend themselves to the current *ontological turn*, and even how Foucault may have preempted it or can be deployed to create a superior account of the *ontological turn*. No matter what one does think of the idea of an *ontological turn* itself – like Tim Ingold and others, I find the constant "turning" a bit nauseous as to the effect that not every shift of perspective needs to constitute a "*turn*" of any kind, and I also agree that many ideas in these turns can be found in much earlier literatures (many decades before Foucault) than the ones that become canonic in these turns – not only have the scholars that redirect our attention thus read their Foucault and on occasion decidedly moved away from parts or even most of his ideas, but one is also left to wonder why some (and especially already well-established) scholars are so hell-bent on connecting Foucault to every new trend in significant ways or even paint Foucault as their origin (precisely counter Foucault's own argument against *zero points* in the process of knowledge production). Foucault did not have something to say about everything, nor do his concepts and methodological discourse lend themselves to every discourse either. Lemke's (2015) proposal of deploying Foucault and seeing in him an origin for a discussion of a "government of things", within the discourse that currently is subsumed under the umbrella term *ontological turn*, stretches Foucault's reach too far.

However, my critique here is far from any overgeneralization of Foucault as a mere "textualist" such as that which Richard Rorty (1981) made in a paper on idealism and textualism. Rorty grouped Foucault among a number of authors who ascribed to the idea that there is nothing but texts – a position Rorty sees as a development from certain Kantian insights. The problem is, as I said and however brilliant Rorty may have been, that he gives in to overgeneralizations – regarding both Foucault and Kant. While it is not wrong to say that there was a general trend – or rather more of a Weberian *ideal-type* – of 20th-century scholars who would underwrite a statement such as "there is nothing outside text",[9] I cannot endorse that Foucault is categorized as such. It is precisely because Foucault made his struggle with the nature of the notions of "practice", "discursivity/non-discursivity", and the "outside" explicit, that we know he took into consideration that there is an "outside of text" and that this outside did have implications for the *subject(ivity)* matters he studied; yet he rarely engaged into an explicit analysis of what these "outside matters" are in detail: So, clearly, I argue that where Rorty neglected to take into account Foucault's consideration of "the outside", for "environmentality-qua-Foucault" accounts to have any purchase, Foucault would have had to pursue the whatness of "the outside" in far more explicitness and with far more vigor. Why are so many scholars now bent on breaking Foucault into shape of the current geopolitical discourse of academia? Well, if you are a member or disciple of a discipline or a group of scholars where Foucault used to be the theoretician then membership is maintained by a kind of disciplinary decadence: White Collar Academia all over again! These people fight to stay relevant and prior association with Foucauldian theory and work means to make Foucauldean thought fit for the current economy of relevance. That's how economies of relevance work, after all. It's funny how Foucault himself, of all people, produced a series of gestures and a way of thinking that allows us to criticize this very function of an economy of relevance itself.

On Our Post-Risk Society

Some 30 years ago, a German sociologist couple suggested that our society is a Risk Society, that the idea of risk has become an integral part of our social structures and a lens through which we experience life? Now risk does not mean that we anticipate something will happen, but to anticipate that something could happen. There is an element of predictability to it, and we can attempt to compensate. Now, I think, we have gone further than that today, that we live in a post-risk society: What comes after risk, when risk is already built into every aspect of social, political, technical life. Our lives are controlled by an increasing number of algorithms that make up digital architectures as well as molecular life of biological organisms and ecosystems that we transform or even create. And these algorithms already have risk built into them, not in ways that merely anticipate risks, but that use these risks to produce new things and forms. The real danger here is that there are interactions between these many new forms and objects, and algorithms that we don't fully understand: For example, how do the super-speeds of financial market algorithms interact with the effects of deforestation and reforestation in rainforest agriculture? How are the development of bacterial resistance to new antibiotics, the use of these in biomedicine and in agriculture, the shareholder value of research-intensive pharmaceutical companies, the future market on food production, and the spectacle of new cycles around the "risk" of an epidemic entangled? Because risk is already a "normalized part" of all these issues, these risks interact with one another very productively. So what we should look at is not the anticipation that something might or might not happen, but what happens because too many things that may likely never happen are becoming part of all the structures of anticipation, of all our algorithms digitally or biologically. What we want to think with is this concept of precarization, which is the process of increasing a condition of existence without predictability, security, or sustainability in one's modes of [re]production. So, perhaps because of the risk society, after the risk society, we have to look for the precarization of life itself, the precarization of everything, really. Risk has become the lens through which we experience life. Today, risk is built into every aspect of life, through algorithms in both our digital architectures and the molecular interventions into biological organisms and ecosystems. In many issues, we are witness to the precarization of life itself, the precarization of everything, really.

The imaginary and the practices of how land and livelihoods are (over) taken by urban/neoliberal/colonial Settler forces that silence the land's agency and silence those living with the land, the animals, and the plants are eerily analogous across the globe at times. In her inspiring, meticulous, and beautifully written study of multispecies relations, *Animal Intimacies* (2018), Radhika Govindrajan tells the story of rhesus macaques that "terrorize" rural villages in India. These macaque hordes, however, seem not to originate in the regions but are troublesome former "city dwellers" that have been relocated to the rural areas by official and unofficial actions to ease the lives of the cities' human populations – killing the animals was not a moral option. Villagers, on the other hand, in Govindrajan's study, seem to be able to account for differences in local and "foreign" monkeys' behaviors, the latter being highly aggressive. These new arrivals even manage to destroy livelihoods of people, which is by the locals often seen in concert (and thus as a deliberate outcome) with some political actions and the actions of land developers, among others, who would drive them off their land by sophisticated criminal strategies, which represents a criminal mentality that urban macaques are set to have acquired from humans. In one particular vignette illustrating the human side of this *urban/metropolitan criminality*, Govindrajan replicates a villager's account of being coaxed into selling his home but being cheated out of money (2018: 110f).

> The developer had paid him an advance and promised that the rest of the payment would soon follow. In the meantime, he suggested that it was expedient to go through with the legal process of changing title lest his partners get cold feet. When Mohan recounted this part of the story to me, he grimaced at the memory of what had followed.
>
> "I was so stupid. I thought such a big man would never cheat me. He was so nice; he brought books for my children. But then after I registered the land in his name, the rest of the money never came. I used to call him a hundred times a day. He never picked up the phone. When I sent him an SMS threatening to call the police, the *gundas* came. He broke my hand and said worse things would happen if I pursued the case. In the end I gave up. There's no point in going to the police; he can buy them easily.
>
> [….]"
>
> Such stories of betrayal abound.

I found myself immediately reminded of the practices the fracking industry deploys in the United States, as accounted for in Sara Wylie's seminal book *Fracktivism* (2018). Landowners she encountered, too, had in the end little chance of resisting that the ground underneath the land they owned was used for fracking and that they, subsequently, were exposed to the toxic chemicals used by the industry. In gaining access, representatives and agents related to the industry used tactics of disinformation, attempted to divide communities, and deployed all kinds of pressure. When exposures lead to severe illness, the industry is in a position to deploy an arsenal of tactics that are helped by the political power they have, quite literally, acquired. Even when legitimacy and legality of their practices may be doubtful, the power they wield can drag legal and policy procedures out near indefinitely and make them costly to the isolated individuals who seek redress, as Wylie's research illustrates. Of course, similar tactics have been deployed as well in regard to agriculture and the use of genetically modified seeds, and efforts to establish seed monopolies and disempower farmers.

To regain power over stories, we must engage in exercises of multi-pole and diverse story-telling and, more importantly, story-listening, and to speculative/fabulative design and fabulation, and imagine a different tactical media and civic science, and to take seriously, as Boaventura de Sousa Santos argues, *artisanal practices*. The Bioeconomy and Digital Culture present the most influential transformations of society and economy emerging from within the Global North. They come together, joined at the hip in *Industry 4.0*: a buzzword for politicians, media, and business representatives that caught the imagination of public worldwide, ever since its introduction at the 2011 Hannover Fair. The discourse surrounding this supposed "fourth industrial revolution" is often presented as a revolution of manufacturing processes through all-out automation, pervasive datafication, and creating an "Internet of Things", which results in the creation of "smart factories", cyber-physical systems, and integrated value-chains. However, this process, based on what is more commonly yet vaguely described under the term "digitalization", is only one half of the major technological-cum-socioeconomic transformation global societies and economies are undergoing. The second (and no less dependent on datafication, cybernetics, and integrated value-chains) process that guides the current transformation is about biotechnologies and ecosystems. Taking note of the ecological destruction caused by human consumption, congealed in a geological era now called *Anthropocence* (or *Capitalocene*), has necessitated debates on concepts such as "sustainability", "circular economies", etc. Technological solutions for these issues are inherently understood to revolve around biotechnologies – regardless of the amount of involvement of digital tools and Big Data needs – and, subsequently, their application needs to be conducted conscious of the measures(!) of ecosystems they interact with, draw from, and/or become embedded in. In order to avoid ending up with a polarized situation – as we find in so many political arenas today – we need to develop a different perspective, which I aim for with the elaboration of aesthetic process in science: With polarized I mean that there are, on the one side, those who believe uncritically in a linear progress of digital and biotechnological solutions for global problems– this stance we find equally among biotech industry and biohackers, although with different stances toward capitalism. On the other side, there are those critical thinkers who are often flat out rejecting biotechnology, often with little understanding of the biological foundations. Art can serve as a bridge between those poles, and the avoidance of these pitfalls can be facilitated by creating what I call "artisanal stance". This aims to realize the call for a *civic science*, formulated by Sara Ann Wylie, that "ties together research and action to improve shared economic, environmental and social conditions" (Wylie 2018: 265).

This is situated in the geopolitics of knowledge and knowing, as we learn from Walter Mignolo, which is worth quoting here at length:

> Geo-politics of knowledge goes hand in hand with geo-politics of knowing. Who and when, why and where is knowledge generated (rather than produced, like cars or cell phones
>
> […]
>
> Last but not least, my argument doesn't claim originality ('originality' is one of the basic expectations of modern control of subjectivity) but aims to make a contribution to growing processes of de-coloniality around the world. My humble claim is that geo- and

> body-politics of knowledge has been hidden from the self-serving interests of Western epistemology and that a task of de-colonial thinking is the unveiling of epistemic silences of Western epistemology and affirming the epistemic rights of the racially devalued, and de-colonial options to allow the silences to build arguments to confront those who take 'originality' as the ultimate criterion for the final judgment.
> (Mignolo 2009: 2–4)

So what do we do, for example, as scholars who want to speculatively design "a different economy"? Generally, economics and other social sciences construct economies along the reification of an ideal-type of "*the* Economy", which is an economy wherein markets exist somewhat necessarily and to the point of also being the sufficient criterion, from which follows the centrality on price (and a quantifiable and monetizable concept of value/utility). However, I propose to look at economies differently, and create another economics, which does know non-market-based economies (wherein markets are not necessary, and often don't exist or are not even possible) and wherefore there is no price(-function) and *value* has a very different meaning. *Mindless* adherence to *the* Economy leads to precarization (of everything).

Notes

1 A nice little summary of Foucault's notion of care – in relation to Aristotle, see: https://politicaltheology.com/foucaults-care/.

2 This is derived from the pervasiveness of *oikos* that emerges with the Athenian law and policymaking.

> [T]he conception of the state as a household is slow to emerge as a fully-fledged political image: only in the late sixth century, the context of the story in Herodotus (5.29) that the Parians chosen to act as arbitrators in the civil strife at Miletus placed the administration of the city in the hands of those whose lands they saw to be well managed, do we encounter a clear if still only implicit analogy between domestic and political administration. The development of the image in the succeeding period is hard to trace and attested only by a scatter of images of the ruler as steward in Pindar (see below); it is not until the second half of the fifth century that we see evidence that the image was firmly established and indeed commonplace in the use of the verb *oikein* in place of *dioikein*, the standard term, to denote administration, particularly when applied zeugmatically to both cities and households in Euripides, Xenophon, Plato and the orators, and in the figurative use of *oikonomia* and its cognates.
>
> [...]
>
> The keynote of the fifth-century imagery of domestic management (*oikein*, *dioikein*, *oikonomia*) had been administration, and indeed that version, with the associated implications of competence, honesty and service, persisted in the democratic discourse of the fourth century (..). However, fourth-century intellectuals shifted the focus to authority and to the figure of the householder, and the analogy between the two spheres is made explicit in the assertion that there is a single art of rule, so that expertise

> is transferable between them. The firm establishment by this point of the lexical usage bridging the two spheres doubtless made it easier to assert that what was in question was essentially the same activity, the only difference being one of scale, but the principal attractions of the householder as the model of the authority figure were, I think, his ubiquity, and the fundamental importance of his role: since effective running of the oikos was a pre-requisite for the survival of the polis and its inhabitants, the authority of the householder could likewise be regarded as fundamental and natural, always operative and not requiring explanation or justification.
> (Brock 2013: 25 and 149)

3 www.washingtonpost.com/opinions/the-gop-is-quietly-crafting-work-requirement-waivers--for-white-people/2018/05/16/fcef4eb8-5928-11e8-8836-a4a123c359ab_story.html?utm_term=.c9ad5a9ce9c0)?
4 It is also a harsh truth that lies in a popular "critical" meme that posits that "if facebook and instagram shut down, many people would suddenly no longer be activists".
5 A decolonial critique would have to point out, of course, that there are notions and ideas that this "the world" rests on, which many people living on this planet, especially in the Global South, might not (wish to) share in the same manner.
6 Which can include but should not focus on appropriation on the expense of continuing precarization, exploitation, and suffering regarding the basics of life.
7 Arne Naess, who coined the term, participated in one of four debates between philosophers organized for Dutch television in 1971; one of the other four debates was the (in)famous debate between Chomsky and Foucault. It would not be unreasonable to assume that Foucault took notice of Naess's ideas at that time.
8 In the attitude of a "happy positivism" that Foucault promoted.
9 Although it should not be forgotten that it was Walter Benjamin who said that "[w]e cannot imagine a total absence of language in anything" in 1916 and, as such, long before (Heidegger,) Derrida, and others that Rorty mentions, and that the reconstruction of the reception of Benjamin's works by various influential 20th-century thinkers is still an unfinished project. While Derrida has, of course, written on Benjamin, we can only speculate at present about Foucault's reception of Benjamin's work (whom he cites on less than a handful of occasions, despite striking similarities, especially in his work on *heterotopias*).

6 The Emergence of Green Precarity

Bioeconomies as Technological Zones in the Imaginary of the Global Rural-Urban Matrix

In this chapter, I broach into the notion of *Bioeconomy* as an imaginary that intersects with practice regimes, discourses, and a transnational policy network (TNP). Here, I am predominately interested in *the* BioEconomy's "techno-social imaginaries", as my research into practices, discourses, and the TNP is still ongoing: We are witnessing a biotechnological change in the general techno-political and socio-economic imaginaries, reshaping interrelations between practice regimes and subjectivities as the intersecting standpoints that enable and constrain experience. Imaginaries, as I have elaborated elsewhere (Stingl 2015), may be described as kinds of fluid, semi-stable reservoirs of meaning that are conceptualized in images. Hence, notions such as *liquid types* and attention to multiple ontologies are interesting additions for the sociological toolbox and imagination to *think with*. In this chapter, I am also guided by a lesson learned for my friend Hans J.I. Bakker, who states that (2015b) "[w]e cannot understand any society when we fall victim to urban normativity."

There is, after all, a relation between the *oikos(/polis)* operating system and the rural/urban dichotomy that *favors* the urban over the rural along the lines of coloniality and Settler Colonialism. These "operational parameters" describe a phantasmatic/phantasmagorical enclosure of the imaginary in play here.

To Imagine Only that Which Is Imaginable

In order to create in our minds and in our shared collective lives any picture of how we will do something, be it that we wish to develop a successful smart-phone app or build a better dairy farm or make a nation's people understand and appreciate the importance of not using plastic bags, it is not enough to merely "imagine" it spontaneously. The "imagination" draws on resources. By resources, I mean the vast set of simple and complex images of available practices, their deployments, and their constraints (Stingl 2011), sometimes loosely, sometimes structurally coupled, but overall what we can call imaginaries. Science, Technology, and Society Studies (STS) have studied these socio-technical imaginaries such as in the case of biomedicine and genetics by Joan Fujimura (2003). Imaginaries are entailed in all performance and enactment of practices, including everyday and quotidian practices:

> Imaginaries demarcate what we think we do, what we think is possible to do, what we think is not possible to do, as well as whether or not and how we can possibly overcome constraints of our possibilities. Imaginaries are not limited to a specific field or sector, but interconnect them: Technobiological imaginaries concern not only biologists, but they come to play an important role for parents considering genetic counseling or in-vitro fertilization while in-vitro fertilization, on the other hand, plays an important role in agriculture, evoking and intersecting with a different imaginary.
>
> (Stingl 2015: 168)

The imaginary I am concerned with here is what is generally understood as *the Bioeconomy* (see https://gbs2018.com/home/; Lewandowski 2018), as outlined, for example, in a larger research initiative by the German Ministry for Education and Research (BMBF) in its programmatic papers on "National Research Strategy for BioEconomy 2030" and "BioEconomy as Societal Transformation", or by other national and supranational organizations.

The (F)Acting of BioEconomy

With Kristin Asdal and Ingrunn Moser (2012) and Christopher Gad and David Ribes (2014) we can make useful the idea of *conceptual-empirical laboratories:* As such, I read together *into* one another Nicole Karafyllis's concept of *Biofakte* (2003) with Stefan Beck's *Technoakte* (n.d.) – in English: *biofacts and techno-acts*, to see what they can do, once we let them lose toward some practice or phenomenon within a practice regime.

Social scientists shun this – at once experimental and normative – *e(qui) vocative* approach somewhat, rather settling for the less risky "concept-applications" that ask what concepts do to describe things or fail to do. But there are things/practices that refuse, resist, obfuscate, obscure, and elude classic conceptual treatment. There is narrative dialectics in play, and we must also ask what practices, relations, and things do to concepts, in turn, as things are transformed into epistemic objects, while at the same time the affective ecologies of practices (aka *political ontologies of production of expressiveness*) labor away at the concepts, inducing a manifold of temporalities – the concept of statehood being one such example (see later chapters). Thus, we arrive at the conceptual-empirical laboratory of *bio(f) acts* and *techno(f)acts*:

Think of urban gardening, of growing your own vegetables in a city like Paris (e.g. Demailly/Darly 2017) for which any urban gardener may assemble and enroll a variety of tools (both analog and digital), soil (what is soil if not itself a complex assemblage and matter of care [Puig della Bellacasa 2017]), seeds, friends, and legitimation of food justice – our aspiring Parisian gardener assembles, enrolls, builds, and deploys to grow a particular type of vegetable in the city in a manner that both "*acts* and *facts*" by co-entangling "*acts* and *facts*".

The contemporary *political economy* as our present *nomos* (rule/law/rationality) of how to hold (*oikein*) – owning *and* belonging to – our house (*oikos*) in and through politics/*la politique* and the political/*le politique* (as in *polis*) can be called, with Bakker, our *global urban-rural matrix*. This matrix, part of the colonial matrix power (Quijano; Mignolo), establishes its own regimes of *techno(f)acting* and *bio(f)acting* of which the so-called BioEconomy is but one.

Governmental elites, leaders of influential politico-economic organizations, and swaths of the "scientifically informed" public associate geopolitical and geo-economic hopes and dreams with the programmatic idea of *the BioEconomy* – as evidenced by recent research initiatives and programs organized by the BMBF, the European Commission, the US-American White House, and the OECD (Organisation for Economic Cooperation and Development). The promise of *BioEconomy* as economic policy and political transformation lies in the idea that biotechnology can improve provisioning for human well-being directly and sustainably while also ensuring profitability. However, what can be realized as *BioEconomy* in lifeworlds and ecologies of lives (*Lebens[um]welten*) as well as in the various social, political, and techno-scientific systems depends on far more than mere biophysical environmental conditions: it depends, above all, on the expectations, ideas, and intuitions of a large number of concerned and affected agents, who negotiate – inter- and intra-actively – the boundaries of this *imaginary*:

> Roughly the last decade has seen the growing interest in a new object of social science inquiry and socio-political imaginary, the "bio-economy". As imaginary, the bio-economy is perhaps best captured by the OECD's aspiration to "turn the disruptive potential of biotechnology to economic advantage" [...] It is also in these frontiers that biopolitical struggles are fought over a new emancipatory "ethics of the bio".
>
> (Haddad 2015)

The practices of frontier struggles and boundary negotiations, whether within a frame of collectively shared motifs or motivations of individual actors, are revealed as being a kind of *normative commitments* that can be described as generated and permeated by "urban normativity". The politics, the economic poetics, the techno-science, and even the social scientific critique are implicitly governed by these commitments. On the other side, we find a line of inquiry into "the Rural"/"the Urban(Metropolitan)", the Modern/Primitive, the High/Low Tech, and numerous familiar binaries that are definitely *not innocent.* A *rural-urban imaginary* wields definitional powers over the social transformations that we can expect under the programmatic of *the* BioEconomy in the *global rural-urban matrix* (Bakker). This matrix, its practical and material determinants, and its political, ethno-economical, and social (justice) consequences, must be revealed in its local, transnational, *and* (de)colonial relations. A *decolonial perspective* focuses on at the very least the following two lines of inquiry in particular: On the one hand, it raises an epistemic question, whether there can be a *BioEconomy* outside of a Western referent; and on the other hand, it engages the geopolitical question, in what ways *the BioEconomy* constitutes a form of neocolonial relations between Global North and South.

As a discourse, it represents the shift we see, for example, in so simple a thing as a plastic bottle, which was once produced by use of petrochemicals and as waste interferes with complex ecosystems. These bottles are increasingly produced as biodegradable bioplastics by biochemical means. In the future, bacterial nano-factories, conducting more and more complex production-degradation processes, will manufacture these products and later process them into animal feedstock.

But how to describe *the* Bioeconomy in general? The easiest way is, perhaps, by three simplified "clichés" (*aim*, *claim*, *task*) about bioeconomy: (a) The *aim* of *the* Bioeconomy is that all of global economy and industry are under a transformation from traditional, fossil-based industry to one entirely based on biotechnology and exploitation of ecosystem services. (b) The *claim* that Bioeconomy is sustainable while global economic productivity is increasing. (c) Its *task* is to increase the efficiency of the use of all biomass.

These three are misconceived. Indeed, we can see that we are currently defining both (1) bioeconomy as its own (social) scientific field *and* (2) a critical and realistic methodology suited to study bioeconomy and its alternative conceptions. Again, current definitions of Bioeconomy differ slightly for every country or supranational organization we are looking at. Bioeconomy stretches from biofuels to pharmaceuticals, to bacteria as factories, to agriculture, and much more. Two concrete examples are given below:

First, in France, the use of "soil-preserving" "alternative permaculture" methods in agriculture are part of a political and legal fight with "soil-precarizing" industrial agriculture. A fight, however, ruled by the logic of financial food futures markets.

Second, corporations use deep sea bacteria to create microbiological-factories qua gene editing. They are chosen because of the indeterminacy of the laws of the seas that allow intellectual property to be asserted easily.

Despite their obvious differences, these examples are equally considered to (discursively) produce *the* transnational Bioeconomy by those stakeholders in biotech-industry, research, and governance who together form *an expert epistemic community*. My (innovative) form of discourse analysis (Stingl 2011), which I apply to the study of "bioeconomy" in all its forms, focuses on how seemingly diverse practices shape and are in return shaped by *the* Bioeconomy. In short, the Bioeconomy is at the same time a lawyer's game, a biologist's game, and a governance game: The rules of these games share and co-produce properties of the Bioeconomy across its different ecological niches. So to speak, *the* Bioeconomy is the forest, the ecological niches (biofuel, bacteria as factories, etc.) are the trees. While traditional STS, sociology, and governance studies research tend to focus on the trees as case studies, metaphorically speaking my focus is on the forest and how it shapes ecological niches without, however, losing sight of "the trees each being a tree".

Two central, intersecting developments have taken place within the 21st-century world politics: (a) *Bioeconomy:* A policy-drive toward supplanting almost all traditional, fossil-based resource-dependent industries *with* new biotechnology industries and markets: In policy, research, and the entrepreneurial sector, this policy-drive is explicitly called "the Bioeconomy", a vague, ambiguous, and often obscure term, behind which hides a complex discourse, almost exclusively conducted by a transnational epistemic community of experts but with far reaching consequences for society at large. (b) *Transnational law*: A global legal doctrine, *lex extractiva*, has emerged in place of *lex mercatoria*, ensuring a global power shift of "corporate capture of normative spaces" by extractive industries. (c) *Intersection*: Both developments intersect in areas where extractive industries produce incalculable risks for populations and ecosystems, increase existing inequalities and ongoing global processes of dispossession, and obscure avenues for democratic participation of citizens in scientific progress and market transparency.

With "Bioeconomy" (which ranges from agriculture, biofuels, urban [re]construction, medicine, biorefineries, biofactories, biomaterials, climate change, etc.), biotechnological industry, researchers, and governments have entered a new constellation, regime, and accelerated phase. This TNP and agenda – as a discourse encountered in policy, funding programs, white papers, etc. – reimagine how we live together in an ecosystem at macroscopic and molecular levels. In other words all kinds of livelihoods in their biophysical, social, economic, and ethical relations are currently put in motion. The questions for future *global governance* are "How is this transition and can it even be regulated, e.g. through the creation of transnational law?" and "Can ways of thinking with alternative environmental legal orders and alternative living systems be a creative and provocative contribution to transnational governance?". "The Bioeconomy" is introduced inside the field by practitioners themselves to serve as an umbrella term and a programmatic agenda; its primary areas of legal contestation are *(intellectual) property*, benefit sharing, and risks and damages. In short, *bioeconomy* is not a term introduced to analyze a discourse of techno-social processes, such as concepts of *biopower, biopolitics,* and *biocapital*. Instead, *there is* an existing discourse on "the Bioeconomy" created and led by members of an epistemic community. This expert discourse is the object of my research. Transnational law is one of the crucial fields where the fate of this discourse and its effects on societies and the global economy in general will be decided by powerful actors. We must take, for example, inspiration from diverse research fields, *environmentality and food sovereignty/security* research, and *critical realism*. We need to integrate disciplinary knowledges from sociology, legal studies, political and governance studies, biology/biotechnology, and food sovereignty studies. Bioeconomy is becoming subject to norms and regulations under transnational law, and the normative spaces of transnational law are no longer founded on the regulative basis of "dead organic matter" of fossil resource economy (*lex petrolea*), but on the basis of intervention, mutation, and creation of living matter (*microbiopolitics*). Our ongoing and future *object of study* is the Bioeconomy discourse as a general phenomenon, whereas for discourse analyses I have been and continue working on the following *objects of observation*: statements, publications, and regulatory efforts contributed by members that comprise the epistemic community, as well as documented reactions and counteractions by affected populations (NGOs, indigenous populations, and their legal representatives).

I critique that BioEconomy, as *(f)acts* in a monocultural and urban-normative hegemony, will lead to a colonial "mining away" off of the very foundations – material and otherwise – for expressing claims to justice and equality, in order to establish what is known as *Green Precarity/ Precariousness/Precarization*.

I have come up, instead, with a proposal of *generative justice* and *generative values* within diverse ethno-economic systems of human provisioning, integrating ideas formulated by Ron Eglash, Enrique Dussel, Nathalia Jaramillo, Angela Mitropoulos, Godwin Y. Agboka, Rainer Forst, Catherine Walby, Sylvia Wynter, and others. An important aspect is to understand that empirically we can encounter *bioeconom**ies*** in the plural, and that bioeconomies of various kinds already exist and have existed historically; this must be understood not in terms of any *doxa* of academic economics but in terms of ethno-economics as complex knowledges and imbrications of *all genre-specific human modes of material provisioning including our contemporary own*.

Conceptual stakes for the "objects" of "*the*" BioEconomy that are subjects-(-and-abjects-)in-waiting to be *precarized* have been laid out in debates within feminist and postcolonial science studies on what "nature" is. Taking cue in Spinoza's distinction between *natura naturans*, as any self-causing activity of nature, and *natura naturata*, the passive product of an infinite causal chain, we land, for example with Donna Haraway, at a notion of *natureculture*. *Natureculture*, as indicated by its name, attempts to forgo the binaries and dichotomies, suggesting that there is no separate ecological relation between biophysical and social processes, i.e. these processes are originally and irrevocably entangled, imbricated, interpenetrative, interactive, and interdependent. Therefore, there is nothing purely natural (nor purely cultural), and there is no Nature with capital N. Thus, the objects of "*the*" BioEconomy cannot be assigned a quality such as natural or artificial/cultural, nor are they in anyway stable but rather are very fluid. This is, in fact, the obvious state of affairs that can be seen in one of "the" *BioEconomy's* most fundamental practices, i.e. patenting.

The history of what is a patentable object the very moment we enter the area of biological life has been well-documented, and *what* is patentable is also still contested. However, the fact *that*, today, there are patentable objects does have real life consequences, since the way that patents, particularly in the influential US-American market, function "does" things in the world. One of the many aspects of patents that "do things" is their disclosure requirements.

Disclosure requirements have ontological function (Carolan 2010), because to be able to enforce your patent, you must be able to specify (as in specific) or actually to "specify" (as in species) "things". The bioeconomic patent becomes an inclusion-and-difference engine. But this presents as Derrida-en "Il n'y a pas l'hors texte", claiming that there is nothing outside the text. But thus it is rendered false by the sheer fluidity of biological realities, which enforce the acknowledgement of what, with Viveiros de Castro, we understand to be perspectival ontologies. And one much remarked on problem here are indigenous cultures and their environments, and how bioprospecting and extractive patenting have repeatedly played out, against the *naturecultural* ontologies of indigenous groups and the fluidity of real biological process and entities. "*The*" BioEconomy, through the practice of patenting, is engulfed in acts of ontological violence by means of kind-making. Patents and digitalization play a major role here.

Big Data is also an important issue for bioeconomy, because objects of bioeconomy are often either Big Data objects themselves or have been originally made possible by Big Data to begin with. While gene-editing methods like CRISPR congeal, on the one hand, around the idea of being able to tailor an edit to an individual (for example in the possibilities of personalized medicine), on the other hand, these biotechnological advances of the past 20 to 30 years have only been possible due to the availability of Big Data sets and the capability to analyze those. But just as with patents, from the point of view what Big Data performs and how it is entangled in different relations and entities, we also must look at the question of how it can be deployed considering dangers, risks, conflicts, and damages – damages meaning the damage caused in the form of destruction as well as in the form of legal restitution. In the context of anticipation, prevention, and/or evaluation of environmental destruction and environmental damages, Big Data plays a dual role, as shown by Alice Mah in a recent paper (2017). As she reminds us, the burden of proof regarding the potential or actual environmental destruction most often lies with victims and not perpetrators. Thus, finding the relevant evidence can be immensely accelerated by Big Data sets and Big Data methods; however, and here lies the crux, the same Big Data was, of course, used to create the danger in the first place; some of the adjoining questions that arise from the use of Big Data are who owns the data and who has the capability, tools, and/or financial or human resources to analyze and evaluate it. Big Data as an archiving technology and an object in the context of BioEconomy is thus ambiguous in terms of power and proves one of my personal pet peeves, namely that one cannot have empowerment without enablement: Mere access to Big Data and, thus, having some power *over* it do not mean one has the power *to* use it – perhaps an instance where further reflection is invited in regard to "Big Data come puissance et pouvoir"? This "present to future"-oriented minimal consensus (in which futures rewrite the present, and, at times, even the past [E. Esposito 2002, 2011]), in which I label here BioEconomy, can be found in white papers issued by politico-economic organizations. Like many techno-science-oriented public-private governance agendas, it is characterized by a conceptualization of progress as "moving ahead and along" versus "falling behind". This logic is in tandem with a rationale that techno-scientific progress leads to more social justice. However, this should not be accepted uncritically: (1) As a transnational political and economic agenda BioEconomy may present an unwarranted reduction of complex ecological and political-economic realities and diversities of both present and past bioeconomies – the issues at stake are: Bioeconomy or(!) bioeconomies; (2) (a) differences between Global North and Global South, while (b) within the Global North agricultural, biomedical, and ecological diversity are at stake – the large number of issues in dairy farming being one good example; and (3) finally, the question "What concept of justice and whose justice?" does bare that the techno-scientific rationale of social progress strongly depends on the underlying concept of justice and its inherent logic of production – the problem of generative justice versus the logic of extraction.

At least two decades ago, cultural theorist Sylvia Wynter argued that the contemporary hegemonic genre of human provisioning at the heart of the conceptions of what it is to be human, what is human flourishing, and how humans relate to their environment rests on extraction and exploitation of all kinds of life, which she named bioeconomy; she called for the possibility for diverse genres of human provisioning. Postcolonial and feminist economists have also argued for different regimes of provisioning such as non-market collaborative practices, subsistence value, and exchange. The realities of production and exchange of value obtain meaning only entangled with living ecologies of different human and non-human agents (both in the macro-world of plants and animals, and the molecular worlds of biochemistry), which rest on complex value and knowledge practices (=cognition) that are shared, organized, and institutionalized among these different agents and in their ecologies (=distributed cognition); this is the reason is why it is important to consider these bioeconomies in terms of cognitive cultures and organizations. However, it seems as though BioEconomy as a cognitive culture is conceptualized in terms of what is currently known as cognitive capitalism with its reductionist concept of value (=monetarized and capitalized) as a mere measure of anthropocentric utility (=usefulness obtained through consumption of good). Classic theories of distributive justice implicitly rest on the latter definitions of value and utility, while dismissing or colonizing ecological relations, non-human actors, and alternative genres of provisioning. As a consequence, in cognitive capitalism, economic theory, and theory of justice, life itself is undergoing a process of precarization (the process of increasing a condition of existence without predictability, security, or sustainability in one's modes of [re]production). Since cognitive capitalism is widely understood as the production of wealth/welfare that occurs largely through knowledge practice (with labor defined by cognitive activity), it is no surprise that many firms in bioeconomy function in accordance with rules of capitalization and exist mainly on the financial market, often not even producing an actual product before becoming defunct (Birch 2016, 2017a,b, 2019).

Against this extractive logic, the idea of *generative justice* proposes

> a universal right to generate unalienated value and directly participate in its benefits in, proposes 1. the rights of value generators to create their own conditions of production, and the rights of communities of value generation to nurture self-sustaining paths for its circulation. Identify what might generate unalienated value; to 2. Identify the threats from extraction of value; to 3. Identify ways in which extraction can be replaced by generative circulation.

The Global North's Productions

The Global North produces the assemblage Green Precarity/Precariousness/Precariate Bringing decoloniality to the bioeconomy, to avoid the precarization of life. I think of this triplet as an assemblage, because various regimes, orders, positionalities, relationalities, and statuses that we consider under the term "precarity" can be separated in analysis but not in the actual experience of the messy reality of precariousness as the subjectivities in play, the messy reality of experiencing precariousness when surfacing into precarity, the messy reality of drowning in precarity; precarity is always relation, interdependency, and interaction with others who already and because of precariousness belong to and participate in precarity. I qualify this assemblage as Green to emphasize that it affects and subjects life, as both bios and zoë, at each level; that it implicates while therein it also intensifies precarization. BioEconomy as optimizing tactics accidentally connects interactively between digitalization, algorithms, and living things, but often doing so in circumvention of what is currently called the human mind. Computational and algorithmic powers in machines we implant, machines that govern lives, begin to interact with our gut biomes and skin biomes, with individual trees and entire forests, without our conscious minds having any part in these interactions let alone having access to the temporalities of these interactions.

In this assemblage, we, as humans or as compost (Haraway), encounter new forms of entanglements of oppression; forms which have traveled and become trans(dis)posed from risk as "the anticipation of catastrophe" to the obsession with the "possibility of risks" as control society, biopolitics, discipline, and technologies of self, to become a new formation: a formation for the production of control of possibilities for risk anticipation – across ecologies, across general ecology – by means of a precarization of everything. Therein, all forms of possible and impossible labor become subjected and affected by the coloniality of power and rendered conjointly exploitable and disposable.

Biotechnological, biodigital, and bioeconomic designs, contemporary or speculative designs will have to answer the following question: Whose lives are improved by these technologies? Investigating and intervening, we will come to know that many of these technologies are actually obtained as discriminatory designs; Ruha Benjamin proposes various strategies from resistance to biodesign, refusing consumption, biodefection, and development of individual and collective modes, as well as alterna-designs emerging from our learning spaces that require not the rigidity of STEM disciplines but the versatility and motion of STEAM – including the A for Arts&Humanities. Perhaps ours can be a steam-boat. I dream of a speculative design for a steam-boat, with steam produced not by extraction but by becoming both cognitively and generatively just, wherein

> cognitive justice encompasses not only the right of different practices to co-exist, but entails an active engagement across their knowledge-systems Recognising peasants and their ways of knowing agriculture, generates 'new collective understandings, meanings and knowledges'
> (Coolsaet 2016)

Ron Eglash et al.'s discusses generative justice by considering three problems mentioned in the following quote:

> 1. Identify what generates value; 2. Identify dangers, risks, and precarities that derive from external value-extraction; 3. Identify ways that allow to replace extraction with generative circulation – and, second, that it is defined as such:
>
> "The universal right to generate unalienated value and directly participate in its benefits; the rights of value generators to create their

own conditions of production; and the rights of communities of value generation to nurture self-sustaining paths for its circulation."
[see Eglash et al. at: https://generativejustice.wikispaces.com/]

In this research area, *bio(f)acts* and *techno(f)acts* play their part as we deploy these concepts methodologically: Biological systems are included qua methodology here simultaneously as both "acts and facts". But in the *trans-scalar* BioEconomy, biological objects (from the nano- to macro-level) and systems (including ecosystems) are subject to the idea of being used, exploited, and extracted: This evokes notions of their supposed rurality, primitivity, exploitability, and Otherness constituted in coloniality: Through urban architectures (which include urban bio-/ecologies) and Settler Colonialism in comparison to rural/indigenous landscapes, they are reconceived as having been transcendentally available for possible human use, as they are as of yet unused, ownable: *terra nullius*.

Only a rural area can become urban, only a rural area can become agricultured to sustain the urban area – regardless of whether urban dwellers (and White Collar researchers) forget about where their food comes from. Only the "unused", wrongly used, or not "efficiently" (qua Western extractive rationality) used land/ecosystem/bio-object can become property. Thus, the (new God) trick is to define these lands/systems/objects not as property but as new (Groys)/unused/misused/malused: Bioeconomy means "to use all biomass more efficiently!"

But this *Green precarization*, as I call it, includes *us*, *for* are not "humans simultaneously using *and* also being part of the biological systems"? This is why it is important to understand that "the rural" and "the urban" are intertwined: The matrix is "*global urban-rural*". The matters are – even if used inefficiently – reconstructed as already *mattered*, as *biological kinds*. Their other/previous relations and multiple ontologies are ignored, suspended, and silenced:

"The" BioEconomy and its logic of extraction silence and/or ignore any other mode of provisioning that might even suggest plurality of bioeconomies as Paul Kockelman's seminal book *The Chicken and the Quetzal* (2016) illustrates in taking us to the Cloud Forests in Guatemala and to the people and creatures who live there. The region is famous for a colorful bird species, the Quetzal, which is – however – endangered by farming activities carried out by indigenous people living there. Thus, an NGO founded by Germans tried to move the local people toward recreating their economy through an ecotourism project, guiding Western/Northern travelers to see these beautiful birds in their natural habitats. Besides being a failure on economic grounds, the other effects of the project exemplify precisely the questions and discontents underlying "the" BioEconomy as well as the concepts of justice, value, utility as well as the logic of production on which it is founded. Villagers enrolled into the ecotourist project ended up not simply adopting a new set of values but rather creating – however in co-production with the "Western" values – a new set of values; they changed the world views that were crucial to their social systemic and ecosystemic niche, and reshaped their ontological and affective relations to the world and their own selves.

One of the key insights Kockelman speaks about the Q'eqchi' Maya people is that their original logic of use value was formed around a logic of replacement that infused activities ranging from "housebuilding, civil-religious elections, vengeance, loans, illness cures, adultery, namesakes, and resource replenishment" (2016: 9f.)

By introduction of a different, non-commensurable logic for use and equivalence, namely a sense of "irreplacability", and, thus, of "scarcity" the local Q'eqchi' people were undergoing a significant transition in all these activities. And while it sounds paradoxical, it is precisely the idea of "irreplacability" that feeds the logic of extraction, because one can only extract that which cannot be replaced, and subsequently this informs our Western ideas and practices of "sustainability": We are quite aware that the reforestation of a deforested area is not an act of commensurable replacement but instead we are looking to find something that counts as an equivalent that can be put in place for that irreplaceable matter that was extracted. The logic of extraction and its subsequent conceptualization and operationalization of sustainability is built around how we negotiate what "equivalence" means. Kockelman's insightful study makes this among other things – such as the lives of Q'eqchi' whose own gendered economy and independence are based on the integration of chicken – perfectly clear.

An aspect that the few sociological studies in this field (that are directed at BioEconomy) are missing is the (historical-)comparative perspective: If the goal is to build Bioeconomy in future, we must understand the past and present of bioeconomies as bioeconomic *alterontologies* (Papadopoulos 2018). Here, too, we observe an artifact of White Collar sociology's *urban normativity* (as identified by Bakker) and metropolitanism (as identified by Bhambra, Go, and Boatcă). Hence, sociologists must allow for a comprehensive understanding of the history of the discipline and provincialize to establish a new sociological imagination. Of course, Bioeconomies of the past stretch further back in time than the 19th century, but the 19th century is where and when the science of sociology was created (Stingl 2008[2010]) and this is where the critique of our concepts and of urban normativity comes into play. Early sociologists, including Max Weber and Talcott Parsons, did not forget the fact that *we are* intertwined with biological agencies and with what they would have called the "land" or land factors. How these notions, now often buried in a hygienically clean narrative of "canon", established not only the discipline but also the effects it had, in turn, is deeply important (see more in Part C of this book), for sociological concepts, too, are playing into what I have called *bio(f)acting* and *techno(f)acting*.

Urban normativity is packaged into the (Euro)modernization/industrialization narrative, which, in and for the Global North, is currently retold: On the one hand, there is the idea that by means of growth in the service sector, digitalization, and emergence of the "knowledge-society/-economy" aka *cognitive capitalism*, we are looking at de-industrialization of the Global Northern economy; on the other hand, automation and digitalization are re-industrialization in terms of the *Industry 4.0* narrative, which supposedly indicates a different type of "white collar" industrial labor, and a rise in jobs that offer services for the workers in Industry 4.0 – however, these service jobs (care, cleaning, deliveries, transport, etc.) are either part of the precariate (uncertain futures, gig economy, etc.) or dead-end (i.e. full-time but with no career improvement) jobs. The fact is that we are witnessing a fundamental change in the political economy of the Global North for some time: Industry is disappearing from the Global North, and seems to have mostly relocated to the Global South – with all the ill effects from exploitation to pollution. Now, what kind of economy can we see emerging in the Global North that will eventually also "colonize" the Global South? How "*green*", how sustainable will this economy be? Not to mention the amount and nature of precarious jobs it will entail.

But with BioEconomy as a global agenda furnished in a logic of neocolonialism, neoliberalism, and neoextractivism, precarization and precarity mean *the precarization of everything.* It is not possible to separate a *Green economy* from *precarity.* Instead, we discover the *capillaries of extractive power* that constitute a *Green precarity*, precisely because the imaginary of *the BioEconomy* that underlies many Green visions is no more sustainable than the system it is intended to replace, because it is steeped in at the root level in forms of discourse, of thought, of speech, and in ways of doing that are inherently extractive and exploitative: Nearly every major document on Bioeconomy asks that we "exploit" biomass more efficiently. Biomass means, here, everything from cows, to bacteria, to human stem-cells, to ecosystem services indiscriminately; all are rendered reducible to a "more efficient use" that is measurable in % of GDP growth in US$. This reduction means that all the doings of these very different biological entities and relations, all their labor is transformed into precarious work, because their "use", the "value" they generate and circulate is one that must be transferable into the ultimate value of quantifiable production in US$, and the "return" received by any living entity/system is just about enough to (or sometimes even less than a) guarantee for the entity/system to at least continue in some form or another… to be of use. This is *precarization* from ecosystems, people, and animals, down to the molecular level of bacteria and cells.

In this line of inquiry, we find that this techno-social imaginary of BioEconomy is instrumental in bringing to bear the question of justice in revealing that the main procedural antagonism of our times is between the *precarization of everything* and a general realization of *generative justice.*

We Need *Ethno-Economics*

Since a key issue is the conceptualization of value and utility, which is qua the *colonial* history of capitalism, neoliberalism, and the (White Collar) canon of economics defined very reductively and narrowly, mostly in quantizable, monetizable terms, we need a scholarship that looks at existing, alternative "systems or systematicities of *generating and circulating value*" (which is a possible definition of economy) and/or theorizes about such. What we need, in other words, is to make a serious effort at *ethno-economics*, i.e. developing ethno-economics as an integrative interdisciplinary platform. While the term ethno-economics has been traded in economist and anthropologist circles since at least the mid-1980s (Mayer 1987; Mazzucato 1997;Griffiths 1998; Sayle 2000; Cavalcanti 2002; Ovchinnikov/Kolesnikov 2006) neither of the two disciplines has gone on to develop systematically(!) this notion, which demands a genuine and generous interdisciplinary outlook. Recent efforts have actually largely come from within the discourses of feminist economists and propose refining models, theories, and methodological discussions for "diverse economies" (exemplary: Gibson-Graham 2011; Gibson-Graham/Roelvink 2011) along with complex systems approaches that introduce recursive depth and generative justice to the study of value exchange systems (Eglash/Garvey 2014). Furthermore challenges have been introduced to the psychological foundations of "Western" economics (Henrich et al. 2010), challenges that make explicit the ethnocentricity of its value-concept and rationality definition (affirming a tradition of sociological critique). Each of these elaborate critical discourses suggest we must analyze and further develop economics in more diverse and diversifying terms; we need to rethink economics as ethno-economics not in terms of economic rationality but in terms of genres of human provisioning (of which [neo]classic Western economic theory presents as *one* viable but not the only alternative option). From the words of Sylvia Wynter/McKittrick (2014: 17), I propose to develop ethno-economics not as

> economics as a behavior-regulatory order of discourse that is [...] *indispensible to the replication of out present economically homogenized world-systemic order*," but instead a "*science of all genre-specific human modes of material provisioning*," this including our contemporary own.

Therein, I understand ethno-economics as a *naturalized economics* perspective – similar to the notion of *naturalized (bio)ethics* (Lindemann et al. 2009) – a perspective that, other than a "principalist perspective", is presented as a parsimonious and minimalist account of the *naturecultural* facts of the world that involved, affected, and subjected stake- and stockholders (as agents) find themselves afforded and in which they constitute them-*selves* in *becoming-with* one another and their world. Economies in this view are understood as different value-creating and value-exchanging systems that differ, however, not only in their system characteristics but also, and importantly, in what is (defined,) generated, and exchanged as "value". This insight and its empirical corroboration are important to any refinement of economic subdisciplines and related fields of study in terms of cognitive and neuro-economics, which often assume a fixed and prerequisite (decidedly Western/Global Northern) demarcation of "value" (using "value" as *explanans* where it should be *explanandum*). Actually existing *bioeconom**ies*** (rather than *the* BioEconomy) offer themselves as exemplary research sites for the study of diverse and alternate systems of value exchange and generation, precisely because they are unexplored empirical arenas to which we are as of yet in the discipline of economics theoretically and conceptual-empirically naïve,[1] a condition partially shared by involved agents: Actual *bioeconom**ies*** are, in this respect, similar to the use of digital devices and their economics, where most users experience these objects as black boxes, the more refined the devices or bioeconomies become, the more they become part of an infrastructure we no longer actively perceive. But precisely *because* (a) *bioeconom**ies*** are not reducible these tasks and outputs and (b) even for the outputs and tasks themselves, which are the "values" that are themselves already more diverse than in the mainstream economic models, *does* a systematic development of ethno-economics seem like a fertile ground as well as present as an integrative interdisciplinary platform where the disciplines of economics, anthropology, and geography meet (and New Political Economy and Ethnographic Theory obtain as guiding methodological consideration). Ethno-economics, as understood here, is the study of different genres, practices, and ecological entanglements of human provisioning.

The working of imaginaries as "political ontologies of the modes of the production of expressiveness" is my main line of inquiry, which leads me to welcome Eglash et al.'s proposal, which integrates with a line of inquiry into political ontology of how justice can become expressed, precisely in a form that does not lead to a "universal" theory of justice: It is about how rights, judgments, and (demands of) justice can become articulated and practiced as rights/judgments/justice, and the ontology for the conditions of their possibility of becoming expressible: This "relational point of view" proposes, also, a shift of focus on *generative values* as criteria for the creation (or imagination) of "the question *for* justice"[2]: *How* can ethics as *ways of knowing an doing* function and *how* can they be used productively? My rather obvious proposition, following this line of flight, is that people make ethics explicit in the form of values, and that values can be studied in terms of their generative potential, which, unsurprisingly, is precisely in terms of *how* they imagine practices follow and interconnect with the values.

How to Imagine the Nature of Phantasm and the Image of Nature: The "Nature" of Rural Imaginary in General Ecology

"You are imaging things." While the ubiquitous deployment of this sentence invokes a judgment that obtains on the derogatory or dismissive, there is yet another, more complicated aspect: We are, indeed, imagining things.

Without delving here too deep into discussions of epistemology and philosophy of mind nor subscribing to a dogmatic view of "social construction", I think most social scientists as well as a majority of researchers in the STEM fields can and do accept that *how* we think of things and *how* we imagine them to be have very practical effects. I think it is acceptable for most researchers and scholars that it is true for both material "things/objects" (for example, a lump of dirt or soil) and non-material relations (such as for example "justice"). It should also be acceptable that for most of us, our differences concerning this question of "*how*" are often rather a question of degree. In the pages that follow, I won't be addressing matters of degree and will presume that we can agree that how we imagine matters to be has real effects, which create difference in the world – differene which can become so extreme that it may even seem that different people live in different worlds. This is why feminist and postcolonial science studies scholarship – such as Donna Haraway – talks about *situated knowledge, mattering*, and *worlding*.

Here, I am predominately interested in techno-social imaginaries, which may be described as types of fluid, semi-stable reservoirs that harbor images, patterns, and phantasms that can be deployed to imagine, and subsequently, to know, to judge, and to perform (social) practices with things, to project one's own life-courses, to coordinate collectively with others, and so on. The talk of "imaginaries" is both recent and at the same time anchored within the social scientific tradition. While some authors (Eaton et al. 2014) credit Sheila Jasanoff and San-Hyung Kim (2009 as largely responsible for investigating the techno-social imaginary in relation to the context of "nation", many others have introduced and used the concept more definitively and in broader contexts both within the Science, Technology, and Society Studies (STS [Fujimura 2003]) and outside (Taylor 2004). A genealogy of the concept in relation to STS would certainly reveal even far more interesting capillaries of discourse. However, before I embark on developing some consequences of recent transformations in the techno-social imaginary of – what for now I want to awkwardly call – "the Rural", I want to unveil some deeper imbrications of both the notion of "an imaginary" and of "the Rural". I want to do this precisely for imaginaries or want to use another term, namely phantasmagorias, which are the very topographies of political economies. A familiar genealogical account links the term "phantasmagoria" from Plato to Marx to Žižek, but my concern is more with understanding operationalizable deployments of the concept, which I do not see in the Marxian account. In general, with regard to political economy, I have never been much impressed by Marxist contributions (thus, I have never been much interested in Marx as a consequence, either, who I consider too deeply rooted in his period of history of science and the context of colonial ideas of race and gender), while I found the direction taken by Max Weber far more fertile and open for boundary-transgressing conversation – admittedly, with the necessary limitations of "provincializing" (Chakrabarty 2008) Western social sciences to their Global Northern context.

Thus, in developing the "phantasmatic" as an operationalizable conceptual tool – or as I prefer to say as *a conceptual-empirical laboratory* (Gad/Ribes 2014) – which can be deployed in empirical rural sociological work (on political economy[3]) to address questions of how anticipatory practices enable thinking and living toward the future by *becoming grounded* in technological means, I propose that we should follow other available accounts.[4]

Pierre Janet (1859–1947) provided, perhaps (see Mayo 1952; Valsiner/van der Veer 2000; Ellenberger 1970), the elementary conceptualizations for operationalizable notions of the imaginary/phantasm in his psychological interests of understanding the phenomena of dissociation and psychological automatism. Among other important channels of dissemination, Janet's psychology became extremely influential in early 20th-century industrial fatigue research conducted at Harvard University under the supervision of Lawrence J. Henderson and Elton Mayo (both well-known researchers in their time, as well as mentors of sociologist Talcott Parsons). Mediated through William James's pragmatism and Jean Piaget's psychology, Ernst Boesch (2002) developed an applicable version of the study of myth and phantasm, genealogically derived from Janet: To deal with what biology and culture afford, myths represent rather unspecific patterns that we learn in childhood. An example that Boesch emphasizes in particular is the "myth of lurking chaos". Phantasms are a more structured way of addressing or compensating these myths, which emerge both individually and collectively as imaged patterns for perceiving, transforming, or anticipating, whereas we should keep in mind that phantasms are also overdeterminative (similarly, Deleuze/Guattari speak of overcoding, and, of course, Lacan should be mentioned in this context as well).

Along similar lines, we find a concise discussion of imaginaries and phantasms of "the city" in the works of Georg Simmel, and explicitly in those terms in the writings of Walter Benjamin, precisely in the *Passagen-Werk* (the *Arcades Project*, 2002; see here especially the accounts of Benjamin's ideas in Buck-Morss 1991 and Weigel 1996). Benjamin's ideas on, for example, *collective wish-images* (Buck-Morss), or *Bilddenken* and *Denkbilder* ("thinking-in images" and "thought-images", Weigel – meanwhile *Denkbilder/Gedankenbilder* is a concept introduced by Max Weber) provide us with some direction. Pensky (2004: 179) describes the dialectical nature in Benjamin's conceptualization that I want to emphasize in terms of the imaginary:

> Thinking in images rather than concepts is, of course, a hallmark of Benjamin's work from its very beginning. Unlike concepts, the claim to immediacy inherent in the graphic image contains the potential to interrupt, hence to counteract modes of perception and cognition that have become second nature. The primary locus of the term "dialectical image" is thus itself the establishment of a (eminently dialectical) tension between two terms which, developed to their extreme, suddenly overcome this opposition.

However, it should be clear that by evoking Boesch, Simmel, and Janet in terms of rural imaginary and its phantasms, I refer to something a bit more abstract than any concrete image, and I pick-up, precisely, on the question of fragilities and imbrications of "nature" within the techno-social imaginary and the rural imaginary alike.

What I am effectively interested in is, therefore, not so much the empirical study of a neatly carved out slice of rural social life, but rather I am interested in the subcutaneous modes of production that inhere in questions of how are statements/enunciations even possible (to imagine) and how do they make possible to re-imagine further such as the following:

> An IT-industry manager blogging on the following subject (Hollis 2012): "Growing your digital business model. Fertile ground helps!"[5]
>
> The technological transformations of agricultural subjectivities emerging in a study on robotification of the milk-industry:
>
> With a robot, cows must voluntarily present themselves to be milked, whereby an ethical norm of (individual) freedom for cows can be seen to emerge together with this new technology
>
> (Driessen/Heutinck 2015: 3)

> The agricultural future evoked for Digital Culture, that transforms it into a *CultureNature* (to play with Donna Haraway's considerations of *natureculture*):
>
> "Games and the Internet: Fertile Ground for Cultural Change"
>
> (Castronava 2014[6])

> Or David Danks (2014) deployment of the metaphor of *"Free Range" Cognition* in his book on graphic representations in cognitive models.

The point here is to walk the fine line between taking the phantasmic connection between these statements (and its *ekphratic/mattering/worlding forces*) seriously, without over-emphasizing in terms of a kind of naturalistic fallacy, subscendental, or fallacy of misplaced concreteness (on some additional forms of fallacies best avoided here, see Weiss/Restivo/Stingl 2014: mainly 55f., additionally: 110–118).

There is, of course, a deep connection between the history of philosophy and the history of science and technology that connects this as an issue of social reproduction, historicity, and transformation of "the Rural" and its relation to the concepts of "Nature" and "the City" within the Western imaginary of the Global North, which takes its course from Aristotle forward (see: Driessen/Heutinck 2014). I am not going to spend time here with an in-depth genealogy, but will *fast forward* to the 20-th century point of condensation of this discourse.

When Martin Heidegger was offered a chair of philosophy in Berlin in the 1930s, he communicated the justification for declining this position with an essay on "Creative Landscape. Why do we remain in the Provinces?". In this essay, he emphasizes and elevates the connection with nature as landscape – as the condition of philosophical work – in a confluence with the "world of the farmer" that the philosopher should share, denouncing, on the hand, "the City" and its urban occupants and their culture (see Heidegger 1994 [1933]). This depiction of an antagonism between the Rural and the Urban has become an explanatory trope for a general divide within not just continental philosophy – for example in the tropes "Adorno v. Heidegger" and "Freiburg v. Frankfurt" – but Western society's general attitude, mentality, and imagination regarding its technological culture, as can be easily identified in the works of Simmel and Weber (both of whom Heidegger implicitly drew many of his ideas from). What seems to me, however, an intricate issue is not only this fascination in philosophy with the binary opposition of Rural/Urban, but also the inherent connection between an idea of (an untouched, original) Nature and "the Rural". This connection also sets up a linear-causal logic: Urbanity, in this imaginary, can only be achieved through "the Rural", i.e. only the Rural can be urbanized, and the imaginary of "the Urban" demands as its logical predecessor an imaginary of "the Rural". In temporal terms, "the Rural" is, subsequently, always narrated as the necessary past of "the Urban". Holding on to this imaginary connection leads, among others, to known forms of romanticized resistance against techno-scientification of agriculture and equally to the orientalism that Westerners deploy when considering indigenous people and their relation to nature and agriculture, i.e. the popular Western idea of what it means to be "indigenous" is largely (although quite inappropriately) constructed in these terms.

Furthermore, the main alternative accounts to consider in the introduction of technology into agriculture seem to be either the notion of destruction and annihilation of nature (see here also the Heideggerian trope of "the ruin") or an oversimplified dialectic of hybridization (which is, by the way, not[!] at all compatible with what feminist science studies, such as in Haraway, talk about in terms of the cyborg). I attempt, as a consequence, to sketch a more meaningful and pragmatic yet also critical realist alternative to approach the "Rural Imaginary", its interrelations, and possible futures.

This is a critical issue of our times and requires to be addressed with (regard to) *care*. Hans Bakker has, in my view, quite correctly made the point that the "matrix" that emerges as the – to use his terms – "global rural urban matrix" must be seen in terms of a *political economy*. I connect in this discourse the three terms of "political imagination", "political ontologies of production (of expressivities)", and "political economy" as an act of mapping and navigating the same topography. We must, I argue, tread carefully, since historical predecessors have spectacularly failed in political terms with well-known outcomes such as in the case of Friedrich Ratzel (1844–1904) who developed a co-productive research program – I am tempted to say *paradigm* – between "political geography" and "anthropogeography", which culminated in the pseudo-evolutionary concept of *Lebensraum*, which was ideologically exploited by the Nazis.

An alternative, and – from a techno-scientific point of view – more productive account, can take its cue from Joachim Ritter's work on "Landscape" (1963, also in Ritter 2002). Ritter reminds us that the disposition to think in terms of "landscape" is an aesthetic disposition. Citing Petrarca's depiction of climbing a mountain in 1335, Ritter embarks on an analysis of the way Nature becomes intelligible as "landscape", how Nature as a whole (*phusis*) is reconstructed in *theoria (ways of observational seeing), episteme (science)*, and *techne (arts&crafts)* as a new relational form of Nature as landscape in a "process of mind(s)", in accordance with Simmel (Ritter 2002: 420), which interrogates the question of "utility". As an aesthetic object Nature becomes landscape as "Nature itself, free of any instrumental use" (421, with reference to Alexander von Humboldt), but thereby nature as aesthetic object is within the boundaries of art, whereas nature as subject to technological intervention and use is clearly demarcated as something else. This is, however, troubled in the example that Baumgarten chooses in his work on *Ästhetik* (1750) as connected to (logical) science, the example of the shepherd in nature, the technology of landscaping in the "art of gardening", and Hegel's notion of freedom. These three imbrications, in Ritter's text, make explicit the opposite absolutes of instrumentalized nature and "total, aesthetic nature", precisely because they interrogate and trouble them, which Heidegger cannot resolve but only ignore in reifying a binary imaginary of Rural versus Urban.

The underlying problem, we understand with Hans Blumenberg (2001 [1951]), is rather that of the relationship between nature and technology, which is for him an inherently philosophical problem, which reaches its pinnacle in the introduction of an idea of a "second nature" (2001: 264), which we find not only repeated in modern philosophy in the reception of Sellars and Davidson in the work of John McDowell (1996) but also in sociology with Max Weber and his reception in the works of Norbert Elias and Pierre Bourdieu. Blumenberg alerts us to the fact that "second nature" may owe its existence to an internal differentiation within the realm of technology between "tool" and "machine" on the one hand (264), and, on the other hand that makes the "machinic" a problem of political economy, the notion of an "organization", which requires as its prerequisite that "the organic is the product of a construction": It is here that matters become more interesting, namely in arguing that the organization – as political economy – is afforded in the very terms of a general ecology; this also means that the "machinic" is more interesting than the tool, *pace* Heidegger, and that an account of Max Weber as a political economist demarcates him metaphorically as a "machinist".

This line of inquiry of *general ecology* has been introduced in its most effective terms by Gilbert Simondon and developed by authors such as Jean-Luc Nancy, Erich Hörl, and Luciana Parisi.

Hörl (2013a,b) goes on to connect here Simondon's notion of the "open machine" to repurpose our history of the sense, rather than merely the history of *our* senses, that is fashioned in a general ecology in the following quote (political ontological principle):

> The phrase "Being is relation" —which Didier Debaise rightly recognizes as a key phrase of contemporary thought as well as an expression of our current ontological state—encapsulates the fundamental principle of general ecology. In other words: in the insistence and virulence of the question of relationality, the core of our eco-technicity is revealed. (2013a: 122)

> Whilst Nancy had previously brought the concept of "eco-technics" (*écotechnie*) into play in order to describe the general becoming technical of the world, it is now, ultimately, a generalized ecology itself that figures as the pivotal moment in our highly technicized sense culture. [...] It is an ecology of a natural technical continuum, which the general environmentalization through technology and the technosciences and the concomitant explosion of agency, schematizes as the core of our current and, even more, of our future basic experience. (Ibid.: 128)

What's additionally and may be more important is that the spectrality of this general ecologicalization, because it is a spectrum (spectral continuum), obtains virally by contagion (Parisi 2007).

Because of the continuality of the spectrum that rural imaginary presents precisely as a future imaginary, it is also imbued with moral ambiguities. These ambiguities can be construed with regard to an orientation in *matters of responsibility and care* such as an orientation on consequentialist ethics, or they can pose moral questions of political economy such as in the alienation that is experienced in technology-induced changing relations such as between (farm-)animals and humans (Driessen/

Heutinck 2015:9f.). It is Alfred Sohn-Rethel, rather than Marx whom he criticizes, who points to the alienating effects of technology as scientific knowledge-production (Sohn-Rethel), which is rooted in the idea that concept-formation itself is abstraction as the source of alienation, but in the very terms of a social-practice rooted in *Realabstraktion* that involves, precisely, "second nature". Here, I argue that the problem is less that of alienation but more the problem of how the imagination of *groundedness* aka "rurality" in terms of socio-technical practices produces subjectivities that can individuate in general ecological terms, and what topologies of political economy this individuation allows us to study and how.

The Land Factor(s) in the Times of Risky Subjects, Late and Near-Liberalism

What I have called *groundedness* deserves some explication. To begin with, *grounding/groundedness* is not the same ground in terms of foundation or first principles. I adhere, in this case, to the ideas proliferated in terms of post-foundational political theory (Marchart 2007), which we have to negotiate on grounds and principles, rather than find grounded principles. However, for the connection of *groundedness* and, thus, location (geopolitics of knowledge [Mignolo]) along with the notion that science and technology (Blumenberg) produce knowledges that have alienating potentials (Sohn-Rethel), we find "fertile ground", indeed, in recent rural sociology and its attempt to connect to science studies. This is exactly the direction that Ashwood et al. (2014) are opening up in their conversation on "grounded knowledge", "that describes participants' ability to recognize and link different grounds—different theories and practices—of knowledge in the participatory setting" (427).

To describe the way knowledges can be grounded by different participants in these settings, political economy can account for some of the positioning of subjects (*subject positions*) *and* the *subjectivities* in-play (for more details on this distinction "subject position/subjectivity", see Oliver 2004). If we want to account for positioning of *groundedness* in political-economic terms, I propose to take a look at a previous work in the history of sociology for inspiration, namely Talcott Parsons's and, subsequently, Parsons's and Neil Smelser's discussion of "land factors" (Parsons 1991 [1953]; Parsons/Smelser 1956). In his 1953 Marshall lectures, Parsons – returning the gaze toward Ricardo – discussed "land factors" as a concept that he wanted to be understood quite explicitly as an adoption of a "classical category" simultaneously extending further to include additional aspects, which would be otherwise subsumed in contemporary sociology under terms such as symbolic, social, and cultural capital:

> The classical category of "land" may be thought of as the economically significant input into the economy from the situation external

> to the society.' The essential criterion is that it is not involved in the reciprocal processes of interaction within the society as such. In the usual economic terms its total supply to the economy is not in any important sense a function of its price; somewhat more generally, it is not "contingent" on economic sanctions. Seen in these terms, "land" should include not only natural resources or the "gifts of nature," but the biologically and the socially given elements of the "natural abilities" of persons within the society, and the "cultural heritage" with which the economy has to work as given, in the usual terms of economic relevance the "state of the arts." This is the set of primary givens from which any analysis of process in an economy must take its departure.
>
> (Parsons 1991: 21)

Both in the lectures – which were met with both ignorance and disdain from the economists that presented its original audience – and even more extensively in the book he co-wrote with Smelser, Parsons pays much attention to the development of "land factors", which include science and technology, as a relational component between the social and economic system, as well as in relation to other aspects (in terms of his [in]famous AGIL-scheme), such as labor/activity, capital/(production/consumption), organization/integration, and the four resulting outcome categories of rent/interest/wage/profit.

I am not saying that I want to deploy Parsons's (largely but not thoroughly outdated) systems-theory and AGIL-scheme here. But I think that his re-conceptualization of "land factors" is an interesting opportunity when considering the formation of subject position and subject formation in terms of the "rural imaginary" that can be operationalized further along the notion of a *groundedness* that, for example, Harden et al. propose to utilize, based on their participatory action research on nonpoint water pollution in two watersheds; they

> found that actors from across the spectrum of local to expert—for example, professor, farmer, activist, resident, and agency worker—were often able to work together to achieve meaningful dialogue in the participatory setting, despite their different social locations. The local-versus-expert distinction did not pose the most significant barrier in our two watersheds. The key in our research was not where actors' knowledge came from (i.e., local or expert), but whether participants were able to understand their knowledge as situated in the ground of their own experience, and able to link that knowledge to the ground of others' experience, building a kind of landscape of knowledge that connects one ground to another. To better capture what we found key to participatory deliberation, we propose grounded knowledge as a phrase that describes participants' ability to recognize and link

> different grounds—different theories and practices—of knowledge in the participatory setting.
>
> (Harden et al. 2014: 428)

Along these lines, Marazzi's analysis of the shifts (in the violence) of finance capitalism develops toward "new land factors" in biocapitalism – and it is worth pointing out that Marazzi emphasizes the temporal dimension being of utmost importance in the analysis of the recent economic developments and crises, precisely because in terms of political theory and economy temporalities play such a crucial role in my approach to their narrative:

> Today's system of production curiously resembles the eighteenth century economic circuit centered around farming and theorized by the physiocrats. In Quesnay's Tableaux économiques, rent represents the quota of the net product, appropriated by the landlord, generated by agricultural labor of wage workers (including the labor of the capitalist tenant where income was considered in the same way as the wage of his workers and not, as it later will be defined by Smith and Ricardo, as profit). In the Tableaux, the physical instruments of production are not even taken into consideration. Quesnay defined the producers of instrumental goods (constant capital) as the active part of the sterile class, that is, not productive of net product. The exclusion of constant capital, instrumental goods, from the factors of production of net product was certainly a mistake, as was later shown by the fathers of classical political economy on the wave of the first industrial revolution. But it is a question of a mistake that is productive of knowledge, if it is true the subsequent discovery of economic value of constant capital and its qualitative difference with respect to variable capital—that is, the discovery of generic labor, the labor abstracted from specific sectors where it is carried out—was at the basis of the epistemological leap that radically distinguished modernity from capitalism. That is, the qualitative-subjective separation between capital and labor, the contradictory relationship between the two "factors of production" as the leverage of crisis-development of nascent capitalism. It is possible to say that the forms of life weakening the social body are equivalent to land in Ricardo's theory of rent. Only that, unlike Ricardo's rent (absolute and differential), today's rent is subsumable under the very profit by virtue of financialization processes themselves. The financialization, with the logics defining it—particularly the autonomization of the production of money via money by the directly productive processes—is the other side of externalization of the value production typical of biocapitalism.
>
> (Marazzi in Fumagalli/Mezzadra 2010: 42/43)

In terms of how subject positions and subjectivities can be grounded through the deployment of the phantasms furnished in techno-social imaginaries, I am particularly vexed by three contemporary *pre-figurations* (Blumenberg) of the political economy that certainly do not leave "the Rural" (dialectically) untouched: *Risky subjects*, *near-liberalism*, and *late liberalism*. Elizabeth Povinelli, for example, deploys a challenge to cultural values for recognition, which I propose to recast in terms of re/cognition that can be thought of in terms of a re/enacting embodied *interaction ritual chains* (Collins 2005; Weiss/Restivo/Stingl 2014) within cultures of cognition, which also demarcates a new direction in sociology through the integration of sociology of culture and cognition with theories of embodied and distributed cognition. Povinelli has put forward a diagnosis, backed up by analyses of policies and political culture, for example regarding the treatment of indigenous people in Australia, queer politics, or the establishment of temporalities in political cultures, which she views as configured by what she calls *late liberalism:*

> By late liberalism I mean to indicate a formation of power—the twined formations of neoliberalism and liberal cultural recognition—that emerged in the late 1960s as a method of solving the crisis of liberal economic and social legitimacy in the wake of economic stagflation and colonial and social revolutions. Late liberalism solved these legitimacy crises by arguing that the population of the nation-state would be secured against harm if it were opened to the precarious character of life as difference. In other words, late liberalism marks a period in which the population would be secured by a new reading of society and individual, a reading that ignored people and their freedoms as a kind of truth-speaking (*dire vrai, parrhesia*), and focused instead on the care of economic and cultural aptitudes and attitudes that enhanced the life of the population but only in so far as it maintained the flows of economic and cultural values from subordinate to dominant groups.
>
> [...]
>
> In late liberalism, the population was secured—and, visà-vis this security, the flow of economic and cultural values regulated—by making every element within it conform to a specific division between the autological subject and the genealogical society and their corresponding social tenses of the future anterior and the past perfect.
>
> [...]
>
> Late liberalism is late in the sense of belated insofar as it never catches up with or gains a hold of itself. It is always internally disturbed by the history of its spatiotemporal dispersion and thus also by the otherwises it finds and produces in this dispersion. The factual spatiotemporal unevenness of the historical and social unfolding of liberalism meant that all spaces drew from a different array of elements at hand.
>
> (Povinelli in Holbraad/Pedersen 2013: 30/31)

The spaces, Povinelli refers to, can be described as "cognitive" and/or embodied (depending on the type of entities, human/non-human animals, or other objects), since they are *geontological* (2016). *Geontology* refers to the "imbrication" – which I prefer over Povinelli's use of "interpenetration" – of biography and geography as well as the power-related bio-geographical obligations across instantiations of governance.

I propose that we consider the instantiation of the various existing forms of this imbrication in terms of their technological condition *and* in terms of the available collective behavior patterns (i.e. distributed cognition) that are enacted as *ethical plateaus* and *contact zones: Geontologies* are to be conceived as sensually concrete spaces of power, where the forms of *governmentality* (Foucault) and the embodied subject collide and the political field assumes a somatosensory gestalt, "where multiple technologies interact to create a complex terrain or topology of perception and decision making" (Fischer 2003:36). This invokes (for the rural sociologists) a need for methodological transformation and reoccupation of the technological imagination in the form of *itinerant, insurgent ethnographies* of *discrepant paradigms* that are *cutting together-apart* translocal and transmedial sites, mobilities, and episodes (Mankekar/Schein 2012), disrupting the

> set[s] of assumptions, concepts, values, and practices that pertains to a specific yet unbounded worldview of an imagined community of actors. [Contesting] the authority of explanatory accounts that reduce radical challenges to dominant orders by narrating such challenges within the parameters of established networks. [...R]ather than foreclosing horizons of possibility, these paradigms suggest the potential for alternative orders that are radically different.
>
> (Juris/Khasnabish 2013: 19)

In terms of these "tangible *itineraries*", I do not only propose the study of past trajectories already traveled but also how itineraries of the future *are* and can possibly be imagined and find – not their expressions but – expressiveness (as in finding their possible modes of producing expression).

A primary expressiveness established in these *geontological* terms, which establishes a subject position, is concretized around, as Geeta Patel describes (2006: 46), the figuration of *risky subjects*, wherein each "individual" as a

> good neoliberal citizen invests in her future by voting, paying taxes, privatizing her ventures, and purchasing safety nets. These ideas of trustworthiness and a future that will happen because of them come partly from insurance and from contracts of various sorts, whether they be financial, legal, or speculative. These ideas are embodiments of how we imagine an ideal state of things, an ideal state of being, and an ideal person.

This "*risky* subject position" is inscribed into the geontological topography in this way, precisely because of the lines of flight created as the permissible possibility of futures that become intelligible *and* communicable through – I am extending Mignolo here – the "*geopolitics of knowledge* of futures". Supplementing Povinelli's *late liberalism* here to accommodate the notion of "futures", Stefan Ecks (in Ong/Chen 2010: 159), in his research with regard to biomedicalization which can be easily extended, characterizes this geopolitical condition of knowing the futures of our times as *Near-Liberalism*, which

> emphasizes that freedom, emancipation, and autonomy are always held out as future goals, even if they are not realized in the present. It suggests that neoliberalism is not a uniform practice, but that it tends toward nearness and local differentiation. It also accentuates that neoliberalism can switch to restrictive modes of governing at any moment. Near-liberalism is not a program to get rid of the social and political from the sphere of economics, but one that seeks to produce new alliances between markets, states, and citizens. These alliances need to remains as flexible as possible, to allow moving in new directions at once, and to allow severing ties form stakeholders that have become liabilities.

The affective dispositions toward permitted possibility of futures, as I have alluded to in my book on *The Digitality of Power* (2015), not only but also shape temporalities through the form of anticipation (as Adams et al. 2009 show), but there is also an uncanny "backdraft" contained in them, which recalls – in the techno-scientific reconfigurations of "digitalization" and "The Algorithm" – the time and temporality of "the Sacred", similar to what Dupuy (2013) developed. This is not so surprising if we consider the temporality of the risky subject position with Jane I. Guyer (2007) as articulated in her influential and much discussed paper "Prophecy and the near future: Thoughts on macroeconomic, evangelical, and punctuated time". Guyer argues that there is a "consequential refocus" in the idea that the orientation of reasoning that is directed toward the near future should be both "responding to immediate situations and [an] orientation to a very long-term horizon". Guyer expresses her intrigue, since this orientation reflects a surprising co-incidence in two geopolitical knowledge venues: The "public rhetoric of macroeconomics" and "evangelicals' views of time". It is here, again, that I wish to also re-emphasize the importance of Max Weber's work, whose sociological oeuvre spans the Roman juridico-agrarian perspective, historical shifts in economic theory, and the Protestant Ethic (including their temporal dimensions).

In similar fashion, Guyer traces

> the optionality and consonance of this shift toward the relative evacuation of the near future in religion and economics by examining different theoretical positions within each domain. In conclusion, I suggest

> that the near future is being reinhabited by forms of punctuated time, such as the dated schedules of debt and other specific event-driven temporal frames. For me, a sense of foreignness in the current present has come to revolve around a strange evacuation of the temporal frame of the "near future": the reach of thought and imagination, of planning and hoping, of tracing out mutual influences, of engaging in struggles for specific goals, in short, of the process of implicating oneself in the ongoing life of the social and material world that used to be encompassed under an expansively inclusive concept of "reasoning."
> (2007: 409)

Moreover, and this is the gist of my argument, the Sacred temporality in-play in the "imaginaries of the future" is the same as is folded into the "Rural Imaginary":

These are socio-technological temporalities that are contained, or rather become *grounded* in the shifting temporalities of "land factors", at once of "the Sacred", "the (secular) Rational", and "the (Evangelical) Algorithm". They become grounded in the same way that Povinelli (2016) proposes that "the biopolitical is not a space but a spacing—thus tense is internal to this spatializing function" – the *geopolitics* of the "Rural Imaginary" is not a ground but a *grounding*. And just as there is in many "rural places" a Sacredness to the place, which is retained in their urbanization as is promoted by digitalization. It is retained the same as there is a notion of "the Sacred" retained in "the Secular" for secularity to make any sense at all, which simultaneously begs the question for the "legitimacy of the Modern age", which Hans Blumenberg (1985) asked both seminally and magisterially; there is the question for "the legitimacy of the Digital age", which I pose in *The Digital Colonality of Power* and which must be reflected in any political-economic line of inquiry into "the Rural imaginary of futures".

We Are All Urban-Rural Cyborgs Now: Imagi(ni)ng Digital Topographies of *Oikos*

"The Urban-Rural Imaginary" works with an "open machine", open in two directions: It (a) spatializes *geontologies* of the digital spaces in rural boundaries and temporalities, whereas (b) qua digital topography "infects" (Parisi 2007) rural fields (both in space or of knowledge) with urbanity as *telos*: In the metaphorologies that are deployed to construct the Internet as a rural space which is "fertile yet not regulated in terms of urban surveillance" such as we find in the following report (NN/London AP 2012):

> The Home Affairs Committee said the Internet "was now one of the few unregulated spaces where radicalisation is able to take place" and

> played a greater role in promoting violence than prisons, universities or places of worship – a pointed rebuke to other government officials who had identified those areas as high risk.
>
> "More resources need to be directed to these threats and to preventing radicalisation through the Internet and in private spaces," said Keith Vaz, chairman of the committee. "These are the fertile breeding grounds for terrorism."
>
> (www.jamaicaobserver.com/news/-Internet-fertile-breeding-ground-for-terrorism---mdash--UK-lawmakers_8252855?profile=&template=PrinterVersion)

The promotion of *risky subject(ivitie)s* in processes of identity formation is equally called upon as "fertile grounds", when youth is reimagined in its rurality and futures, since "[t]he internet is fertile ground for the mosaic of allegiances out of which teens build identity" (www.theguardian.com/commentisfree/2015/jan/12/internet-mosaic-teens-build-identity), a transformation that is reflected in the Western constitution of the Global South in the Digital age: "Brazil is Latin America's most fertile online market" (www.digitalcommerce360.com/2013/08/09/brazil-latin-americas-most-fertile-online-market/).

This brings us face to face with the notoriety of "white Settler Colonialism/capitalism", for even in the neo-"new (political) economy", there are rural moments of colonization when "[n]ew seed stage investors are finding fertile ground in Internet start-ups" (http://archive.boston.com/business/technology/articles/2011/01/16/new_seed_stage_investors_are_finding_fertile_ground_in_internet_start_ups/).

Of course, all of these statements are minor and benign symptoms, and their individual importance should not be exaggerated. But my point is that as symptomatic expressions, they are salient reminders of their shared conditions of possibility, of imaginaries of the future(s) which contain a decidedly urban-rural phantasmatic (and ekphratic) force. How the "land of youth, the Global South, of the internet itself" will be seeded, settled, and colonized, how it will become a landscape, become rural before it can become urban, is up to the temporalities and tactics of retransforming space that is imaginable: This is a struggle over the question on how our knowledges can become grounded. As a geopolitics of knowledges of possible futures it is a struggle that defines how lines of inquiry can become lines of flight, i.e. a struggle about itineraries and discrepant paradigms.

In constructing the Internet as a fertile, rural space, it is proposed that it is waiting to be seeded? But what then are the plants that will grow there, what fertilizers and what pesticides will be deployed? Should we be surprised when copyright and patents laws, software updates, and licenses online resemble the strategies deployed by corporations like Monsanto regarding GMOs and seeds (Kinchy 2012)?

The *technological condition* that Erich Hörl describes in terms of general ecology and the question of individuation/milieu and "of struction" (Nancy) obtain elevated relevance in the making of contemporary "landscape": These are landscapes of the future, future landscapes, as much as landscaped futures – often in deploying means that creates a reiterative adoption of a past: This is what I understand from the analysis of Monica Brannon (2013, 2017) who in her research on satellite imaging and standardizing space engages with the concept of a geopolitical gaze that she describes as "statistical-mechanical geographic vision". She emphasizes (Brannon 2013, 2017) that imaging by satellite both simultaneously discloses what was previously invisible and constructs visibilities/visible things, while invisibilizing others. Her concern, whether looking at Cold War deployments of satellites of Google Earth, is with the techniques of quantification and algorithmification of the social as well as the ordering of information and of archives, and their interrelation. What is hidden, for example, is the rural nature of the infrastructure of the Internet, as artist John Gerrard has attempted to reveal; see here the report by John Jones on Gerrard's attempt to make Google's "data farm" visible:

> The internet, says Gerrard, "doesn't not exist. It is physical. There is a great cable running under the Atlantic ocean from Ireland to America. There's a new set of infrastructures which are great information railways being put into place. I became interested in asking: what does the internet look like?"
>
> [...]
>
> Farm is a stately and hypnotic portrait of our time that gradually becomes more disturbing and melancholic as you watch. It is as if the internet has congealed into metal and concrete, our delusions of technological liberation brought, literally, to earth.
>
> For Gerrard, all our beautiful gadgets create a "distractive dynamic". But distraction from what?
>
> "Maybe it distracts us from just how dilapidated the natural world is becoming."
>
> I look up from my screen. I get ready to send this story. To you… Via the farm. A farm, suggests Gerrard, "which is eating us as we eat it".
>
> (www.theguardian.com/artanddesign/2015/feb/04/google-data-farm-john-gerrard-internet-art)

The trope elicited in this depiction coheres with the tropes of fertility, seeds, and Settler Colonialism, evoked by Povinelli and others in their conceptual deployments as well as invoked in the statements that circulate in the empirical realities I have cited here: In conceptual-empirical imbrications they collide/collude[7] (Mbembe in Mbeme/Comaroff 2010: 663), instantiating the capillaries of the political economy Hans Bakker describes as the global rural-urban matrix. If the data farms eat us as we eat them, is ouroboros image not inscribed in the stakeholder/stockholder relation of debt and credit, and the runaway rent, interest, and profit economy that presently "eats away" the future "land factors"? "Debt/credit relationships tie up and encumber the future with present obligations, and these obligations are literally carved into landscapes and subjects" (Povinelli at www.e-flux.com/journal/27/67991/routes-worlds/).

If we think of "data farms" as "data mines" instead, since in regard to Big Data the talk is of "data mining" – which is a no less rurally phantasmatic – the question becomes,

> *Who/what is the (domesticated and disposable) canary that signals to the (data) miners, that the air is no longer safe to breathe?*

And here a more perverse dialectical conflagration occurs: In the re-orientalization of indigeneity, indigenous people's modes of existence, cultural practices,[8] resources, and knowledges become recreated, or rather carved, in their landscapes, bodies, subject positions, and, eventually as a consequence, subjectivities as "Indigenous": Indigenous people are redefined in the terms of a "Rural imaginary" first, which then allows them to be reimagined as *urban subjects*, and thus also radically, neo-/late/near-liberally individualized: Are we witnessing here a transformation into the possessive individual that C.B. MacPherson (1962) critiqued? This question pertains to imaginaries that set up the practices of digitally archiving cultural practices of indigenous peoples (Christen 2011, 2015; Fitch 2013); it pertains to the introduction of Global Positioning System (GPS) and personal digital assistants (PDAs) so that the (imagined) "local bushmen" can cybertrack wildlife in their role as "poachers or conservationists",[9] and so forth: The phantasms, (re-)orientalizations, and neo-coloniality of the global rural-urban matrix can be observed at work in these constructions of indigeneity. At the same time, the absorption of the supposed "goodness" of the "simplicity" of "indigenous" living, a connection with a better nature, a rural rather than urban lifestyle, is promoted and proliferated in the Global North, in the deterritorialization/abduction of a concept such as *buen vivir*, which demarcates the re-ruralization by technology of the urban? But is the actual reality of *buen vivir* not something else? Can it be had without the *vivir en comunidad*?[10]

The *Urban-Rural Imaginary of Futures* that is rooted in the Global North with its focus on the "possessive, individual, and risky subject" is both a landscape differently carved and carves landscapes differently than any genuinely "rural" community would.

Land, I should add, is something that, today, can be consumed, too: Every year Northern governments disclose land to be consumed, which is constructed in an economy of scarcity (see on the German example of *Flächenverbrauch:* www.zeit.de/wirtschaft/2015-03/flaechenverbrauch-nachhaltigkeit-umweltschutz).

Technological Objects and the Urban-Rural Imaginary

Bakker's New Political Economy, and his discussion of the global rural-urban matrix, has its merits. However, still rooted in the Euromodern project of a (Weberian) comparative sociology articulated from within the methodology of symbolic interactionism, it needs to be seen from the positionality of its articulation and, thus, provincialized, lest it turns into just another colonial academic venture of a White Collar provenance. On can begin by reading Bakker through the lens of the postcolonial and feminist STS and apply it to the black boxes of (social) technologies and non-human agency that are busy with their own *bio(f)acting, techno(f)acting,* and *ethics as ways of knowing* (Stingl/Weiss 2016; Stingl 2017). Emerging in the Global North, I will briefly present three interrelated issues that – qua *the* BioEconomy as a global, transnational agenda – will waft geopolitically toward the Global South propagating in terms of the global rural-urban matrix: Urban Food Justice, the robot-induced autonomy of dairy cows, and the digital ecology of infospheres.

Generally, food justice (Gottlieb/Joshi 2010) has become more than a mere human rights issue, a part of sustainability studies, or environmental ethics. It is important to stress, for example, that the related issue of food security has been found wanting unless the discussion of food sovereignty is included, which addresses the entangled power relations: With the complex sociological as well as science and technological issues involved, food justice scholarship that is rooted activism has begun to re-evaluate its own foundations (Daigle 2019; Levkoe et al. 2018), but the broader discourse of social scientific research itself, for example the so-called "sociology of food" or political sociology, has to make those steps, too, and do so with the necessary respect for scholarships that take positionalities, multiple ontologies, and radical contextuality seriously (cf. Blaser 2009a,b).

This becomes especially evident in the ways we change the face of our cities, because we can see that due to one of the *(f)acting* effects of Euromodern *urban normativity* (which is not only a term describing an academic mentality but a general one) urbanization eats up much more space (and minds). Ensuring food justice (including the dimensions of food security and sovereignty) for those living in the city – and especially if they want to and need to "retake the city" (Saskia Sassen) or exercise their "right to the city" (Henri Lefebvre) – therefore requires new strategies: Urban spaces need to be (re)viewed as and remade into gardens, forests, and fields. Several contributions take note of this necessities and already existing initiatives such as Agyeman/McEntee (2014) on urban political ecologies or Poe et al. (2013) on the use of materials that urban forests illustrate: But the danger is, as even my resorting to the terminology of "urbanization" shows, that we end up re-inscribing the rural-urban divide along with, for example, the global-local divide and an inherent center/periphery narrative. In the end, these efforts run into the danger of failing to communicate that to enable food justice, different actors would have to overcome the bias attached with these distinctions and accept that for their needs, normative ascriptions (especially along these established binaries) are a problematic hindrance – or as I stated in my contribution to Bakker (2015a):

> Instead of uncovering the capillaries of power that exist in linking the rural with the urban, the task would be to uncover how the rural/urban distinction and the rural imaginary establish power-relations in the first place by turning 'the rural' into practice, for example through making and enacting of (imaginary) technological and biological/ecological boundaries between the rural and the urban.

In another example, my interest wanders toward agriculture and animal husbandry, where automation and robotification are constantly taking over, producing a series of issues that range from ethical questions to matters of theory and conceptualization (Carolan 2019). It means a confrontation with different types of extra-human agencies – perhaps the title of this book should, then, rather have been: Cows, Plants, and Algorithms.

Milking robots are, indeed, taking over dairy farms as Michael Carolan (2019) or Clemens Driessen and Leonie Heutinck (2015) show. The seminal work of these researchers illuminates not only *that* but also *how* political economies, attentions economies, and knowledge economies of agricultural production are subject to *cyborgification* where biological and the algorithmic worlds merge and remake, as a consequence, human-animal relations in the "two-way influence between technology and society/culture/morality" (Driessen/Heutinck 2015: 5).

What we are witnessing is that any dairy cow's future(s) are being re-imagined as that of as a neoliberal, risky, and quasi-autonomous subject. In a techno-phenomenological turn to techno-scientific intersubjectivity, we witness that *shared* and both increasingly and decreasingly embodied *experience* of cow and farmer is becoming an *urban-rural* cyborg. Say Driessen/Heutinck,

> [t]his conception of experience makes for an approach to ethics that allows for a self-evident role of both farmers and—so we argue—also animals when considering ethical concerns on the farm. Not as ideal liberal subjects: rational, autonomous, atomistic individuals, but relational characters immersed in practices. The thick description of

> everyday experiences, and how farmers and animals communicate these, is then an appropriate way to promote ethical reflection and debate
>
> [...]
>
> In exuberant styles, these authors have crafted alternative forms of relating to "companion species" and "intimate assemblages of corporeal becoming." The ethical call in these approaches amounts to learning to be open, to tune in, and to seek to encounter "mindful others." Not by extending humanist models of subjectivity and rights to a select group of higher mammals, but by becoming aware of the various ways in which our language and material practices contain species hierarchies, make us overestimate our human specialness and prevent us from more truly encountering and imaginatively responding to a variety of living beings.
>
> (Ibid.: 6)

What becomes necessary to problematize are the individuality and autonomy of dairy cows: As milking robots or "automated milking systems" (AMS) emerge, so do unique (con-)figurations of "robot cows and robot farmers", troubling temporalities, autonomies, individualities, and groupings of cows, farmers, and robots:

> Thereby the co-evolution of ethics and technology is not merely a matter of marketing rhetoric accompanying a new device, but is part of a (to some extent) sincere process of reinterpreting and rearranging practices in the light of emerging norms and behaviors.
>
> (Ibid.: 17)

This is embedded, however, in complex epistemological, ontological, and ethical histories (or, to speak with Foucault, histories of the present and, to speak with Hannah Landecker, biologies of histories) such as for example the complex history of *milk imaginaries* (Atkins 2010; Otomo 2014; Cohen/Otomo 2017; see on epigenetics, breast milk, and race, also Mansfield/Guthman 2015; Mansfield 2017). Techno-social phantasms supplied therein are contained in the "urban-rural imaginary" or BioEconomy, and are, thus, subject to a complicated genealogy imbued with entanglements of oppression (Nibert 2002), racism, Settler Colonialism, the coloniality of power and Being, and precarization. We are dealing, of course, here, in *matters of care*. We need, I claim, new methodologies, new kinds of scholarship – especially within the social sciences – to disclose the ontologies and the cognitive policies in-play, to reclaim the normative appropriateness, and to create windows of opportunity that make explicit entanglements of oppression and open up entanglements of liberation; the question is, again: How to create cognitive and generative justice?

We should, once again, take note of the deployment of agricultural/rural metaphors when it comes to digitalization, digital information and communication technologies (ICTs), the Internet, and digital culture, including with regard to ethical questions. It is no surprise that renowned information philosopher Luciano Floridi, since the late 1990s, carries concepts that seem to come from discourses on ecology and agriculture into his digital ethics: To think ICTs in terms of an ecology, as an infosphere that elicits ethics phrased in the terms of "regimes of care" that we'd expect in biomedicine, environmentalism, and (Green) agriculture: Information Ethics (IE) are thus defined in several of Floridi's works by combining the languages of medical care with metaphors of "land" as

> [...] an ontocentric, patient-oriented, ecological macroethics (...). An intuitive way to unpack this answer is by comparing IE to other environmental approaches. Biocentric ethics usually grounds its analysis of the moral standing of bio-entities and eco-systems on the intrinsic worthiness of life and the intrinsically negative value of suffering. It seeks to develop a patient-oriented ethics in which the "patient" may be not only a human being, but also any form of life. Indeed, Land Ethics extends the concept of patient to any component of the environment, thus coming close to the approach defended by Information Ethics.
>
> (Floridi 2006: 25)

From BioEconomy to Digital Culture, both work on a myth,[11] aka are metaphor(olog)ically speaking run by an operating system, founded in (the prefiguration of) *oikos*.

Notes

1 In the same sense medical trials require "treatment naive" populations for research.
2 A rare but, therefore, all the more important alternative is found in the work of Sabrina M. Weiss who has, for example, conducted some inspiring and highly original directions in creating a fertile and opportunity-oriented conversation in the field of ethics (bioethics, eco-ethics, food ethics) between Eglash et al.'s generative justice and Rainer Forst's (2002, 2013) ideas on relational justice and noumenal power (see also the works of Sabrina M. Weiss).
3 If we understand "theory", with Donna Haraway and Marilyn Strathern as tools to "think with", then political economy is, in my view, just such a theory.
4 See also my earlier attempts to demarcate the different interpretative avenues on my blog: (a) https://alexstingl.wordpress.com/2010/11/09/what-is-a-phantasm-first-approach-to-tackle-this-travelling-concept/; (b) https://alexstingl.wordpress.com/2011/09/21/what-is-a-phantasm-second-approach-towards-tackling-this-travelling-concept/.
5 https://chucksblog.typepad.com/chucks_blog/2012/10/growing-your-digital-business-model-fertile-ground-helps.html.

6 www.technologyreview.com/s/532331/games-and-the-internet-fertile-ground-for-cultural-change/.

7 Before I became aware of Mbembe's resourceful combination of "collision/collusion", I have coined the term *implification* to disclose figurations wherein an implication (in the juridico-political sense) occurs which simultaneously functions as an intensification of the trajectories so implicated (see Stingl 2015, forthcoming, and Stingl at: http://installingorder.org/2015/03/04/the-genomics-of-race-implifications-for-digital-cultural-health-capital/; last accessed March 13, 2015). This is similar to Mbembe's proposition, although differently useful. However, had I known about Mbembe's combination, I would have used it instead of creating the conceptual-empirical laboratory of *implification*.

8 Should it not already strike us as odd that we think of them more as cultural practices than cultural techniques? (see on cultural techniques: Winthrop-Young 2013).

9 www.nbcnews.com/news/world/poachers-conservationists-use-drones-gps-wildlife-battle-n287611. www.wired.com/2004/01/learning-to-track-like-a-bushman/.

10 I am indebted to my good friend Cristina Blohm (Leuphana) for alerting me to that aspect (among countless other things). Furthermore, I want to express my gratitude to the students at Leuphana University who came together for our discussions in the four events we organized under the title "Decoloniality/Decolonialidad/Dekolonialität- Liberation and the Possibility for Community", as well as the co-organizers (Nicholas Rowland and Sabrina M. Weiss) and all panelists for the three "Decoloniality and the Social Sciences" panels at the 2015 Meeting of the Eastern Sociological Society in New York, NY.

11 Not just as "Arbeit *am* Mythos" but also "Arbeit *im* Mythos" – prefiguratively (cf. Hans Blumenberg).

Part C

(As in Creative Misunderstanding, Care, and Commoning) – Contact Zones and Imaginaries as Permeable Borders of Empirical Arenas on the Road toward *BluesCollarship*

7 Nomadic Statehood

Toward a Theory of the Question "When are States"?

The Dancer and the Dance

The following chapter has as its subject matter the political imagination. In it, I propose the notion of *nomadic statehood*. The state, as a concept and as an abstract and general entity, has often been attacked by scholars and activists, formerly from the Left and, increasingly, from the Right. I find this, however, problematic because I have come to believe that if "the state" is a problem, then only one concrete form of it – the Euromodern nation-state form (Richter 1996) – is, and, instead that the concept of "society" may be the actual problem, and that the concrete form of the state that may be problematic is dependent on Euromodern and colonial ("fixation on") *society*. In this chapter I will propose only a step or an idea toward a different conception of statehood, again not wishing to provide a readerly, magisterial, and hegemonic text here. To do so, I take inspiration in a quote from French psychologist Felix Guattari:

> What is now on the agenda is a 'futurist' or 'constructivist' opening-up of fields of possibility. The unconscious remains bound to archaic fixations only as long as no assemblage exists within which it can be oriented towards the future; and in the future that faces us, temporalities of both human and non-human nature will demand just such an existential reorientation.
>
> [...]
>
> Our objective should be to nurture individual cultures, while at the same time inventing new contracts of citizenship: to create an order of the state in which singularity, exceptions, and rarity coexist under the least oppressive possible conditions.
>
> (Felix Guattari *The Three Ecologies* (1989))

Key in this earlier quote, for me, is the issue of *temporalities*, and the idea that we need to ask about the "*temporalities* of statehood", which is a rarely (if ever) asked question. I was once presenting parts of this chapter at a conference (as a guerilla paper, for in protesting the nontransparent policies of the association conducting the conference, I had originally retracted the paper and refused to pay for the conferences, but was invited to give the paper by the panel organizers anyway in order to help along the protest). I had planned to ask attendants of the panel: "Who of you can dance?" and to have asked one of the responders to come down to the podium and made him or her dance to the music, so to speak, along with me. I aimed to illustrate that, when talking analytically about the State (and the same goes for body-mind, as the two are connected in mutual intra-action through the political imagination [Hengehold 2007]), the question we really have to ask is: *How do you separate the dancer from the dance?*

Another reason why I do not present a "final word" on a "theory of the state" here – *pace* Foucault – is that it is and must be a discourse in becoming: We do not yet and will not for quite some *time* have a theory of the State for our *times*.

That doesn't mean that there aren't things we can't say or use for analysis or intervention – indeed, there has always been a strange relationship (or should we rather say: *strange attraction* or, with Hong/Ferguson [2011] *strange affinities*) between political theory and practice. Our task is to appraise/assess theoretical contributions we have, let go of those that don't hold up, and graft useful tools onto our current discourses: So, let's bring the State back in – as if it had ever left.

My fascination with this topic began through my deep involvement as a Magister student in Germany with the works of Kant and of Foucault – in particular the latter's rejection of a theory of the state –, and my parallel and early-onset fascination and eventual frustration with Jürgen Habermas's cosmopolitanism. I think Habermas's[1] work is indispensable for any serious discussion in and around Euromodern political philosophy, theory, and practice. But his reading of Kant's political philosophy is just like his interpretation of Weber, and his refusal to understand Foucault's premises and goals – despite his literature of Jaspers and Arendt (Habermas 1987) – flawed and biased: In his (1999) almost Dewey-like conflation of democracy and republic into one, monolithic and quasi-absolute category, Habermas' conception of sovereignty surrenders crucial territories in the discussion of the concepts citizenship, integration, and political intervention. Furthermore, and as Enrique Dussel has of course shown, Habermas must be provincialized; his typically Euromodern, explicit, and implicit *claims to universality* must be rejected.

Emboldened by my ambiguous reaction to Habermas's own "philosophical-political profile", reading literatures in this field – a field that is an imperative zone of engagement for me in the issue of technoscientific governance, and the question of the political imagination and its double imbrication with the body and the State –, I found that these literatures were pointing to a deeper issue in the past few decades that needs to be made explicit and theorized properly at the core of political theory and for which, despite its many accomplishments, the still colonial (white, male, center/margin-focused, textual-linguistic, and territorial) matrix of even post-fundamentalist political theory has not yet found a way to catch up to. Nonetheless, the literatures of Marchart, Soerensen, Slaughter, Sassen, Florini, Zuern, Fischer-Lescano, Teubner, Peters, and Ong[2] among others proved to be interesting in the aspect that they do pose the question for the state in different ways yet pointing at the same emptiness – absence – at the center, that there is no state at the center and that everyone's fascination with this absence – why isn't it there, where is it hiding, should it even be there, is still there, has it never been there to begin with, etc. – marks the problem that what is at stake is not the state but our theory of state, for in our theory of the state – how, where, why and when we ask about it – lies the clue of how to deal with this absence in practice: All there *really* is this absence, and with regard to the human condition our question must be if in full view of this absence humans are now *homo compensator* or *zoon politikon*. Obviously, I am saying they are neither; instead, I'd argue they are dancers in a world of many kinds of music (perhaps Habermas is "notoriously" un-musical in more than one way).

Recent scholarship in international law and systems theory (Fischer-Lescano/Teubner 2006; Teubner 2010, 2012) focused on *regime collisions* that are considered the result of functional differentiation on a global scale. This reinforces the trend of challenges that the modern State is facing both empirically and theoretically. But which is that "state" that is so modern?

The Euromodern state, on its surface level, is a product of a long history of secularization and continuous codification of governance and mechanics of rule into (positive) laws, at least, that is, in Western political ontology and epistemology, and the social structures they touched. Beginning with the theologico-political and *oikos*-ecclesial state (and not the Roman reinterpretation of the classic Greek philosophical state – *pace* Agamben), we should point out that the more pronounced influence of neo-Platonism and neo-Aristotelianism – and the reinjection or, should we say: reboot of *oikos* – came into play later with the canonization of translations of manuscripts obtained in the Middle East and hand-written – often immensely divergent copies – and later the dissemination of printed copies. Many Latin scholars were appreciated differently and diversely over the lengthy process of the emergence of the state concept we'd recognize today: Largely, the history of the state in the West tends to be viewed through the lens of the history of law as governance and the emergence of the secular state from the ecclesial state: Boeckenfoerde (1991: 92f.) suggests the "state" concept – while not per se "universal" – both designates and describes a form of political order that owes its existence and specifities to developments in the Europe from 1300 to the late 1800s but is now widely disseminated across what Boeckenfoerde calls the "civilized world". In his view, therefore, there is no more possibility to talk of different "states" in specific historic contexts or as different concepts by scholars from Plato and Aristotle to Aquinas.

Beyond any "folk heuristic" of the meaning of secularization, Boeckenfoerde's approach – an offshoot of the influential German Aristotelian-right-Hegelian *Ritter School* founded by Joachim Ritter and carried forward by intellectually diverse scholars such as Hermann Lübbe, Willi Oelmüller, Robert Spaemann, and Odo Marquard – brings together the "two sides of the historical process", the constitutional history with a process I would call a re-worlding of the previous politico-religious cosmology that dissolved the umbilical-like chord between political order and a religious teleonomy. The beginning of the "modern state"[3] (aka the Euromodern state), thus, is often dated to occur beginning with a series edicts leading to the Huguenot Wars in France between 1562 and 1598 and with the Peace of Westphalia (1648) which effectively ended the Thirty Years' War. However, these events only were the *registrature*,[4] making explicit and establishing as formal sets of practices that had their beginning in much earlier friction and rupture. The beginning of the process of secularization on the level of institutions and networked actors that gained traction throughout (European) history was found, as Boeckenfoerde diagnoses, in the *Investiture-controversy*, which resulted in decades of war during the 11th century: The emergence of the state and its concept as the modern state in European history is bound to the relation between violence and governance, as Norbert Elias and Charles Tilly have argued in their distinctive flavors, as a competitive monopolizing process (Elias) or in the metaphor of organized crime (Tilly).

Connecting with the history of the Roman Empire, the early European order yet installed a political order that wasn't "political" but "politico-religious" in every aspect of governance and, moreover, even in the acts of everyday life: Says Boeckenfoerde (1991: 95), *it was an order not fused but infused and informed by "Christian historico-theology and expectation of the coming apocalypse" (christlicher Geschichtstheologie und Endzeiterwartung)* which installed the *populus christianus* and *res publica christiana* both as *in (oikos-) ecclesia*, in practice, one could go as far as saying every institution, office, or locus where acts of governance were practiced counted de facto as agencies that were at the same time *ecclesia in ecclesiolae* as well as *ecclesiolae in ecclesia*, and could not occur outside of this (dialectic) relation.

The pre-history of the Investiture-controversy lay in the open problematization of the relation between political office and clerical office in the 1075 *Dictatus Papae*, issued by Pope Gregor VII, asserting the Pope's sole and universal right and authority to dethrone the emperor, and to name and depose bishops (the *investiture*).[5] This right was challenged by the German king and Holy Roman Emperor Heinrich IV, and ended in the (in)famous *Walk to Canossa*, which only briefly ended the process the controversy had caused as the ensuing series of wars shows.[6]

Again, Boeckenfoerde (1991: 96) is an interesting source for a noteworthy, if not always, critique-free account of history. He muses that the sort of revolution presented in these events exceeds a mere "desacralization" of the position of emperor and the political order, with the latter sent on an autonomous path of development (which we can consider to establish the colonial, normative politico-historiographic idea of path-dependency) describable as "worldly".

The papal strategy in the Investiture-controversy to devalue the emperor's claim to power over the realm of ecclesia had an opposite effect, and separated *oikos/polis* from *ecclesia*: an emancipation in the irrevocable dialectic of historical processes. The political order of *oikos-polis* now was render independent, self-reliant and self-evident. Devoid of a spiritual connex, the political order was now seeking for worldly statement of grounds, which other discourses on moral anchors offered, including a new kind of discourse on natural rights. In the 16th and 17th century, on the religious-political territorial level as well as the level of the mechanics of governance practices, the introduction of the idea and practice of the *raison d'etat* changes the playing field.[7]

To reiterate in short, the (academic) history of the Western matrix of the concept of state can be described in the following stages:

> The religio-political unity (Roman Empire to Middle Ages), the rift in this order with the Investiture controversy (11th century), the *registrature* of dissolution of the chord between political and religious governance with the Huguenot Wars (second half of the 16th century), the dissolution of the art of governance as *arcana imperii* with Machiavelli, the institutionalization of the new political in the peace of Westphalia (1648), the discovery of new political technologies in Industrialization (19th century), and in the late 20th and early 21st century the 'hollowing out' and 'fragmentation' of the state vis-a-vis the emergence of the transnational statehood/governance.

Because these differentiated regimes cannot call on an original foundation of practices of conflict resolution such as would be found in the constitution of a nation-state, it is argued that the state of and practice of law is now a fragmented affair (*the state of affairs*): Global legal fragmentation is supposedly the mirror result of global social fragmentation, and as such fragments are only "weakly compatible". But I don't buy that, actually. Something doesn't sit right with me, regarding the concept of "the state" in these discourses. I turn to Passoth and Rowland, who begin their seminal paper on the *actor-network state* (2010) with the very one question that really seems to matter in this discourse, or that, at least, should have matter: *What are states?*

What Are States?

In looking at what it means to view the State as an actor and/or as a network, Passoth and Rowland begin their argument by taking away the monolithic character of statehood, instead discussing the *performativity* of statehood. I agree with them for the most part. Further, I am inspired by a quote from Manuel DeLanda:

> I criticize the use of the concepts 'The Market' and 'The State'. Not because they are a duality, but because both are reified generalities that do not really exist. Adding a third term, like "The People," would not help. What we need is to replace the reified generalities with concrete assemblages: many bazaars, many regional trading areas, many national markets… each with a date of birth and (potentially) a date of death.
>
> (Manuel Delanda in Dolhijn/van der Tuin 2012)

Rowland/Passoth's account is progressive and inspiring and highly useful. But many who read it will be tempted to reduce its theoretical premises to the ideas of Bruno Latour and the specific limitations of his actor-network theory (ANT) version in the realm of political theory: ANT does inherit a certain obsession with spatiality from its discourse with the theories of Michel Foucault and Algirdas Julien Greimas. The problem with spatiality and language – which precedes ANT, even though it must be noted that ANT does go beyond the issue in some areas while even in Foucault there are potentials to connect to an outside perspective – lies in the "colonial matrix of power", which Walter Mignolo[8] describes with reference to the historico-practical dimension of secularization and, more importantly, in the theoretical realm to linguistics and Emile Benveniste – my depiction is basically just repeating Mignolo's superior one.

Benveniste, in view of the semiotics of Meillet and Saussure, formulates the "formal apparatus of enunciation" in the form of a discussion of the pronominal system of any language, and temporal and spatial deixis or markers, as Mignolo says: This "pronominal system is activated in each verbal (that is, oral or written) enunciation" (2009: 5). Thus, a reference address is produced as a spatio-temporal location, here and now.

This is a problem for the enunciator loses the ability to narrativize, aka. Account for different temporalities:

> The enunciator can only enunciate in the present. The past and the future are meaningful only in relation to the present of the enunciation. And the enunciator can only enunciate 'here', that is, wherever she is located at the moment of enunciation. Thus, 'there', 'behind', 'next to', 'left and right' etc., are meaningful only in reference to the enunciator's 'here'. Now let's take a second step. The extension of linguistic theory and analysis from the sentence to discourse prompted the introduction of 'discursive frame' or 'conversation frame'.
>
> (Ibid.)

Benveniste's concept is paradigmatic for the linguistic framing of theory in the 20th and early 21st century, which Elisabeth Grosz (2011) has emphasized that we need to "get back to the Real", to "become undone" or as I argue to get "beyond *and* before" the text. The colonial fixation on the linguistic frame (which includes even movements such as Derrida's Deconstruction) is nothing but a set of constraints that limits our understanding of the discursivity of and concepts as practices. Text, even if equipped with history and time, is nothing to do with temporality as I speak of it, then![9] The colonial or Benvenistique discursive practices are only textual (linguistic), thus spatial in being only here and *in*(!) the now: that is how I tentatively define the "textual" as constraint to mean "linguistic plus pronominal" practices: The state is enunciated so: spatial, pronominal, linguistic in time as understood through spatial metaphor and as a somewhat "static" concept of present time.

All of this notwithstanding, Passoth and Rowland manage the impressive feat of, at least, partially *decolonizing* the state concept to a certain point; meaning, it is not necessarily bound to the quality of actor-hood described in nation-stateness, but it can be captured in richer concepts beginning but not ending with Foucauldian *governmentality*[10]; however, therein the governed, the techniques, the measurements, and histories are still geometrically and geographically bounded. This has to do with Foucault's conceptualization of language in a dialectical relationship with spatiality.[11]

As they themselves say about key concepts in ANT (826f.) such as the concept of *translation*[12] that has "a geometric meaning that refers to attempts to mobilize human and non-human resources".

The socio-technical assemblages that ANT deals with refer to markets, bodies, states; these are not abstract entities but "concrete localities", which, by way of their assembled relations in borders, taxes, pink forms, actual bodies, etc., constituted state actor-hood.

States *have* sizes as actor-networks; size is also another spatial concept. Let me emphasize: I am not against analyzing spatiality. What I am trying to say is that we cannot focus on spatiality alone, because we would be missing something. Harold Innis made this point, in reference to the use of the "fallacy of misplaced concreteness" (Whitehead) in his "Plea for Time" (1951), arguing that a previous monopoly of time was followed by a monopoly of space. In political theory and practice since Innis "plea", 1950, we have not seen much of a pendulum swing away from spatiality. Moreover, I argue, we do not need another monopoly, but we need to stop the reification of these generalities and to think of space *and* time.

So, what are states? States are topological, a relationscape *in* space, local, in situ. They are *relationscapes* (Manning 2009a,b): of course, this is just another technical term, but it is intelligible and useful for the future development of this issue:

> Topological spacetime refutes this dichotomy between the abstract and the concrete. Topo-logical spacetime is not 1 + 1 but n + 1, always more-than.[...] Topology refers to a continuity of transformation that alters the figure, bringing to the fore not the coordinates of

> form but the experience of it. Topologies suggest that the space of the body extends beyond Euclidean coordinates to more abstract space-time. In topological geometry, I am both here and there, actual and virtual, real and abstract. Topology potentially deforms linear progression, rendering the concrete abstract.
>
> (2007: 142)

Being in space *is*, therefore, not all there *is*.

You can guess what I am getting at: I want to talk about something to do with time but *more than* (the monopoly of) time. It's also not just about something so linearly thought of as scales. Our scales ultimately fail, as Karen Barad (2017: 63) perfectly demonstrates for us on the example of nuclear force:

> What is the scale of nuclear forces? When the splitting an atom, indeed, its tiny nucleus, destroys cities and remakes the geopolitical field on a global scale, how can anything like an ontological commitment to a line between 'micro' and 'macro' continue to hold sway on our political imaginaries?

When Are States? (How Are States?)

Again, Manuel Delanda paves the way:

> The idea that we know already how all past discourses have been generated, that we have the secret of all past conceptual systems, and that we can therefore engage in meta-theorizing based on that knowledge is deeply mistaken.
>
> [...]
>
> I also reject the neo-Kantian thesis of the linguisticality of experience. To assume that human experience is structured conceptually is to dehistoricize the human species: we spent hundreds of thousands of years as a social species, with a division of labor (hunters, gatherers) and sophisticated stone tool technology. Language is a relatively recent acquisition. Are we to assume that those ancient hunter gatherers lived in an amorphous world waiting for language to give it form? That's Creationism again, you know.
>
> (Delanda in Dolphijn/van der Tuin 2012)

States can be *translocal* or they can exist because of a translocality. This, *translocality*, is my first suggestion for the deterritorialization and transformation of the discourse on the State.

Translocality[13] means broadly that stakeholders and stockholders do not share the same space of cause and <u>e</u>ffect; however, they share the temporality of <u>a</u>ffect; only resolutions are usually localized in specific sites – in this fragmented constitution lie both the discontent and hope for a future grounding of an epistemic and communal concept of democracy and the State: States perform the negotiation between stakeholders and stockholders; they are brought *into play*, into existence through the performativity of this negotiation. Yet, the problem is not spatial but the fact that negotiation is necessary lies in the double separation of stake and stock, and cause and effect. And the events enacted are a narrative of sorts. I say that because stakeholders and stockholders have separate *narratives*, conflicting narratives aka conflicting of practices that need to be negotiated.

Practices, even conceptual practices, are not exhausted by linguistic practices. Linguistic practices are practices, too. But I use the notion to mean more than linguistics aka textual practices. I see practices as discursive and non-discursive, and among discursive practices there are textual and non-textual ones. The theory of semantic agency that I have partially outlined across several essays and books (e.g. in Stingl 2011, 2014a,b; Stingl in Weiss et al. 2014) refers to the comprehensive notion of practice not to the reduced one.

I like things simple, so I'd like to go here with the simplest concept of *narrative* possible and sufficient, which is Barbara Tversky's concept[14] which states that "narrative is the sequentialisation of (at least two) events in time."

So this is my problem with the focus/monopoly on spatiality; in the monopoly, the focused gaze through a singularized lens;[15] there is no space, pardon the pun, for temporalities.

Temporalities[16] are the sequentializations of events in time. Various temporalities coincide and need to be aligned. I make an analogy here: What if we think of states as we do of cooking a pasta dish.

At first, it's simple, boil some water, cut[17] mushrooms, heat pan, add cream, etc.; well, "assembled", we get a tasty pasta with shroom'n'cream-sauce when it's all done – wanna have omelet you gotta break some eggs.[18] It is that simple and yet it is not. Making it, as well as making *tasty*, does not simply come about like this. There is a lot of knowledge and practices involved in making this, a heat source, the pans had to be produced, things to be grown – all of this is infrastructure. Just think of the mushrooms that take time; how and how long they grow depends on environmental factors that also affect their taste. This technique is used for all the other ingredients, including the (variously composed, for it's not just H_2O) water you boil your pasta in, depending on things like where in the world you are (on a mountain of 4,500 meters, we're talking about 85 degrees). We can include the materialities of the pan (non-stick pan? Who paid for that to be developed? And how does that affect cooking times and textures?). Most of us, who are cooking, just do so as if it's nothing, but all of the above and more go into the cooking, including our knowing when to put in what, and how long to leave it in there. My point is all of these steps carried out, i.e. knowledge, practices, different stuffs and normativities of cookery, are the infrastructure of each dish you cook - in analogy when talking about human beings participating, belonging, provisioning with, for, among, and even against each other, the infrastructure is *the state* – and an integral part is temporality, which is my second suggestion for the deterritorialization and transformation of the discourse on the State.

We can see this problem, not of time but temporalities emerging both at a level that is deterritorialized, such as in the question of nomadic statehood, and inside territorial nation-states, where the different statehood modes, such as social state, l'état du droit (*Sozialstaat*, *Rechtsstaat*, etc.), are called on.

One example is found in the empirical distinction between civil society and political society, which Chatterjee had diagnosed in the postcolonial development of India's peasantry since the 1990s:

> The critical difference, as I have pointed out elsewhere, has been produced by a split in the field of the political between a domain of properly constituted civil society and a more ill-defined and contingently activated domain of political society Civil society in India today, peopled largely by the urban middle classes, is the sphere that seeks to be congruent with the normative models of bourgeois civil society and represents the domain of capitalist hegemony. If this were the only relevant political domain, then India today would probably be indistinguishable from other western capitalist democracies. But there is the other domain of what I have called political society which includes large sections of the rural population and the urban poor. These people do, of course, have the formal status of citizens and can exercise their franchise as an instrument of political bargaining. But they do not relate to the organs of the state in the same way that the middle classes do, nor do governmental agencies treat them as proper citizens belonging to civil society. Those in political society make their claims on government, and in turn are governed, not within the framework of stable constitutionally defined rights and laws, but rather through temporary, contextual and unstable arrangements arrived at through direct political negotiations. The latter domain, which represents the vast bulk of democratic politics in India, is not under the moral-political leadership of the capitalist class.
>
> (2008: 57)

So we can say that capital, with Chatterjee, comes in two flavors: *corporate* and *non-corporate*. For instance, *information orders* that are assembled under each flavor can sequence time-scales quite differently: For in *corporate capital* the logic of profit and the logic of extraction are necessary and, probably sufficient as criteria of ordering, whereas for *non-corporate capital*, profit is certainly not sufficient, and while it can be accounted for, it may not even be necessary.

In his seminal exploration of the *Politics of the Governed* (2006), Chatterjee shows *that* and also *where* and *when* the relation between citizenship and the (nation-)state is becoming contested in practice through the existence of a political society of people whose creation and maintenance of livelihood occurs "on the margins of legality", and, in turn, their constitution as political participation occurs while they are or despite them not being regarded "by the state as proper citizens possessing rights and belonging to the properly constituted civil society".[19]

The issue of citizenship is an important dimension for the discussion of statehood and of temporality, neither of which, I think, has yet found satisfying accounts: Just consider the recent discussion by Butler and Spivak (2011) of the problem of citizenship and statelessness that was introduced by Hannah Arendt. In Chatterjee's model, members of the political society do not have rights in the way members of civil society have, but they enjoy – based on conceptions of tacit knowledge and expectation – entitlements (even if these are never going to become rights). As such entitlements are often associated with functions or functionalities, leading to the question of the (temporal) relation (and its emergence) between function and state in political theory/practice and how it is philosophically accounted for in its own production, i.e. "etiological accounts" in the question of "What is entitlement?"[20]

Quasi-States

In the recent discourses on neoliberalism and post-democracy, Colin Crouch has created a vital argument by pointing out that ideologically informed comments and analyses of late have been blind-sided by their focus on two actors, market and State, while there should be a minimum of three actors under consideration: States, markets, and corporations. I do agree with Crouch that with regard to a plethora of political and social theoretical questions, ranging from power and influence to structuration and (social) agency, we need not only bring the State back in, or need to accept that it has never left; we also need to accept that the *corporation* is in and may have been in for a long time.[21] Second, we need to address the concept of corporation in the same way as we did the State, beginning with the question what are corporations and moving toward the questions of when.

Think about it[22]: *Are* corporations their corporate headquarters (a location), or are they their corporate identity, are they management or the stockholder? Are they the performativity of the relations between management and stockholders? What/who are stockholders and who/what are stakeholders? What are events that are sequenced/sequentialized in time? Is there a mono-causal and linear logic of production and location?

Herein, temporalities matter. Nation-states, as we know them, *do* intervene in these logics and causalities; they create conditions, reactions, potentialities, and agencies. So yes, territorial states *do* matter. But there is more to reality than just this: Corporations – like all organizations – do need well-educated staff and workers that embody a particular form of companies' memory through *tacit knowledge* they both construct and acquire *in performance*, and corporations do need to hold on to them, and bind them and their competencies for lengthy periods of time. Sure, they also have an FQu-economy styled fun time in firing people, but they simultaneously have a hard time hiring as well: labor versus skilled labor!

And yet, specialization has that kind of price-tag: skilled laborers need incentives to come and work for you, and more so to stay.

At the same time, corporations expand into more and more markets, not just through a monopolization mechanism,[23] that too, but also by using up more and more previously "public" aka state functions (health care, education, research, infrastructure, etc.) by way of public-private partnerships and privatization.[24] It is not unusual anymore that corporations offer care services or education and training to their employees directly or their families. They offer services to their (often internationally migrating) employees that nation-states used to offer to their citizens; sometimes they must circumvent a nation-state's influence because the bureaucracies of health care, education, and pension systems of nation-states just have found themselves unable to process internationally mobile professionals. At the same time, corporations help turn these services into products and structure them as markets, generating new precarities. For example, most Ivy League universities are now major corporations, and corporations buy into or fund universities, such as in Britain and the United States – or try to influence the way "universities think" by introducing and promoting the ideology of "creative destruction" (sometimes in the guise of "innovative disruption" [see also on the history of "creative destruction" from Wundt to Schumpeter: Stingl 2008/2010]).[25] Corporations offer services that were once deemed functions[26] of the State. Corporations offer services to their employees and recreate them as "markets" while they seem to "take them away" from the state – i.e. nation-state – through privatization and lobbying. In doing so, I argue, corporations are becoming (already have become) *quasi-states*. They are not states in the "conventional", territorial sense; but in assimilating more and more state functions, they become more and more "like states". As a consequence, are they actually becoming states? Can this process be reversed or, at least, controlled? Are they maybe already states?

Some colleagues and friends have responded over the past eight or nine years since I first stated this idea that corporations are states – without me originally even getting into the personhood and transnational law question, and how this translates into the colonial issue of sovereignty, which comes with Euromodern conceptions of personhood and the colonial-legal provision of corporate sovereignty and the complexities of issuing identification documents by corporations. One response has been, for example, to point out that nation-states – which I propose are only one form of statehood, historically relatively recent and emerging from Western epistemology (and false a universalizing claim) – have a monopoly on violence, e.g. regarding the rights to carry and use weaponry. But many multinational corporations have their own armed security forces and even small armies, which can operate for them across the globe in different locations. Corporations have territories in different places in the world, and they increasingly manage to make the rules, like states, on these territories, but the point is that they do not depend on each single territory, the way a nation-state does, while yet exercising similar powers as states do. My point is, however, that we are not per se looking at some new phenomenon; instead corporations become state-like because the state is not and has not been the same as the nation-state.

Nomadic Statehood

Erhard Eppler (1926–2019) in *The Return of the State* (2009) says,

> The German term Rechtsstaat – meaning "constitutional state" – carries a positive connotation, and Sozialstaat – social or welfare state – is likewise seen as a positive thing by most. But *der Staat* – "the state" – is a very different proposition: a sinister system, impenetrable, all-powerful, dangerous. Why would it not sacrifice even the innocent? Above all else this system is remote and alien.
>
> [...]
>
> We, the children of a technological civilization, in which everything rapidly becomes obsolete and consigned to the scrap heap, find it hard to imagine that there are areas of life that do not obey the laws of the throwaway society, because a better way of doing things is not apparent, or perhaps even possible. It will always be necessary to adapt the democratic constitutional and welfare state to new circumstances – or in other words, to reform it. But it cannot be replaced by something better. Because a better way of doing things has yet to be devised.
>
> –Erhard Eppler in *The Return of the State* (2009)

As my suggestions go, I propose that the concept of statehood itself must be reconceptualized – or brought under the spell of *the b'n'b* (to go before-and-beyond), with the (Euromodern) territorial state being merely a special branch. Any future theory of *the State* must conceptualize statehood as *nomadic statehood*. The challenge that quasi-states and the question "when are states?" pose in the issue of *temporalities* will lead us to the insight that we must embrace *nomadism*, i.e. we must learn to think in *deterritorialized* ways (Deleuze and Guattari being only one possible yet provincializing way), and this is hard.[27] Perhaps, we begin by dancing, and ask how to separate the dancer from the dance? Perhaps we can't, and perhaps our question must be? How do we dance the State?

Let me explain this point; unlike Immanuel Wallerstein, Kenneth Pomeranz, or Andre Gunder Frank,[28] I do not think that this dance/dancer, that this temporality, can be made explicit or intelligible by pointing out "the dance system" in its history. Wallerstein's *world-system* was and is an impressive and thought-provoking historico-academic enterprise. But that is not what I mean by deterritorialization or temporality. I think postcolonial and subaltern studies in the way of Dipesh Chakrabarty have begun pointing out the crucial aspect that I see unfolding empirically, too:

> So long as one operates within the discourse of 'history' produced at the institutional site if the university, it is not possible to simple walk out of the deep collusion between 'history; and the modernizing narrative(s)[29] of citizenship, bourgeois public and private, and the nation-state. 'History' as a knowledge-system is firmly embedded in institutional practices that invoke the nation-state at every step – witness the organization and politics of teaching, recruitment, promotions, and publication in history departments,[30] politics that survive the occasional brave and heroic attempts by individual historians to liberate 'history' from the meta-narrative of the nation-state.[31]

In other words, we need to engage in a provocative enterprise that upsets the Western mode of thinking, particularly in academia:

> There is a tendency in Western scholarship and science to accept the post-/de-colonial mode of thought into their (academic) discourses and the subjects that enunciate post-/de-colonially who are or were located geo-/bio-graphically situated in these discourses into the Western institutions (universities) – however, they do not accept post-colonial modes of thought or epistemic divergence/difference from those who were not 'originally' located so geo-/bio-graphically. Thereby, they merely tolerate the 'others', and tolerance becomes, again, just the act of violence of othering, two-fold: in othering the geo-/bio-graphically differently situated ones and in excluding those who think differently; thereby they only reify the patriarchichal, parochial, racist operation of the colonial (center-margin) matrix of power that exists in economy, authority, gender and sexuality, and knowledge/subjectivity). To genuinely become different, comprehensive, integrative and genuinely innovative, Western academia must accept difference in their own ranks and stop tolerating and instead comprehending post- and de-colonial modes of thinking, indeed, a n entirely different mode of thinking is required that manages to radically historize, radically de-colonialize and radically immigrate/integrate in practices of epistemic disobedience.

To talk about *nomadic statehood* can constitute an act of epistemic disobedience toward the hegemony of the spatial notion, aka the colonial matrix of the state as a territorial sovereignty.

Even within Western epistemology, however provincial, it is possible to think about de-spacing the state (Beck, Deleuze, or Sassen) and to implicate the diagnostic despair of the recognition of time differentials that we find in Luhmann or Elster, that remain stuck with a monolithic concept of a universal totality and absoluteness time, and why they don't suffice. One could discuss the difference between topology and spatiality, by showing that *topos* and *basho* can be much richer concepts, because they can account for temporalities without the problems of absolute time, in making for example a connection of *topos* with Walter Benjamin's passages and thresholds, Michel Foucault's concept of heterotopia, and the folding ontology of multi-*pli*-city of Gilles Deleuze, and Kitaro Nishida's reflections on *basho*. Classic topography is territorial and non-temporal, whereas actively provincializing topology to open it for decolonial options can be de-territorializing, and that there can be room for spatiality and/or process. This would also make possible to ask on the question of "boundaries versus relationscapes". We need to make, in the years to come, the case for a concept of state that is nomadic, to re(ad) dress the issues of displacement versus statelessness, violence, sovereignty from a different angle (Arendt, Butler, Spivak, Chakrabarty, Chatterjee), and based on the difference between polity, political, and civil society rooted in the distinction between textual and non-textual discursive practice. Finally, instead of *defining* a concept of nomadic state, I would say that we need a circumnavigation of the concept, or to make process-maps like micro-/macro-climate dynamic maps, and I would argue that we have always been living with nomadic statehood and that the state of (national) territoriality, as a transitional regulative, was nothing but a bad dream, a phantasm.

A crucial aspect, before I begin with the folding of idea of *de-spacing*, is the question of the governing regimes – I am picking up a point I have investigated, inspired by witnessing several presentations by Patrick Carroll,[32] which led me to reformulate his talking points into a meditation on the "old" question of seeing like a state,[33] and asking who or what is it that actually sees in case of diverse regimes. James Scott's metaphor of "seeing like a state" (refracted in Bruno Latour's "thinking like a state"[34]) – despite my critique of monoculture that possessed academic minds and practitioners of the monoculture of optico-centrism – makes me think of the problem of lensing and visual construction of statehood and of state's subject. In thinking "seeing like a state", the state is either an entity (I am not even saying actor, but stay on an ontologically simply level, not defining "what" the state is here, because the consequences of this in even more fundamental level may have grave consequences for the "what" of statehood) that does the seeing as if through a lens, or it is the lens.

We will quickly encounter the problem that the idea of diffraction applies here: In optics, diffraction is the set of possible consequences that applies when a wave hits an obstacle. An interesting phenomenon on a macro-scale is (gravitational) lensing. Perhaps, we can say that from a diagnostic point of view, states function like obstacles on the micro-level or gravitational lenses on the macro-level. This is one metaphoric strategy that would be promising, while another is based on a more "phantasmagoric" image that is oriented on what happens to actual people in actual reality with regard to how they encounter the state, which is through bureaucratic/administrative regimes.

Carroll and others[35] have shown that these practices are also deeply imbricated with, and even constituted by techno-scientific practices. A consequence of these practices is that not only could they be seen as being the mechanisms of state-formation (Carroll,[36] Mukerji, Jasanoff and others) but they also overlap, conflict, micro-construct, deconstruct, malconstruct, reconstruct, the subject that is seen and, thereby, pose the question for the entity that (actively) sees, which is, supposedly, the State. Now, this brings with it a twofold question:

a What if there is, indeed, no entity on the other side of that lens that does the seeing? What if there is only the lens and the subject that is seen through it but no one that/who sees?
b If the lens is made from these (spatially, temporally, etc.) differing regimes (of bureaucracy, conflict resolution, etc.), does that not mean that the (ground) lens is more of a kaleidoscope?

The consequence of this thought-provoking little experiment of metaphors, my own *laterna magica* – to speak while wearing a *media-arché-ologist's* hat – is that the State is a kaleidoscopic lens that is made of itself in the very event of seeing diffractively like a state, because of the two-sidedness of the event. I cannot, in this chapter, explore this into further detail, explore this conceptually and media-arché-ologically through to its intra-actions and infra-actions of the two-sidedness of the event of seeing as a kaleidoscopic gravitational lens in the political field. I hope, however, that this little exercise will be enough to stimulate inspiration, passion, debate, critique, aversion, or even disgust to bring to mind that no concept, not even that of the State, no metaphor, no event that of "seeing like a state" possess the kind of innocence that any (legal) positivism suggests:

In truth, States are just as much kaleidoscopic as they are leaky or nomadically mobile.

De-Spacing

It is not an entirely novel idea to argue that the spatial condition of the state has run into discontents in the empirical reality of regime collisions that have led to reflections and revisions of the state concept and how its spatial condition is constituted. In recent times scholarly talk, in particular among sociologists, political theorists, and contemporary historians – one is tempted to emphasize: in the West, or confined in the colonial matrix – has addressed the different questions of borders/boundaries and the reach of the state (*power-variable*[37]), citizenship (*constituency-variable*), national economies (*resource-variable*), fragmentation of regimes (*legal-variable*), mass-media and information and communication technology-induced social differentiation and fragmentation (*digitalization*), global finance and trade (*capital-variable*), among others.

In each of these variables, we can show discontents that lead to a challenge of how we understand the concept of the spatiality of these variables and how this altered spatiality leads to a challenge of the spatiality of the state. Western/colonial science and scholarship have been answered mainly using two types of strategies, either a strategy of re-phrasing spatiality that remains within a territorial mind-/conceptual-frame or by one of two tactics of de-spacing; the first tactic of de-spacing means to argue that the notion of space must be completely ignored in the discussion and should not be made explicit at all, while the second tactic suggests that some specific variables need to be conceived as unbound by any spatial (and territorial) considerations, whereas others remain bound by spatial (and territorial) considerations. And while neither strategy, ignorance or de-spacing, nor, as a consequence, the two tactics of de-spacing count as genuinely "deterritorializing" modes of thought – not to mention that they are in no way acts of epistemic disobedience and remain in or resort easily to the colonial matrix, some offer interesting analytic ideas that are worth looking at, because some of their results at the very least corroborate deterritorialized thinking and re-emphasize that even a deterritorialized mode will need to account for aspects of spatiality even if in a different way than a territorial/colonial matrix proposes. This also re-emphasizes why the question "what are states" still needs to be asked and why analyses by Rowland/Passoth are so important, for the question "what are states" cannot be reduced to "when are states" or even reduced to being deterritorialized; lines of inquiry are complementary even if they are not necessarily commensurable (but then again, that is just the point of deterritorialized thinking).

One particularly impressive and useful approach to de-spacing,[38] I want to "make some room" for here, is presented by Ulrich Beck in his book *Power in the Global Age*.[39] Beck's approach does not manage to achieve a deterritorialized mode of thinking for the sole reason that while he creates a genuinely good approach to de-spacing, he focuses on understanding the state as the actor of and constituted by politics/policies and analyzes its relation with the economy and economic actors.

He does not come close enough, in my view, to framing a genuine spatial critique of actors that would qualify as what we have defined earlier as *translocality*.[40] Let me begin by saying that I see Ulrich Beck, even in de-spacing, as a social scientist whose frame of mind I tend to associate with what has been called the "spatial turn" or "topographical turn"[41] in humanities and social sciences. Space itself became a new analytical category during the 1980s in social sciences, because *space* – illustrated by Beck's *Risk Society*[42] – became problematic the very moment that it could no longer contain its consequences and its solutions. In the act of being transcended, *space* is problematized and rendered into *place* as an analytical category, perhaps we can tentatively say that an analytical category is a concept/category that is no longer satisfying epistemic needs and warrants. I agree with Sandra Harding's[43] assessments of Beck's work as impressive – I would go as far as saying that he was the most consequent German sociologist in the Weberian tradition and the sociology of institutions –, useful, and important, but also with her criticism regarding the lack of a genuinely postcolonial perspective, and limited perspective on the issue of class in the face of the complexities of postcolonial globality.

I think Beck (2002) is correct when he says that there is *no way around a redefinition of state politics* (*Es fuehrt kein Weg an der Redifintion staatlicher Politik vorbei* [29]), in his discussion of the transformation and *debordering* (*Entgrenzung*) of the state. Beck remains a *political* sociologist in his diagnosis when he argues that those who believe in the humanism that characterized civics and civil society think that the transformation that is called globalization is just another act in the play of the process of civilizing humanity, and that it would be feasible if not easy to take the *crises, discontents, and collateral effects of this transformation* (*Widersprüche, Krisen und Nebenfolgen der laufenden Zweiten Grossen Transformation*) and turn its global and globalized actors into proponents of the same civics and humanism. This, he says, is a trope from the archive of the unpolitical.

Instead, he argues, a reversal of this line of thinking is necessary, and we should consider how Beck equates state here with the reach of politics (30):

> *As unpolitical as the image may be that global civil society will replace the renewal of state politics, as much a novelty and as untested is the idea that civil society takes power. Such a symbiosis between civil society and the state I all 'cosmopolitan state'.*
>
> [So unpolitisch die Vorstellung ist, die globale Zivilgesellschaft vermöge die Erneuerung der staatlichen Politik ersetzen, so neu und unerprobt ist die Vorstellung, dass die Zivilgesellschaft die Macht ergreift. Eine derartige Symbiose von Zivilgesellschaft und Staat nenne ich 'kosmopolitischen Staat'.]

In this discussion it is clear that even though not a right-winger, Beck never loses the actor perspective that has existed in the West since the discourse of governance as a legal problem can be subsumed under the label of Machiavelli and the state as a product of secularization.[44] In order to save the state, Beck de-spaces the nation-state, while leaving the territorial construction of its political agency (of dyadic inter-actors) intact, by creating a symbiosis[45] between state and civil society[46] and introducing a concept of transnationality. His diagnosis is certainly insightful: In the international system, nation-states are no longer actors in this system, while statehood continues to exist in the symbiosis, which means that while nation-states – slowly but surely – are evacuating (or: evaporating) the stage, states in and of themselves remain "in the game" but the fact that they are not the sole actors but "actors among other kinds of actors" is constitutive of their participation and agency, and this "constitution" is "transnational constitution" of states…. as actors.[47]

Beck argues that we must be careful, because in the empirical reality we find two types of transnationalism happening: authentic and inauthentic.

Inauthentic transnationalism occurs when nation-states exert power through the back door in claiming key positions in globalized institutions for their personnel, who operate on cognitive strategies infused by national agendas, at times even to play rail-shot intra-national politics.

In his chapter "State strategies between renationalization and transnationalization" (engl. 2006: 166ff.; German 2002: 253 ff.), Beck created an impressive phenomenology of various strategies that would fall under the category I have described as de-spacing. However, the alternative he constructs ends up with having to resolve the following diagnosis:

In the transnational space (*transnationaler Raum*), authority (*Herrschaft*) is *translegal* because it emerges in the tension that exists between the construction and expansion of governance (*Herrschaft*) on the one hand, and continuous decline of (democratic) legitimacy (*Abbau demokratischer Legitimation*) on the other hand.

I agree that Beck's transnational space, as constructed by the *cosmopolitan gaze* (*kosmopolitischer Blick*), is indeed a de-nationalized spatiality, but it is not deterritorialized nor does it make temporalities intelligible:

- Beck still needs actors; in other words, his *cosmopolitan gaze* still *produces* actors and not networks.
- Regime collisions and their resolutions are *translegal* but not *translocal*, because they still are being coordinated, even when realized and diagnosed as untimely and non-synchronous, in a time-frame that operates on an absolute totalitarian (colonial) monolithic concept of time.
- It still adheres to the colonial matrix: It does de-nationalize but it does not de-colonize.
- It still conceptualizes agency in anthropocentric and anthropomorphic ways, and constructs ecologies and nature from the point of view of human agency.

Antiracism, Decoloniality, and States

In their 2005 landmark paper "Decolonizing Antiracism",[48] Bonita Lawrence and Enakshi Dua make a number of important points about the fact that much of antiracism – understood as the discourse consisting of critical race, feminist, postcolonial, and other (theoretical) lines of inquiry – is largely oblivious to if not even actively obscuring the differences among indigeneous people as well as occluding the ongoing struggles against Settler Colonialism and the suffering under Settler Colonialism experienced by indigenous, aboriginal, native, or "First Nation" people. Beginning with the fact that already a collective naming such as "First Nation" people tends to homogenize different groups of peoples and, thus, erases crucial differences, the paper identifies a number of areas where and ways in which antiracist discourse has been complicit in the ways that "settler colonial *society*(!)" has denied, obscured, negated, and silenced (different) indigenous peoples.

> International critical race and postcolonial theory has failed to make Indigenous presence and colonization foundational in five areas. First, Native existence is erased through theories of race and racism that exclude them. Second, theories of Atlantic diasporic identities fail to take into account that these identities are situated in multiple projects of colonization and settlement on Indigenous lands. Third, histories of colonization are erased through writings on the history of slavery. Fourth, decolonization politics are equated with antiracist politics. Finally, theories of nationalism contribute to the ongoing delegitimization of Indigenous nationhood. Though often theorizing the British context, these writings have been important for shaping antiracist/postcolonial thinking throughout the West.
>
> (Lawrence/Dua 2005: 127/128)

While in the past one and half decade since their paper there were some developments that one can hope may be qualified as positive and for which the wider acceptance of decolonial literatures and the emergence of differential indigenous scholarships and notable voices (such as Zoe Todd, Kim Tallbear, and others) are certainly indicative; nonetheless it needs to be stressed that their diagnosis has still not lost its bite.

However, for the present purpose, I am focusing on one point they make, namely that "theories of nationalism contribute to the ongoing delegitimization of Indigenous nationhood". One of my key points considered in this section is, of course, that coloniality and Euromodernity introduced not so much the idea of the "state" as critical, but that the critical issue is the introduction of the notion of society. Thus, *pace* Pierre Clastres, but it is not that we need to think "society without the state", but "state without society". Thus, I find it interesting when Lawrence and Dua argue along similar lines:

> Finally, theories of nationalism render Indigenous nationhood unviable, which has serious ramifications in a colonial context. The postcolonial emphasis on deconstructing nationhood furthers Indigenous denationalization for those targeted for centuries for physical and cultural extermination, and facing added fragmentation through identity legislation (….). Such deconstructions can ignore settler state colonization (…). Or they theorize, from the outside, about how communities "become" Indigenous solely because of interactions with colonialist nationalist projects (…). If the epistemologies and ontologies of Indigenous nations do not count, Indigeneity is evaluated through social construction theory. More problematic still are works that denigrate nationalism as representing only technologies of violence (…), or a reification of categories that can degenerate into fundamentalism and "ethnic cleansing" (…). There is also the simple dismissal of "ethnic absolutism" as an increasingly untenable cultural strategy (…), which calls into question the very notion of national identity. None of these perspectives enable Indigenous peoples in the Americas to envision a future separate from continuous engulfment by the most powerful colonial order in the world, or their continuous erasure, starting with Columbus, from global international political relations
>
> (Lawrence/Dua 2005: 131/132)

I will, of course, have to diverge slightly and emphasize that my own point is about the state and not the nation or nation-state. But, again, we need not be afraid of equivocation. I assume that what Lawrence/Dua talk about in terms of indigenous nationhood does not immediately and for obvious reasons equate with the Euromodern idea of the nation (state), but refers to different equivalents and analogs, which are yet on par: the point is, precisely, that their addressing Indigenous nations refers to (therefore certainly in part preempting) what I want to address in terms of statehood. This should by no means be misunderstood as me proscribing indigenous authors how they should address, conceptualize, or operationalize the matter. My presentation here is one targeted at one specific attempt at provincializing Western (social) theory and thereby opening it up in the first place to enter into genuine conversations with for decolonial options such as presented for example by Lawrence/Dua.

In this context, Joanne Barker (2015, 2018) provides an interesting reference to the fact how national sovereignty and, thus, statehood, is subsumed under the epistemological Western and Global Northern geopolitical conceptual device of "society". Her analysis, which shows how the 2008 financial crisis, its genesis, and aftermath are tied into and continue the history of land dispossession experienced by indigenous peoples, identifies how a series of treaties, laws, and US Supreme Court (SCOTUS) decisions helped defraud Indian tribes (among others by maintaining that nomenclature of "Indian" and "tribe") in a pattern that has become quite characteristic and foundational for the operations of corporations, which precisely in this history of Settler Colonialism obtain their legal fiction of [corporate] personhood, and, thus, capitalism and neoliberalism – besides she illustrates how the supposed "resistance" to capitalism fails to recognize and include indigenous people.

In one of the landmark SCOTUS decisions that Barker cites (2015: 249f.) in her seminal paper "The corporation and the tribe", SCOTUS decision *Johnson's Lessee v. McIntosh* 1823,

> The question before the SCOTUS, as Chief Justice John Marshall framed it, was what kind of title the Piankeshaw Nation held in the lands. But before deciding, the Court had to address two facts: (1) the US Con-gress had acknowledged in its ratified treaties with Indian tribes— as had all European nations before it— that tribes possessed a land title that they could treat upon; and (2) the treaties themselves referred to Indian tribes as sovereign nations with all commensurate jurisdictional rights over and within their territories. While not missing the import of treaty language, SCOTUS sided with McIntosh on the grounds that Indian tribes had never been recognized as equal "sovereign, independent states."
>
> The uniform understanding and practice of European nations, and the settled law, as laid down by the tribunals of civilized states, denied the right of the Indians to be considered as independent. communities, having a permanent property in the soil, capable of alienation to private individuals. They remain in a state of nature, and have never been admitted into the general society of nations.

The SCOTUS evocation of society in the formulation "society of nations" under which sovereign states thus must fall is emphasizing my point here: The invention of a Euromodern concept of "*society" as a (Northern Atlantic) category is a decisive colonial operation that renders statehood in its variety of forms, which I summarize as nomadic statehood, null, and void (terra nullius) and allows* only *those societies to have states that can be considered under the terms of* "national sovereignty" as part of a "society of nations" – later: "global society" (for posterity, see also a less traditional but different and still Euromoderncentric account by Thornhill 2019).

That Global Dilemma?

There is, I think, at this point a geopolitical dynamic ongoing, which has historical roots from *oikos* to colonialism to neoliberalism, and has in its grip minds, groups, communities, collectiv(iti)es, institutions, organizations, social systems, nations, societies, global society, if you will. It has led us to a point today, for which one quite plausible account of how it can be described poses before us a number of conundrums that are neither well understood nor studied in their current being and effect nor in the plausible and possible trajectories – although speculation and projections would look largely dystopian.

The deft diagnosis is this: At this point global society is as good as totally governed by corporations. And while this general diagnosis may not universally hold for each and every country or, at least not for every territory or place, it is, *translocally* speaking, an effective description of the geopolitical pervasion of *extractive* and *corporate capture*. For starters, many of the ways of thinking, seeing, and doing of *corporate culture/mentality* – that can rightly be said to owe themselves, non-surprisingly, to the *coloniality of power and Being* – can be identified throughout the world. Many papers and studies written on the so-called postneoliberal states, such as for example Ecuador, show that despite the rewriting of political constitutions by including concepts like *buen vivir, sumac kawsay*, little has changed in extraction and precarization still being practiced and accelerated. Sure, they are being *implicated* in media, often by the same politicians or local communities who will then *intensify* the situation by making or supporting deals with the same extractive industries – the reasons are often complicated: such as while leaders were elected to fight corruption, bring transparency and sustainability to the region, re-election, both (!) so-called leaders as well as their voters believe, depends on job creation, which, in turn, depends on GDP growth of the country, and so on, and "there is no alternative"; there is, at least, no alternative that is currently thinkable, let alone viable. In *The Digital Coloniality of Power* (2015), I have called this curious and perverse seeming double-bind of "implication" with "intensification" by the neologism *implification*. It is a concept that is helpful for certain kinds of analysis, certain kinds of thinking with this concept. I think it is a strange development, but we can identify an increasing number of cases in recent decades where a constellation, action, dynamic or person was implicated of something negative, but it seemed that not only despite the implication but rather in some kind of yet ill-understood conjunction, it intensifies. The most visible, and to some ironic degree one would have to say "ridiculous", case is, of course, the election of a politically inexperienced *media personality*,[49] Donald Trump, as US president: The more Donald Trump became implicated – think alone of the revelations concerning his predatory attitude and behavior toward women – the more support he received.

When I say governed by corporations, I don't mean that national governments have no influence or power at all. But the account I am interested in shows a very complicated situation, a situation that we will have to study critically and in more detail in the future: The persons that are corporations – a personhood that some of them obtained through US legal practice and extended internationally and then transnationally through US foreign policy on the one hand and the pervasion of the logic of the extractive Empire on the other hand – have wielded and increased their governing power over, within, and across societies: Not always are they dealing with nation-states or supranationally bonded nations (such as the European Union (EU) is supposed[!] to be). The United States, which I have characterized in *The Digital Coloniality of Power* as an *anocracy*, is a very problematic case. This is true especially in terms of the revolving door between the government agencies of the US States and the Federal Nation State on the one hand, and (multinational) corporations on the other hand. And this is, of course, independent and irrespective of whichever US party – Republicans or Democrats – we are talking about; examples can be found on either side, although given the overall ideology of the Republican Party, it should not be surprising to find more and more extreme cases on their side. But in the end it does not matter if – to use a bit of hyperbole: – qua being "pro privatization of nearly everything" many politicians of the Republican Party happily sell parts of and hollow out the US government, or whether a Democrat US general attorney used to represent the very banks he is supposed to negotiate their punishment with, only to return to his old job representing these banks after his tenure in government. The fact is that these "internal" problems of the United States do have consequences for many other countries, societies, and communities across the world (and even across temporal scales), because of the international political, corporate finance, (political) economic, and transnational legal intertwinings, interdependencies, and entanglements (of oppression) across the globe. In the same light, both China and Russia are similar but very different in their structure but no less intertwined and entangled. And I am not sure we have as of yet – in the Western social and political sciences – an adequate understanding of these, and I believe we will still need quite some research, discussion, and debate before the accounts we can give allow us to understand matters necessary and sufficient for pragmatic sense- and meaning-making. But the best embankment toward an account is that in a sense Russia as nation-state could be seen pretty much a conglomerate of corporations, whereas China is, in economic terms, itself a corporation. Thus, for both it would then be true that their economic interests take the form of corporate interests, and their geopolitical actions are very much in-*formed* by that form. Meanwhile the United States is in-*formed* in a highly fragmented fashion, shaped by the structure of its anocracy, i.e. certain parts of the government do things one way, others differently, leading to the highly incoherent geopolitical actions of the United States. Countries that make up the EU are, at this point, in a volatile state of transition wherein the result is, as of yet, unclear, although the most likely outcome is further fragmentation along old national societal boundaries leading to increasing loss of power while dealing with quasi-states and nomadic states, above all *corporate actors*, including what can be considered in through the following metaphorical lenses: *the corporate nation* China and *the corporation-conglomerate* Russia.

In short, a lot is currently empirically cooking for the future "theory of statehood", but whether it will be a tasty stew, a bitter broth, or just a burnt, charcoal lump we have yet to see.

Notes

1 Habermas's essay on "Kants Idee des Ewigen Friedens" explicitly puts "democracy" into places where Kant refers to Republicanism – a sleight of hand not untypical for members of the Frankfurt School.

2 See here specifically: Marchart, Oliver, *Die Politische Differenz*, Frankfurt aM: Suhrkamp, 2010; Soerensen, Georg, *The Transformation of the State*, New York: Palgrave MacMillan, 2004; Slaughter, Anne-Marie. *A New World Order.* Princeton: Princeton University Press, 2004; Florini, Ann, *The Coming Democracy*, Washington, DC: Brookings Institution Press, 2005; Zuern, Michael, *Regieren Jenseits des Nationalstaates*, Frankfurt aM: Suhrkamp, 1998. Albrow, Martin, *The Global Age: State and Society Beyond Modernity*, Stanford: Stanford University Press, 1996; Sassen, Saskia. *Territory, Authority, Rights*, Princeton: Princeton University Press, 2007; Ong, Ahiwa, *Flexible Citizenship*, Durham, NC: Duke University Press, 1999. Bernhard, Peters, *Der Sinn von Oeffentlichkeit*, Frankfurt aM: Suhrkamp, 2007; Derrida, Jacques, *Schurken*, Frankfurt aM: Suhrkamp, 2006; Not to forget always highly entertaining Fukuyama, Francis, *Statebuilding*, Ithaca: Cornell University Press, 2004.

3 This needs the following disclaimer: I do not propose an original account of this process, nor a complete and comprehensive, or a very rich and detailed one. I am just trying to provide a short summary overview of how Western scholarship has, and often still does, conceived of the emergence of the Western modern state, which serves as a benchmark for concepts of the state, state development, state failure, the end of the state, and the return of the state… from the Western point of view. My summary is informed by the literatures referenced here.

4 On the concept *registrature*, which is similar to but still different from and developed in slight ignorance of Hannah Arendt's concept of "revolution" (1965): Stingl, Alexander, *Aufklaerung als Flaschenpost oder Anthopologie der Gegenwart*, Saarbruecken: VDM, 2009.

5 David Rothkopf hits the nail on its head, when he writes that

> [t]he power to appoint bishops was especially important to the emperors because, while top positions in the nobility were hereditary, top clerical posts could be filled by emperors with loyal supporters or those to whom they owed a debt to from whom they might someday want a favor. The result was a system in which the line between interdependence and too much dependence, between secular and clerical centers of power, was very blurry. Later, and repeatedly, we will see just such a fuzzy and contentious relationship evolve between the centers of public and private power.
>
> (2012: 55)

6 Miller, Maureen C, *Power and the Holy in the Time of the Investiture Conflict*, Bedford: St. Martin's, 2005; Gushing, Kathleen, *Reform and the Papacy in the Eleventh Century: Spirituality and Social Change*, Manchester: Manchester University Press, 2005; Blumenthal, Uta-Renate, *The Investiture Controversy: Church and Monarchy from the Ninth to the Twelfth Century*, Philadelphia: University of Pennsylvania Press, 1988.

7 The concept of the *raison d'État* is an interesting, crucial, and notoriously difficult concept, because its meaning – tied to governmental practices that are shifting with time and space – is vague. In Anglo-American contexts, it is often translated as "national interest", which I take exception to. Indeed, for a nation-state, its national *raison d'État* is its national interest but that does not mean that it is an accurate translation because the nation is one form of spatial political order, but it is not the only one by far, as I argue here, but only one of several forms of state.

A second issue is an important relation, which I unfortunately cannot discuss here in detail but which connects this part of my work – the theory of the nomadic state – to my other work – the (historical) ontology and epistemology of the body, and the agencies and ecologies of vital materialities –, that is between the *raison d'État* and the *raison être*. Regardless of whichever discipline I am engaging with or am being associated with by others, at the center of work is a "strange relation" between "*the* body and the State". I have long sought to understand how both can, in an actual empirical reality, be ontologically multiplicities (Deleuze) while they are all in the case of *the* body a *heterotopia* and in the case of the State a *dispositif* (Foucault). ["*The* body" means the idea of what it is to have *a* body that exists before the text; this is a very difficult concept, because that does not negate the importance and irreducibility of politics of the body, but it is at the same time a necessary (but not sufficient) condition for them; yet it is, at the same time, not an essentialism.]

I have learned from reading Laura Hengehold that there is a name for the connection between body and State in the empirical human reality (while keeping an eye on Walter Mignolo's work, I am not yet sure if that refers only to *anthropos* or to *humanitas* [Following in the footsteps of 1930s debates surrounding the work T*he Racial Myth* {1934} by Franz Boas's student Paul Radin, recent postcolonial discourse has taken up this distinction: Osamu 2006; Mignolo, 2009; Sakai 2010]). Before engaging with Hengehold's *The Body Problematic* (2007, see my review in *Foucault Studies* Vol. 15, 2013), I did not have a good conception of how to really talk about the relation outside of concepts such as "the social" or other vague intermediaries. Hengehold constructs the mutually interdependent dialectic relation as follows:

Body <> political imagination <> State.

I would elaborate this in light of the aforesaid to exclaim the following as the Western model of this intra-relationship:

Body)<*raison être*>(political imagination)<*raison d'État*>(State.

This would also be an excellent model to account for the dependence of Western theory of body and State on linguistics and text, for raison (reasoning) is conceived in the West as an exclusively textual affair *and* linear causal logic.

8 Mignolo (2009: 5–6).

9 Nor as Chakrabarty or Mignolo do. They both emphasize this point very strongly.

10 I do disagree with their passing note, taken from Thomas Lemke, that linking "govermentality" to mentality is a misconception. Indeed, many interpreters who have done so have created this account as a misconception and

misconstrued fact. However, to deny a certain anthropological/semiotic link in Foucault's concept between governance (State), techniques of self (imagination), and mentality (psyche/mind) is problematic: Mentality can be translated into the German concept *Gemueth* which is prominent in Kant's anthropology and political philosophy (Enlightenment), the life-long and recurring reference point in Foucault's work. I feel as though Lemke et al. forsake the anthropological dimension for the political one. However, I think, you cannot have one without the other.

11 See my paper "How to map the body's spaces" Myth-making, myth-breaking in history, Bucharest (2011).

12 I do not have time here to discuss sociological concepts of translation in other schools of thought such as in the fictional accounts of social phenomenology. I needn't stress that I find social phenomenology a la that Austrian banker and systems theory a la Luhmann of very limited use, if any.

13 This concept of *translocality* was originally developed by me, before I had become aware of Patrick Carroll's works. Carroll (1996: 142f.) establishes a concept of translocal practices which is similar but at the same time distinct of my own.

14 Tversky (2004).

15 Like the single lens camera, the single lens of the spatialized (and actor) state suffers from problems: It "blacks-out" the image during exposure. The reflex mirror's movement takes time even though it does not to sequentialize events but describes time-holes between events that limit the reaction speed, while, at the same time, the mechanistic components result in noise and vibrations.

16 My concept is similar to Schrader's but not identical. I had begun to develop my approach when I was still oblivious to her work, but am very happy and grateful to having discovered her insightful studies. See Schrader (2010):

> Pfiesteria are not only context-, but also history-dependent, in which case we are dealing with an entanglement of two indeterminacies – between 'bodies' and 'environment' and 'past' and 'future' – that cannot be resolved at the same time. As soon as a 'time' as an external parameter that *automatically* orders events chronologically can no longer be presupposed, the differentiation between 'bodies' and 'environment' depends on what I call temporalization– the establishment of a relation between 'past' and 'future'. (293)

17 Although, already, we can complicate, when remembering how Deleuze conceptualized the event: There are two sides to it, such as in the event of cutting, there is the knife that actualizes a potentiality, that of cutting (instead of murdering) but the actualization is an event in the knife cutting and the mushroom being cut – cutting *and* being-cut are the events.

18 Or break some unquestioned certainties. Certainties are eggs! Instead of certainties, perhaps we need confidence(s) and plausibilities! I am hardly ever certain, but usually quite confident.

19 Chatterjee (2008: 57).

20 See Casullo (2007) and Graham (2012).

21 See for example the popular but insightful historical analysis by David Rothkopf (2012).

22 The debate between Michael Callon and Daniel Miller is a very insightful exchange that makes a good head-start in this discussion: Miller (2002) and Callon (2005).

23 Norbert Elias *(Process of Civilizations)* described such mechanisms for monopolies on taxing and on violence; as a king-making mechanism, we see a similar

pattern with corporations on markets. Since Elias did also write *On Time*, it'd be interesting to completely re-theorize Elias's works through ANT and semantic agency theory (SAT), see my "Truth, Knowledge, Narratives of Selves".

24 On this point, see Crouch (2011).

25 www.nytimes.com/2012/09/16/magazine/teresa-sullivan-uva-ouster.html?pagewanted=all.

26 On the issue of functions, I follow here a philosophical consideration of *etiological accounts*. That would in and of itself require an in-depth discussion, how political theory is affected by this way of conceptualizing function through selection, warrant, and entitlement, originating the complex discussion of Tyler Burge's original arguments by various scholars, including – for my purposes – Graham (2012).

27 It requires, to borrow from Walter Mignolo, an act of epistemic disobedience (Mignolo 2009).

28 Wallerstein (n.d.); Pommeranz (2000); Frank (1998).

29 And I would add: territorial.

30 Although the same could be said for political science and sociology.

31 Chakrabarty, Dipesh. *Provincializing Europe*. Princeton: Princeton University Press 2007: 41.

32 Carroll is one of a small but growing number who execute a different account of state formation theory, providing an alternative to the standard (national-) territorial account and the (German) legal-historical account of the *Allgemeine (Rechts-)Staatslehre* (cf. Hegel, von Stein, Jhering, Jellinek, Kelsen, Carl Schmitt, Boeckenfoerde, Stolleis).

Patrick Joyce and Chandra Mukerji have argued that the modern national-territorial state is derived in a more recent historic phase as a result of technologies of power that combine military and engineering power as a way of mobilizing the political agency of nature – not unlike recent discourses on vital materialism, but, of course, intentionally anthropocentric. I am very fond of Mukerji's ideas on the subject matter in particular, in part because she is one of only a few Western sociologist/political theorists who have been willing to entertain the potentials of Michel Serres's works. However, I have two points. Less importantly, I think we should pay more attention to the resemblances/differences of her argument and Michael Mann's earlier concept of infrastructural power (Mann 1984). Or at least, I have not been able to find this discussed in the works of Michel Serres that I have seen. Second, and most importantly, I think that her criticism of Foucault is slightly mistaken. While I am happy to agree that she does cover aspects that Foucault does not address, in part because he may not have been interested in diverting his focus on these issues, her interpretation of Foucault's concepts is different from those I find more plausible and I would argue that a less Americanized and more actual Foucault would be an important asset in this discussion.

Indeed, American and German interpretations of Foucault (still) largely tend to read Foucault's concept of power as a Marx-Hegelian or Freudo-Marxian concept of domination, ignoring the fact that not only was Foucault's concept inherently different from these concepts but he also quite vehemently and explicitly rejected the concept of domination. This problem becomes quite apparent in her recent and brilliant work on state formation (Mukerji, 2010), where she seems, I think, to conflate Hegel-Marxist domination theories with Foucault. Given that her adoption of the ideas of Foucault's student Michel Serres as well as Dorothy Holland's (et al., *Identity and Agency in Cultural Worlds*, Cambridge, MA: Harvard University Press, 1998) ideas on *figured world* resonates completely with the theory of SAT, which I unfold elsewhere ("Truth, Knowledge, Narratives of Selves" in: *American Sociologist*, Vol. 42, 3/3, 2011), and that SAT is originally derived in my continuous interlocution with the works of Foucault, my reservations regarding her nearly complete rejection of Foucault's conceptualizations are well justified. I see no major difference between her, SAT, and Foucault, as illustrated in her acceptance of Holland et al.: "A 'figured world,' [...] is a politically infused culture that shapes cognition as well as action. The culture consists not only of a constellation of ideas, but also physical forms systematically infused with meaning". Neither in Foucault nor in SAT is meaning reduced to the *logos* of text. That is precisely the gist of Foucault's stance against Derrida – shared by Deleuzeanism, recently exemplified in the work of Elizabeth Grosz.

Patric Carroll's work, on the other hand, is interesting for his decisive stance on the connection between state-formation and techno-scientific practices.

See here: Joyce, Patrick, *Democratic Subjects: The Self and the Social in Nineteenth Century England.* Cambridge, UK: Cambridge University Press, 1994; Joyce, Patrick, *The Rule of Freedom*. London: Verso, 2003; Joyce, Patrick, "The Potency of Things: A Critique of Cultural History" in: Niall O'Ciosain, ed., *Explaining Change in Cultural History.* Dublin: University College Press, 2005; Bennett, Tony, Joyce, Patrick, eds. *Material Powers.* London: Routledge, 2010; Mukerji, Chandra, "Material Practices of Domination and Techniques of Western Power," *Theory and Society*, 2002 31: 1–31; Mukerji, Chandra, "Intelligent Uses of Engineering and the Legitimacy of State Power", *Technology and Culture*, 2003 44: 655–676; Mukerji, Chandra, "Tacit Knowledge and Classical Technique in Seventeenth-Century France: Hydraulic Cement as Living Practice among Masons and Military Engineers," *Technology and Culture*, 2006 47: 213–233; Mukerji, Chandra, "The Territorial State as a Figured World of Power: Strategics, Logistics and Impersonal Rule," *Sociological Theory*, 28 (4), 2010: 402–442; Mukerji, Chandra, "Unspoken Assumptions: Voice and Absolutism in the Court of Louis XIV," *Journal of Historical Sociology*, 2, 1998: 283–315; Carroll, Patrick, "Science, Power, Bodies: The Mobilization of Nature as State Formation," *Journal of Historical Sociology*, 9, 1996; Carroll, Patrick, "Tools, Instruments, and Agents: Getting a Handle on the Specificity of Engine Science, *Social Studies of Science*, 31, 2001: 593–625; Carroll, Patrick, "Material Designs", *Theory and Society*, 31, 2002: 75–114; Carroll, Patrick, "Water and Techno-Scientific State Formation in California, *Social Studies of Sciences*, 42, 2012: 489–516; Carroll, Patrick Papers presented at 4S 2011 and 2012.

33 The notion of "seeing like a state" was famously introduced by James C. Scott's, *Seeing like a State.* New Haven, CT: Yale University Press, 1998.

34 Latour, Bruno, "Thinking like a State", Lecture Delivered, November 22, 2007.

35 Above citations, in particular: Carroll, "Science power, bodies,"1996.

36 So the state, I suggest, is realized not by a power that is located in a center and spreads out in concentric circles, but in multiple locales in which socio-corporeity and socio-materiality are invested, and through the networking of all of these by a government which claims a unitary and univocal voice over "life". (Carroll 1996: 142).

37 I leave open for moment what is meant by "reach"; this can mean legal, military, etc. This constitutes the variable summed up under power, because the concept of power is vague in the very same dimensions, as can be seen in the many "theories of power" that exist between historians, political theorists, and sociologists from Marx (domination), and Weber (the forms of ruling)

to Parsons (power as a medium to fulfill the function of goal attainment), Luhmann (success medium), Mann (infrastructural versus despotic techniques of power; military/economic/ideological/political [re]sources of [social] power), and Foucault (capillaries of creativity) among many others. Each case opens a different analytic dimension; however, I argue that even in their individual analytic prowess and even in their individual usefulness in a theoretical framework such as SAT [Stingl 2011], (almost) all of these concepts of power remain stuck in spatiality and/or a monolithic concept of time.

38 I will not have opportunity here to enter into a discussion with the other few interesting – mostly (Marxist) relational – approaches such as, for example, Bob Jessop, Mouffe and Laclau, or Nicos Poulantzas (Jessop, Bob, *State Power: A Strategic-Relational Approach*, Cambridge, UK: Polity, 2007; Poulantzas, Nicos, *The Poulantzas Reader: Marxism, Law and the State*, London: Verso, 2008, Mouffe, Chantal, Ernesto Laclau, *Hegemony and Socialist Strategy*, London: Verso, 1985).

39 Beck, Ulrich, *Power in the Global Age*. Cambridge, UK: Polity, 2006. I will quote (in paraphrase or translation) from the original German edition: *Macht und Gegenmacht im globalen Zeitalter.* Frankfurt aM: Suhrkamp, 2002.

40 Contrary to any assumptions, *translocality* is not the same as Beck's concept of (world) risk society, which is still inherently territorial and rest conceptually on territorial boundaries between populations, territorial states, and economic actors. At the same time, understanding Beck's concept can be a first step towards learning to think outside the nation-state box.

41 It has also sometimes been called the "topological turn", which I refute based on the clear distinction that I make between the static/geometric concepts of topo-graphy/geo-graphy and the dynamic Deleuzean ideas of topology/geology.

42 Beck (1986).

43 Harding (2008), here: 49ff.

44 So is the widely accepted argument made most prominent by the Ritter-School; see for example Boeckenfoerde.

45 When we should be going even further and discuss the issue of parasitism (Serres) and thirdness.

46 This is exactly the point in the colonial matrix that Chatterjee begins to attack in distinguishing civil society, political society, and the state.

47 This leads us back to the critique provided by Rowland and Passoth.

48 I am deeply indebted to the students of my 2018/2019 workshop-seminar *Decolonial Options* at Leuphana University Lüneburg to alert me to this important essay that had previously escaped my attention.

49 At the behest of the publisher's editor, a footnote on the difference between the legality versus the legitimacy of 45th presidential mandate (which is a similar argument as the EU's legitimacy deficit issue) was removed.

8 Covfefe that *Matters*

Toward a Different Sociological Imagination

This chapter explains how in public discourses on truth and facts regarding political news and science we are supposedly living in an age of "post-truth" that coincides with the age of digital culture. This chapter discusses whether the role of "digital culture" in knowledge discourses presents a new development that has fundamentally restructured public discourse. It is considered here that, instead of presenting a new stage, the antagonistic discourse of digital culture is merely an extension of the longue durée of what decolonial scholarship has identified as the *coloniality of power and Being*. Furthermore, an important issue in contemporary public discourses on politics and science is not the "problem of truth", but the fixation on "relevance". Taking inspiration from efforts of provincializing knowledge, a social epistemological line of flight is projected on the basis that an agonistic discourse in political news and science requires a culture of epistemic responsibility based on a general requirement for a "right to being justified *to*".

The following newspaper headlines will serve as illustrations for the stakes entailed in the empirical philosophical thoughts that follow after:

> "Wo warst Du, Gott?"
> *Where have you been, God?*
> – March 12, 2001, BILD
> (Germany biggest yellow press tabloid)
>
> "Trump refuses to rule out opponents being executed for treason: 'They tried to take down the wrong person'"
> (May 24, 2019, www.independent.co.uk/news/world/americas/us-politics/trump-treason-comey-mccabe-strzok-page-white-house-a8928556.html)
>
> "Die Wahrheit über kriminelle Ausländer."
> *The Truth about criminal foreigners.*
> – January 4, 2008, BILD
>
> "Ausländer dürfen über deutsche Regierung abstimmen."
> *Foreigners can vote on German government.*
> February 6, 2018, BILD

"Deutsches Wochenblatt zum Kampf um die Wahrheit"
German weekly paper in the struggle for the Truth.
Subtitle of the anti-semitic journal *Der Stürmer.*

"Chaos! ALLES noch VIEL schlimmer."
Chaos! EVERYTHING so much WORSE.
– *Fictional(?) tabloid headline.*

The Surreality of It All

Four times in my life have I been met with an event, which defied my own belief: that kind of event, which even upon revisiting it in memory, and despite the undeniably lasting and life-course altering consequences for my life and other people's lives, still seems utterly unreal. The first was the coming home from school and learning of the unexpected death of my father more than a quarter of a century ago; the other three are more recent ones. One was standing in the middle of Place St. Michel in Paris, close to where I lived at the time, and seeing Notre Dame burning – regardless of the debates that followed about its renovation, the massive donations by rich people, etc., if you were there and had coin-size pieces of glowing ash raining down on you, the eerie unreality of this would also never leave you. The other two having happened in 2016; the first is of a personal nature and I will not recount it here, whereas the second is the morning after the US-American election, upon learning over sharing a breakfast with two academic friends who were both like so many others also teaching and living precariously at Lüneburg's Leuphana University at the time, that a "business"-man and *yellow press* personality had won the US presidency; a man with no discernible political qualifications, who had over the past years and particularly months escalated the bitter divide not only within US society itself but also, thereby, helped along the already ongoing rise of authoritarian populists in the Global North, the rise of anti-intellectual sentiments, of mysogynist and anti-*LGTBQ…x* sentiments, of racists and white supremacists, and, of course, of neoliberal, neocolonial, neonationalists, radical capitalists. Explanations are, since then, aplenty for the why-s and how-s of the election of Donald Trump as US president, as well as for the kind of support and applause both he and the kinds of opinion-groups that he seems to find among certain sectors and even in segments of the populations of other countries – utter *implification.* Of course, there is – in actuality – not a single *one* explanation; instead there were (and still are) a number of factors in play simultaneously. But among them range with determinative force a certain mix of resentment, status anxiety, and self-centeredness (especially of *white privilege*) that are directed against those that are undeniably Other, following a certain logic of how one perceives oneself as a member of a deserving majority of "meritocratic normals", regardless of any "facts" that seriously question the majority status, the assumed desert, the illusion of "normality" of one's in-group membership categories, the "Otherness" of Others, let alone their supposed lack of desert or their supposed social destructiveness, etc.. It is little wonder that Theda Skocpol and her students found long before the launch of the 2015/2016 Trump campaign that even in places in the North of the United States, where the factual numbers of Latin and Southern Americans were largely negligible, supporters of the so-called Tea Party (a reference to the historic Boston Tea Party, although a later myth credits the slogan "Taxed Enough Already!") perceived these (and other "illegal" immigrants) as a direct threat to their safety and their jobs. The Tea Party was presenting a kind of social movement which Skocpol came to understand as becoming the fledgling power-basis of resentment for the Trump movement (as Skocpol reported in her keynote speech at the Eastern Sociological Society meeting 2016; see also Williamson et al. 2011;): "For the Tea Partiers, the major intellectual distinction isn't between black and white — although that is the color of most of them — it is between deserving, hard-working citizen and unauthorized, foreign freeloader" (summarized in Freeland 2011).

Similarly, we see the success of the German right-wing extremists of the Pegida movement and the party *Alternative für Deutschland* (AfD), which is running on an anti-immigrant platform, particularly in places with decidedly small migrant populations: In 2016, in the German state of Mecklenburg-Vorpommern, the AfD received 20.8% of votes, when, however, only 3.8% of people living in the state were foreigners.[1]

It has been noted that the rise as well as the continued success of right-wing populists is partially correlated with the Internet and social media (see the introduction by Engesser et al. 2017 for a special issue on the topic of the journal *Information, Communication & Society*). For instance, what I have called in a short passage in my book *The Digital Coloniality of Power* a "cold civil war" in the (*anocratic*) US is partially led as well as co-constituted via these "digital sources". However, these processes are not specific to what has been deemed the "novelty" of our *Digital Age* or *Digital Culture*. On the contrary, I think that the current purely *antagonistic* political discourse across digital realms as well as across (what is among digital and game studies researchers being called) "meat space" is merely an extension of a much longer running process, wherein those who are Other(s), in opposition to some vague construction of "we"/"us", are imagined and constructed in a way that renders them, ultimately, "killable, rapeable, disposable" (Maldonado-Torres 2007). Taking inspiration from scholars who have shown that today's "new racism" is anything but "new", such as most recently analyzed by Zine Magubane (2016a,b) among many others, I have looked at knowledge production and the constitution of political discourses in "digitalizing worlds" and reflected on the "newness" of their productions; being a Foucault-inspired scholar, the connection between knowledge and power is of particular interest to me, especially because power is an ambiguous concept in social scientific thought. The "political"/ "public" discourses and their involvement of "knowledge(s)" writ large, I find, have a genuine connection to what Chantal Mouffe or Oliver Marchart understands in "post-foundational" terms as pure antagonism to which Mouffe counter-proposes that we recognize the ubiquitous existence of inherent antagonism as the very source of what the "political" is; *but* in order to have a political discourse that does not end being characterized by (eventually assured mutual[2]) annihilation, we should adopt a culture of *agonistics* instead. With reference to the example of "theory" debates within the sociological profession itself, I reiterate once more Peeter Selg's (2013) short and accessible account of what that means, addressing three options of *antagonism*, *deliberation*, and *agonistics*:

(i) *Antagonistic Exclusivism* means that one perceives alternative views to one's own, but effectively denies the possibility of consensus, and even more than that: any interaction or engagement with these other views is denied[3] and their very basis of even being capable of making an (intelligible) argument is rejected. Instead, other views must be seen as a target for annihilation. (ii) *Deliberation as consensual mainstreaming* recognizes the differences of opinions and views, but this type of recognition is bought at a price: the price of abandoning the possibility of actual dissent – consensus as a result of deliberative process states that the result of deliberation must be adopted with finality through the "forceless force" of the better argument "winning" (the theoretical interlocutor here is, of course, the Habermasian discourse ethicist), which results in the end in a form of exclusion of dissenting views.[4] Finally, (iii) *Agonistic pluralism* accepts the existence *and* importance of *dissensus*; here, *dissensus* does not lead to an inability to act, but it allows us to retain the potential to transform persisting or new problems into opportunities.

In the book *The Digital Coloniality of Power* (2015), my analyses of "the Digital" and related practices were focused not only on the (political) theory issue of antagonism but also on the attempts made at identifying exemplary forms of antagonism, and show where and how digital was (and where it will be) practiced, especially in relation to knowledge, science, and technology. The argument of the book, speaking (primarily but not only) in terms of social epistemology, was precisely that what we are witnessing in the antagonisms of "Digital Culture" is not something new but merely an extension of something much older: In "Digital Culture", we are witnessing an extension of what has been identified and described as the *coloniality of power and Being* or the *colonial matrix of power* (here there is extensive literature, which is, however, rarely read in the social scientific "mainstream": Quijano 2000; Maldonado-Toress 2007; Dussel 2013; Grosfoguel 2013; Lugones2008, 2010;, 2011, 2014; Mignolo/Tlostanova 2006; Mignolo/Vázquez 2013; Mignolo/Walsh 2018;Weiner/Carmona Báez 2018).

While talking about *coloniality*, decolonial[5] authors refer to the fact that even after any geographic decolonization, a regime of aesthetics, practices, and ethics still lends itself to thinking of and (re)constructing Other(nes)s through colonial forms, which include "dismissability", "killability", "disposabilty", etc. The fields that I subjected to this kind of analysis in the previous book were sociological theory, the politics of democracy in US-American discourse (mind you, that was already way before the election of Trump), the interrelated concepts of class society and upward mobility, the technological evolution of classroom technologies and cognitive learning development, online reader comments on issues such as in the GamerGate controversy and others, and, finally, future trajectories of technoscience from the role of biobanks and Big Data Health as producers of inequality to the potential of biomedicine to serve as surveillance technology.

However the need for identifying and *resisting* the workings of the *colonial matrix of power*, and the form of *resistance* which Mignolo and others call the conduct of *epistemic disobedience* (which I try and conduct here myself to some degree), requires that we also further continue the inquiry into coloniality's extension that I identify as the *digital coloniality of power.* I will perform this here by focusing on co-productive processes of knowledge, e.g. regarding the role of "truth" in public discourse in general, i.e. in the *News*, and with regard to the general discourse of science. I will, in the remainder of this chapter, take some steps toward developing an inquiry into the production of "knowledge(s)" through concepts such as "relevance", "fact", "opinion", because we need to better understand how these relate to the current media ecology – aka the topology of which is being constructed by the *digital coloniality of power.* But it should be clear that this is a project that is only in the beginning stages and that it will precisely require that as social scientists we must work together on it for the foreseeable future.

I will have to make another short comment here, regarding any (and especially my) subsequent use of the scholarship of several "Western" authors, to avoid any misunderstandings, since the reaction of my colleagues within the academy to my "conversations with decolonial options" has been met with mainly two types of skepticism: One group of critics tends to generally bemoan that any postcolonial or decolonial contributions, just like feminist theory, are not necessary because we already have all this critical potential covered by Marxist theory; the other group tends to argue that if I am so taken with "decolonial authors", why don't I use them exclusively, and why do I still pay so much attention to Western scholars – this second type of criticism comes in two versions: *either* it is being argued that in using Western scholars I am being inconsistent *or* in using Western scholars I end up re-appropriating or re-occupying postcolonial/decolonial authorship. Of the two types and the second type's two versions, I think that only version two of the second type has real critical purchase and, indeed, points out a genuine danger that I must constantly navigate and can never be entirely sure in advance (i.e. before readers have responded with their accounts) if I do not end up wrong, which requires me precisely to be *care*-ful and response-*able*. I have discussed a few things about the first type and its fallacies in the last book (2015) and on numerous other occasions have pointed out that in the very process of creating theories, analytical categories, gestures of resistance, it is necessary for diverse subject positions *and* subjectivities *and* ontologies to be included from the beginning and should not simply be "theorized" afterward from the point of view of an existing (Western) theory; in addition, this has already been a point of contention with the strategy of arguing that everything can be reduced to class – it cannot! – or the argument that Marxist theories have taken account of women's points of view[6] (which they never satisfyingly have as many feminist theories have asserted, despite the role that Marxism did play in the history of feminist theory, especially in relation to second wave feminist theories (see Gayle Rubin in: Nicholson [1997]). Furthermore, as I am also considering myself a post-foundationalist (and being of the opinion that this position is reconcilable with decolonial options), I do think that this "Marxist" critique falls precisely into the fallacy of the *deliberative mainstreaming*, whereas the version one of the second type – declaring the proposed line of argument inconsistent – has an antagonistic expectation, i.e. assuming that if one argues with postcolonial/decolonial scholarship, one must fully reject (aka annihilate) Western knowledge. The very point is, however, that Western knowledge must be *provincialized*.[7] In deploying "provincialization",[8] especially talking from within the geopolitics of Western social sciences and the geographic location of the Global North, I mean the following.

It means to recognize that Western/Northern (social) theory and the categories, concepts, methods, and classifications it creates and deploys have been (and still are) complicit in the Politics of Empire and in *coloniality* through their claims to universality; however, these Western categories were created by Westerners for the Northern context and cannot achieve or claim universality as other locations and cultures cannot be understood through these concepts/categories/methods, and their deployment (as universals[9]) is an act of colonization that operates geopolitically, epistemically, and often even ontologically[10] (as *ontopower*). However, this does not mean that these categories do not have the potential to help us understand the context in which they were created, or rather they can do precisely that *iff* the context of their creation is included reflexively into the *care* of their use. In the same way, non-Western concepts cannot (and should not) simply be imported (let alone pirated or appropriated [Pham 2016, 2017]) to be used by Western scholarship – neither to "explain" Others nor to better "explain" ourselves. Provincialization here means precisely the reflexively uncovering and careful deployment of these restrictions; it does not mean that all concepts must be abandoned or annihilated. Provincialization here means, thus, the deployment – without simply appropriating Other concepts – of the so provincialized concepts/methods in reflexive awareness of the restrictions and cautious awareness of the consequences.

In the following steps, I will attempt to inquire into what forms of witnessing and giving testimony are considered credible in today's media ecology, as the context for political news as well as science and which are subject to "epistemic injustice" (Miranda Fricker), and why? I will argue, as far as possible in this space, that the "currency" and the "social imaginary" of news have for a substantial amount of time been "relevance" rather than "truth", "truthfulness", or "factuality", which lead to a change of meaning for credence, temporality, and warrant of (public) information and data. Subsequently, I will propose, through the inclusion of a distinction in two different types of *warrant* (*entitlement* or *justification*), originally proposed by philosopher Tyler Burge and compatible with ideas by sociologist Laurent Thévenot (2001, with Boltanksi 2006), that a decolonial ethic can both expose where "epistemologies of public relevance-production" rest on (colonial-type structures) of knowledge entitlement, and where they can be required to cohere with genuine justification. As I have argued in previous work on the context of the public use of medical images, there is a possibility to establish a "right to being justified to" (expanding on the proposal by Rainer Forst on basic rights and the right to justification), which emphasizes that the problem is not found in a theoretical discussion directed against "anti-realism" and post-moderns as strawmen held responsible for the emergence of post-truth discourse. Instead, I claim, in stepping even further, that we should focus on transdisciplinary methodological perspectives (Lewis Gordon) that allow for concrete operationalizations of these different types of *warrant* to understand the role of facts and relevance along the issue of justification, and how the recent "post-truth" climate in public discourse can be accounted for by understanding the workings of the coloniality of power having extended into recent process of transformation in the realm of public discourse. In this vein, my argument here is in congruence with my proposition that a conversation between decolonial options for the social sciences and post-foundationalist views on *the political* that manages to operationalize within the social ontology of media ecology a "right to being justified *to*" may yet realize a more comprehensive and progressive form of epistemic responsibility.

"Anthropocene Does Work." What Is this Thing Called "Truth" and Whose "Facts" Are It, Anyway?

One of the more bizarre explanations for the election of Donald Trump as US president was that we should blame it all on Foucault, Deleuze, Derrida, and Lyotard if not already and first and foremost on Nietzsche. Anyone and everyone who could even be remotely attached to the label "*postmodern*" was, in this account, immediately complicit in the "making possible of" the post-truth, post-fact, fake-news presidency.[11] The supposed willful attack on "truth" of these *postmoderns*, we are meant to understand, made it possible that the American electorate, worse than the electorate of many if not most Western democracies, was now prone to accept both that anything is "fact" and nothing "is true": "Darn those relativists!" Strangely, these anti-Trump-cum-anti-postmodernism voices, occasionally, have strange bedfellows on the political Right in those who are against "feminist theory"/ "feminism" or who blame "political correctness", and also supposed outcomes of postmodernist scholarship – although "political correctness" (along with the so-called "social justice warriors" [SJWs] that are often invoked in Internet commentariats by *trolls*) is, of course, largely a strawman invented by the US-American political Right.[12] In this "shooting the postmodern strawmen ad hominem from the hip" I am strangely reminded of Neil Postman, who certainly was not the first to drive that same horse, but who certainly had been a great popularizer (if not populist) of such sentiments when he summed up in his 1999 book *Building a Bridge to the 18th Century* the reason why we should discard postmodern thinkers for they built on the ideas of Nietzsche and Heidegger: Of course, qua Postman, one was "a lunatic" in his final years, the other "fell in love with Adolf Hitler". If only it were that simple, but the blaming the "postmoderns" for Trump is often portrayed, at its core, even simpler than that: *They said there is no such thing as "truth", now we have Trump!* Well, if this wasn't supposed to be an "academic" publication that I am writing here, I would probably break the fourth wall and get into profane vulgarity and respond: "Boo-fucking-hoo!"

Of course, you can see what I did there, cleverly saying what I won't do but therein perform the very act of vulgarity that I claim I wouldn't. How very postmodern or populist of me, is it not? But the question is why would I perform such a stunt here? Is it just as a gimmick? In a sense, yes. Because one of the main issues in today's "public sphere" and "public discourse" – two very Habermasian terms, which would need to be in themselves *interrogated*[13] regarding their adequacy for our times – has become that the "gimmick" often matters. And it truly *matters*, as in it *matters* matter: For it matters what we *matter* matter with, with what worlds we *world* worlds, with what stories we tell other stories, to paraphrase a lesson we were gifted by Ursula LeGuin, conveyed to us by Donna Haraway (2016). Interrogating the history of political rhetoric in Western countries, we do not have to go far to find that there was an ongoing shift toward political extremes and margins even among mainstream conservative parties, which resulted in the presentation of marginal positions as "gimmicks" over political substance (aka "political facts"). But the problem is, of course, that one day deeds *must* follow words, gimmicks must be not only made to matter, but gimmicks must *matter*. The drama (or farce) that Germans could witness in late June 2018, regarding their minister of the interior Horst Seehofer, the leader of the conservative Bavarian CSU Party, came about due to the party touting for years with increasingly shriller tones, that they should be the only democratic voice of the people on the political Right and setting up policy demands they could never fulfill – in part, because they never had any large enough majorities in parliament, because there were never going to be resources available, or because the ideas would have immediately landed in front of constitutional court. But in seeing their majority dwindling in their home state of Bavaria and the right-wing extremist party AfD gaining ground, political commentators saw the CSU leaders[14] seeking to escalate their rhetoric into policies – regardless of the danger of breaking their ruling coalition with their sister party CDU and chancellor Merkel as well as with the social democratic coalition partner, regardless of the question of international agreements and negotiations among the EU states, as well as that of constitutionality. Numerous political circles were puzzled by the ferocity of what was largely considered a detail question about asylum and migration, but that is precisely the point: The gimmick has come home to (be made to) matter, and it had done so at around the same time that Donald Trump tweeted in line with his anti-immigration stance that Germany's rising crime statistics should be a lesson to US Americans, when in fact crime statistics in Germany showed a significant decrease in crime.[15]

It may sound strangely paradoxical, but I claim that the "post-truth" era is strangely oriented on the idea of "truth": While it often seems like there are sharp divides between political leftist and right-wingers that are construed as "your truth" versus "my truth" and seemingly make "truth" itself seem arbitrary, if we posed the question to either, would we truly be surprised if most people on either side were to claim that their truth is more true, more real, or rather: "the real truth"? In effect, the point about accusing one side as producing fake news means to claim that they effectively do not report something true. But let's be serious here: The cherry-picking of facts, the existence of conspiracy theories, the shopping for arguments or laws, the prevalence of personal interests or collective ideologies over judging the value of information, data, facts, and the "truths" these make are nothing new. Take for example the role of the "new" media discourse and "Yellow Press" in late 19th-century US-American politics, culminating in the 1898 Spanish-American war, which is often called the first "media war" regarding both the role that media played during the war and the role attributed to Yellow media in inciting the war in the first place (see Carey 2016).[16] And it isn't to say that the role of the press and the making of "truth" haven't come under scrutiny in the first half of the 20th century, long before the "postmoderns" showed up. Toward the end of and after World War I, specifically the British government and press were criticized for what would be called "atrocity propaganda" (Robertson 2014), which Holocaust historian Henry L. Feingold partially blames for the slow reaction of the US government during World War II, given that reports on atrocities (specifically those committed against the European Jewish population) were reviewed in the context of World War I reports of atrocities that were deemed exaggerated for propaganda purposes (Feingold 1995).

The academic discourse at the time, again(!) before the postmoderns, was equally attuned to the problem that social action could "construct" what would become a "truth", as can be seen in the work produced by sociologist Dorothy Swaine Thomas and (her later husband) W.I. Thomas from 1928[17] (see here Merton's classic appraisal from 1995 of both the Thomas Theorem and his own 1968 proposal of the Matthew effect), where the basic sociological tenet is stated that "if men[18] define situations as real, they are real in their consequences".

This does not mean, of course, that certain facts can be altered: Just because someone were to define as real that there is no such thing as gravity does not mean all of a sudden that someone would just lift off the ground. It is, however, precisely the error to take this insight into the opposite direction and mistake as a *realist* attitude toward empirical research, a naively simplified empiricism. Michel-Rolph Trouillot effectively put out a warning against this tendency, describing his own approach as

> an epistemological position that is unabashedly antiempiricist and that should be brought forward for the purpose of this exchange. To put it most simply: can the object of study be reduced to the object of observation? The extreme empiricist answer to that question tends to be a naive yes. This answer reduces most matters of methodology to matters of research techniques and mistakenly assumes all empirical studies to be necessarily empiricist in one form or another.
>
> (2001: 135)

In the same vein as we should avoid "fallacies of misplaced concreteness" (Alfred North Whitehead) and the "reification" of our concepts (Raphael Morris Cohen),[19] Trouillot insists that we must analytically strictly separate objects of study and objects of observation and not conflate them or substitute one for the other.

The objects we deal with in science – and here I mean not only the social but specifically the "natural" sciences also (Esfeld 2002) – are often quite vague (Esfeld 2008, 2013; Esfeld/Lam 2011) and it is this vagueness that makes them so *interesting* and *interested* (see here the late Delia Graff's 2002 seminal study of an "interest relative account of vagueness"[20]; also Barnes 2010; Schoenfield 2016).

Scientific "truths", if we understand, such as the knowledge we have about "things in the world", can be surprisingly dependent on a number of practices, which can often involve even the physical body of the researcher herself, as Natasha Myers illustrates in her seminal study of "ontological choreographies" of crystallographers *Rendering Life Molecular* (2015): Crystallographers, who model proteins – a now indispensable area of biotechnological research – have come to create a "haptic" way of learning to understand the structures of proteins, which is akin to dancing. This aspect that is found in other areas of research as well has even led to a competition titled "Dance your PhD". The competition was ridiculed,[21] and the point here is that, precisely, from a public's point of view there are certain expectations about what "science" is supposed to be, what is considered "serious" and "relevant". To put it a bit superficially: A very "common narrative" about science is that there is *a truth* underlying our reality, and there is a strict, serious, nonsense-free (aka objective) way of getting to this truth. For example, anything *considered* non-serious, nonsensical, or, most importantly, irrelevant is subject to anything from ridicule to purge. But Myers's study (like others before her, such as for example conducted by Karin Knorr-Cetina [1999] or Hans-Jörg Rheinberger [2010]) illustrates how important these supposedly non-serious and/or non-relevant elements can be in the "chains of inferential steps" (Adina Roskies) that lead to the *elaboration* of a fact. I want to emphasize here the word *elaboration*[22] and its multidirectional meanings: To elaborate a fact can mean the (a) clarification of the fact as well as (b) its production; thus, it can mean (a') to carve away laboriously that which obscures the fact as well as (b') laboriously constructing a fact. And it is no accident that therein lies a relation to theories of aesthetics and works of art: One can, of course, see the work of art as an artificial creation by the artist or as something that comes closer to the truth, because it allows us an unburdening of concepts and categories; thus, it can involve the creation of an essence of its own and getting closer to the political essence behind a curtain as well as the true nature, depending on which aesthetic theory one finds more convincing. But in the Western philosophical history, aesthetics is not about art as much as it is about perception, and the Enlightenment discourse on aesthetics from Alexander Gottlieb Baumgarten (1714–1762) to Immanuel Kant (1724–1804) reviewed aesthetics in the context of both art and science, and their philosophical discourse always had a genuine political effect: In Kant's case the role of *sensus communis* was rooted in the genealogy of physiology which was itself bound to the affordances of the physical world; it played an important role in moral and aesthetic judgments, in the latter case both important for judgment of taste and in the emerging field of biology, and it is political not only in practice of political judgment and political imagination but also in the question of judgments being made "in public" (see here especially Arendt 1981 *and* 1990). In short, knowledge cannot be had for free, so to speak, and there is no immediate access to any "truth" that can then be commonly shared (neither the truth nor immediate access can, precisely because the sharing of such access would be a sharing of the means of access and means are, obviously, mediate).

To return to Trouillot's anti-empiricism distinction between *objects of study* and *objects of observation* in the context of Natasha Myers's research, Myers refers us to a quite useful conceptual distinction introduced by Hans-Jörg Rheinberger who distinguishes between

> "epistemic thing," the phenomenon that the experimenter aims to describe, and a "technical object," something that is known well enough

> that it can serve as an instrument or component of an apparatus in an experimental system. Before the experiment, an epistemic thing remains elusive and unknown; it is not quite yet within the grasp of the scientist and it has not yet solidified into an object. A technical object may at one time have been an epistemic thing, but once elucidated it could then be put to the task of determining other epistemic things.
> (Myers 2015: 78f.)

Objects that scientists deal with are, thus, not simply there before the knowledge of them is made, but they depend, for example, on experimental settings and individual practices. That ontology and epistemology are not free of material interests and practices can be found and described already in a 1985 book *The Dialectical Biologist* by Richard Levins and Richard Lewontin; but that science and the theory of science would do well to accept that both "knowledge about things in the world" *and* the "things on the world" depend on the entanglement of ontology and epistemology via the means to gain that knowledge and ontology and epistemology are, thus, never "closed" off from one another has been variously addressed by Gilbert Simondon (1924–1998), Heinz von Foerster (1911–2002), or Karen Barad as *onto-epistemo*logy. One of those means are, of course, concepts, and it is important to understand – for science and for the political discourse – that concepts actually do work, that they are working – concepts are, generally, conceptual practice(s). In expressing her reservations about the concept "Anthropocene", Donna Haraway has stated that now that it is in the world, it "does work". It changes how people understand the world, not only in the present or the future, but also in terms of the past. Effectively, the past is being written in the present by means of the Anthropocene concept: Facts that were not facts yesterday suddenly are yesterday's facts, and they go on *facting* our lives and that of the planet… at least until we replace it with something else like Plantationocene or Chthulucene.

However, this does not mean that "facts" are not "facts" anymore, even if there is no such thing as *the* Truth. Now return briefly to the example of gravity. It is a fact that I didn't just lift off the ground as I jumped and flew way, but when we ask for the cause, we tend to say "because of gravity". But what is the truth of gravity? In a way, there is, precisely, none. We have theories of gravity, which *generally* help us understand how this effect we call gravity might work, but gravity is not something we can know the truth of; we can only know its facts – such as I did not right now lift off to fly away. Our theories can, perhaps, seem more true or less true, but they are never *the* Truth of gravity. They give us guidelines on how to, for example, overcome gravity, e.g. while building a plane or a rocket and thus give rise to new facts or connect facts that were previously separate, and so on. Or take another example, the longwinded discussion in the United States on teaching "evolution". Often in this discourse it is said that "evolution" is but a theory. But the funny thing, it's not: Evolution is a fact – we can show it from fossils to a petri-dish. What is a theory, however, is the proposal *how* it works in detail, how the facts connect. And the proposal that is still largely influential today, the "new synthesis", marries some of Darwin's hypotheses (such as on competition and selection) with genetics. Now, proposals from the religious Right – involving God or an intelligent designer – are not "competing" theories, precisely because they are not theories, because God is not a fact.[23] But the controversies and the confusions between what is a "fact", a "theory", a "hypothesis", an "explanation" make a number of necessary corrections in science itself very problematic. The new synthesis, for example, with its focus competition and its methodological individualism needs correction, as more and more data points to symbio(gene)sis as being more important. But this is not an argument against evolution; it's an improvement on understanding how evolution works. But critiquing the common understanding of how evolution works could be perceived in antagonistic public discourses, such as in the United States today, immediately as though it critiques the fact of evolution. We see the same happening with the debates on climate change or vaccination. *Science news* report, for example, on any single study that seems to defy generally held views: Coffee is good for you, coffee is bad for you, coffee is, again, good for you. But these *science news* are not really interested in the details nor do they acknowledge that one single study is not immediately knowledge. *Science news* tend to ignore that it is only across several studies, which are connecting facts, that increasingly reliable knowledge is being elaborated. *Science news* look not at what's fact or looks true, but what seems somehow relevant or interesting. Of course, this introduction into theory of science is a bit unfinished and deserves more elaboration itself, but this is not what this essay is about.[24] What I am trying to point out is this: One can argue that even those who deny climate change or deny evolution in favor of creationism believe in the existence of *a truth*: their point is not that "truth" doesn't matter or does not exist, but that "science" is itself a system of belief that is not free of the influence of interests and ideologies, and that the claim of science to be objective is either a false belief or an outright lie. So, the era of "post-truth" in this regard is not so much an era after "truth" as it is an era in which the "modern" social systems or systematic approaches to producing knowledge about any truth (which includes both science and [political] news/journalism) have become questioned, distrusted, denied, or lost discourse hegemony (in the terms of "post-fact", interlocutors would assume that the only "facts" that are *relevant* are the ones that matter to them, or their filter-bubble/in-group). This can take different forms: Science (or journalism) can be distrusted and, subsequently, denied as institutions in-themselves or it can be questioned whether scientists as individuals are ever able to live up to the demands of the institutions, thus making the institutional/systemic demands simply unrealistic; or if science is seen as just one beliefs-producing "ideology" among others, it is suddenly in a kind of market-competition with other belief-/knowledge-producing systems such as Christianity.[25] But beliefs need to be warranted, and even the belief held by a televangelist that he needs his followers to donate him money so he can buy a new private jet[26] is usually provided some warrant such as "God told him so". It's simply that this kind of warrant is pure entitlement and not justification.[27]

Witnessing and Giving Testimony in Today's Media Ecology

"Fake news" is *the word* in the word-of-mouth today. Not long ago, it was the quasi-Habermasian "truthiness", coined by US-comedian Stephen Colbert.[28] Perhaps we can see even bit of a "fake-news" fatigue the present public discourse – not only among those sharing the currently quite en vogue anti-intellectual sentiment, but even among intellectual people themselves. Who still wants to talk about "fake news" or "post-truth, really? But let's run it one more time, shall we. Several months ago a "news" item caught my eye: *Sean Hannity Flips Out After Getting Busted Sharing Fake News*.[29] Perhaps, this is actually "meta-news", after all this is news about a news-maker, or rather a so-called pundit: A pundit[30] falling for fake news, and then being enraged over news about his falling for fake news – it's a meta-news item of about the third order. If you ask me, personally, this story isn't "news", or it shouldn't be. However, the fact that it sort of is, today, illustrates the status of current media discourses and the role of "fake news".[31] The mentioned story deals with a series of tweets by Mr. Hannity. He has tweeted a link to a story by someone else, which he commented as "Wow if true".[32] Ok, let's stop and think: What is news or what is it that makes news "the news"? We're looking for criteria. I think that a category such as "important" no longer is what qualifies news as news; I think it doesn't even qualify, for example, why this or that research project is conducted or even funded. "Important" has somehow lost its property of being generalizable: To say "this is important" immediately begs the question "for whom?", and that is a "who(m)" which is immediately suspect in that that "who" is not "me". It may be a bit of a hyperbole, to give this dismissive account of good, old "importance", but I think I am on to something here. Ask yourself, truly, when did you (if you are in academia) last describe your own research – especially in a research grant proposal – as "important". Truth be told, I shun that word like Christians do "the devil", and so does everyone I know – and believe me, I asked.[33] "Important" sounds like one either is a bit self-aggrandizing in their own importance for the world *or* selfish and solipsistic. "Important" hasn't been part of the "grant writing bullshit bingo", ahhh, pardon me, the "proposal prose", for quite long time. Instead, today, for an "item" to carry information that is considered news-worthy or funding-worthy, it must have "relevance" (or "impact"); mind you (and *pace* Habermas), it requires not a *truth claim* but a *claim to relevance*. Relevance, I might add here, channeling Tyler Burge, is a kind of warrant, but what kind of warrant? It is a warrant of the kind of *entitlement* and not a warrant of the kind of justification (we will come back to this important distinction below in more detail). Tentatively we could say that this makes for two kinds of possible understandings of what "news" is: "News" is either current information about the world, which is already partially warranted and seeking further warrant through justification (this is, pretty much the ideal-typical as well as the idealistic idea of news, unless, of course, we opt for a somewhat naive "search for truth" explanation), or "news" is what is deemed relevant qua an entitlement structure. This structure is like a kind of moral or political economy, a "economy of relevance", so to speak. "Attention economies (and ecologies)" (Stingl 2010; Weiss 2012; Stingl/Weiss 2013) are a particular aspect of *economies of relevance*, as we can learn in the work of philosopher Bernard Stiegler, especially in his book *Taking Care of Youth and Generations* (2010).

What is interesting about this Mr. Hannity's story is (a) that this falls into the "relevant" category of what "news" is, or rather has increasingly become (in and of itself, this turn to relevance and warrant qua entitlement instead justification is nothing new itself, but more a cascading expansion of tabloid press: the *tabloidization* of "the news", if you prefer to call it that. Boris Groys's book on the question of what is "the new" [2004, orig. 1992] would be an interesting read here as well, as Groys deals with what has caught commentators' eyes as a description of our "vagabonding attention" with an air of "Fuck authenticity!"); and (b) that any "news"-person, such as Mr. Hannity, who is working for an organization of the very size and very means as his employers could offer, should(!) have editorial staff at their disposal who would check the claims made in any source – this could sometimes be called "fact-checking", but I would refrain from calling it that, because sometimes even newsworthy items in the "justifiable news"-category cannot be fully verified as to their facticity or factuality[34] in a given moment, so I would prefer to call it plausibility-checking.[35] But "fact-checking" is for major news organizations on both sides of the Atlantic: *The Washington Post* has its fact checker, same as the German "political" talk-show *Hart aber Fair.*[36] And it is not as though those on the (more or less extreme) political Right don't make claims to the relevance of "facts" as well or enjoy when a fact-check exposes one of their political enemies having made a blunder. Claims to facts and to those facts that are relevant are, however, also made in cases that "a truth" is, well, inconvenient – on all political sides, by the way. The enemy's claims are not claims with facts or not with all the facts, and definitely not the relevant facts – or so some of the many variations go, with which the *antagonistic* game is played. It's not only or simply about "post-fact" but that "facts" can have different weights or, rather, different relevance. So, perhaps, we do not live so much in a post-fact era as we live in an era of diversely relevant facts, where relevance is determined less by the clear cut rationalities of a Habermasian discourse but by a *geography of affects* (Anderson 2010, 2016; Levina/Silva 2018; Silas-Brügge 2018). What has gone out of the window with the emergent dominance of relevance-driven news-cycles is not so much the question of checking "fact versus fake" in terms of truth; what has really gone out of the window is to check for plausibility. Facts are states of affairs in so far as facts present (as) truth-claims about states of affairs (for which justifications can be given). And in practice, news organizations (regardless of their political *couleur*) when confronted with an item that might be news-worthy, they should at the very least do a plausibility-check – even if the fact-check in the traditional understanding of truth-check cannot be conducted to its fullness. But this doesn't happen (as often) anymore for two reasons: There seems (a) to be less spending on editorial staff in both numbers and competence, there are less staff available to work on a news item, and there is an ever increasing number of interns and other low-paid, overworked (precarious) workers doing the job that a host of well-trained, well-paid, and well-rested editorial staffers should be doing (just think of the increasing number of typos, grammatical errors, which is not only restricted to online-outlets anymore); (b) *plausibility* has been replaced with *relevance*, or at least, we seem to value relevance more than plausibility. Why do I think so? Because even "respectable" news organizations nowadays report on issues that have less and less plausibility. They do so in order to stay relevant: Why did Hilary Clinton's "scandals" receive more coverage than those of Donald Trump[37] in nearly every news-outlet? Because in today's media(-attention) economy it is better to appear relevant, even if utterly wrong, than to be right but appear as not being able to present what is currently and in the moment deemed (most) relevant. In the same vein, currently US comedians jump on everything the Trumpites do and try and make it funny – comedians thrive on relevance, too. They even become the main source of news for more and more US viewers. But one can ask, is this "comedic" contribution to the economy of relevance – which gives itself a guise of "criticism" – not contributing to the problem? What if these comedians from Colbert to John Oliver and so on stopped being funny about serious things and actually got serious for a while, because what they talk about is serious? For late night shows in the United States, their economies of relevance come in two layers, too: What is relevant in the news cycles, and what is relevantly funny. But if, what is relevant were truly serious and truly serious to them, why make it funny? If the matters were so serious, for example with real lives being (plausibly) at stake, should they not stop being funny to also convey the seriousness of the matter[38]? The answer to this question is not an easy one, for certainly many people wouldn't even listen to the message if it wasn't presented in a funny way, and that way they at least here the message – though, what is the/a "message" here, one could ask? But it also seems, presently, that comedians as well as many news outlets, after years of decline in audience, have profited from the current political situation. They appear more relevant, again. In the world of news, so many economies of relevance have become entangled with one another that it is nearly impossible to make sense of them with any analytical clarity. What I wanted to point out here, though, is that two main trends over a long time have contributed to the current fray: The substitution of plausibility with relevance, and the decline in quality editing. And these two are dimensions, I claim, we seriously need to talk more about in the future. I think that this notion of *economies of relevance* could be developed into an effective analytical tool: In the ecologies of media or research it provides insight into the affective value that explains concepts such as "impact", "usefulness", and so on. So, the types of questions we need to ask are, for example, "What does it take to seem relevant today?" or "Who is relevant?" It is in this context that questions of "Who counts as a *witness* and whose and what *testimony* matters?" come up, along with the creation of biases and privileges (to speak and to be heard, to write and to be published, among others). It is here that questions of "how do we construct *the Other*" and, thus, the *coloniality of power* come into play. The issue of relevance and of testimony not only plays out in social media but increasingly includes the necessity of social media. In order to be considered a relevant witness and give relevant testimony in public – even if implausible – it must exist through social media. But social media are just a new medium that works with old mechanisms; hence, this is merely an extension of the *coloniality of power*: *Digital coloniality of power*. For example, those whose testimony is not heard in digital culture follow along the same old fault lines of already existing social inequality and injustice, as I have shown elsewhere (2016).

To reiterate, Mouffe and others propose that antagonism constructs an "Us versus Them" situation. In antagonism, one's own opinions are absolutized and construed as "completed", i.e. *dissens*/disagreement is impossible. As an effect the Other means a construction of an antagonist, an enemy, and it becomes a necessary adversary[39]: An enemy must, subsequently, be construed as "destroyable", but therein it becomes constitutive of one's identity. To be destructible and to even invite destructibility are necessary to the function of coloniality. Categories by which the Other can be destroyed or excluded *must*, therefore, be *produced* – even the deliberative mode of discourse can end up being a possible outcome of antagonism, for both modes sharing a negative attitude toward *dissensus*, except that antagonism codes dissensus as its anathema and is thus *destructive* toward the dissenter as its enemy, whereas deliberation is "merely" *excluding* remaining dissenting voices.

Economies of Relevance and Other Epistemic Problems: Warrants and Responsibilities

The following proposal was originally developed with regard to my interest in doctor-patient interactions as well as neuroscience publications and their audiences as particular examples of lay-expert interactions; I looked, in particular, at the role of medical images such as *functional Magnetic Resonance Imaging* (fMRI) scans (Stingl 2014a, 2017). I noticed epistemic gaps that derive from the fact that epistemic production happens in steps (distinct as well as flowing) that comprise chains of actions, but which differ in *inferential distance* between beginning of the chain and its epistemic end result or epistemic product (including knowledges or judgments) with regard to the distance between inferential steps that is assumed by the audience or end user, and their actual distance – and given that the steps have generative epistemic effects, thus effectively changing what kind of belief should be underlying an audience's/user's knowledge.

Situations of this kind are the domain of *social epistemology*. For example, Lorraine Code (2010) in reference to a seminal essay by Richard Moran, titled "Getting Told and Being Believed", explicates that at issue is the *relation* of believing "not *that-P*" but that there is a "person who tells you *that-P*", and that this person is "taken to know *that-P*"; thus, it follows, for Code, that Moran opens up the philosophical interest in "[t]he relationships, the modalities of epistemic dependence and [...] of epistemic vulnerability and risk, that follow from reliance on testimony as a source of knowledge" (Code in Haddock et al. 2010: 29ff.). There are (2010: 32ff.) a few caveats to Moran's (and others') accounts of "the social" in such a moment: Between the "person who tells *it*" and the "person who might listen to *it*" there exist standards of determination of credibility. It stands to reason that our social imaginaries have something to do with these standards. However, these standards are not always fair and just. Therefore, there exist forms of epistemic injustice, to cite the title of the famous book by Miranda Fricker (2007). Along those lines, Lorraine Code's work addresses – with her eye on *epistemic responsibility* – "the implications of prejudice, incredulity, stereotypes, situated knowledge, and ethical-epistemological issues of trust, testimony, and social conditions for establishing the status of a credible knower" (Code in Haddock et al. 2010: 32f.).

Our knowledge acquisition hinges on practices (as stated earlier). Knowledge practices are, like most other things such as judgment or imagination (Kant), subject to practice as in: we have to exercise and train them. Take for example wine, whiskey, or tea connoisseur, who needs to learn to taste and even, sometimes, learn to even appreciate differences. Not for nothing do we speak also of "acquired tastes", because taste must be exercised before it can be used properly in an expert context: Everybody can drink wine, but not everybody can taste it. In some extreme cases, "acquired taste" refers to the fact that not only does a practice require a kind of learning by experience, but even an overcoming of an initial dislike – Matsutake mushrooms (Tsing 2015) or certain cheeses (Paxson 2014) being a good example. Even something as deceptively simple as "seeing" needs to be practiced in order to achieve not only an "expert vision" in a respective field, but more importantly that it becomes "second nature". For a radiologist, for example, seeing cancerous tissues in a technologically produced medical image is a *second natured expert gaze.* Patients as well as their treating physicians do not possess such a near "intuitive grasp" of this seeing practice. And yet, patients, treating physicians, and radiologists are entangled with one another through these medical images. Between the "being shown an image" and "being told it shows" a tumor on the one hand, and on the other hand "seeing" the tumor in the image lies an *epistemic gap* because on both sides, actors have a concept of "tumor" and understand in a "more or less"(!) meaningful way that one of the parties "has a tumor", but their concepts and meanings are different: Herein, there is an *epistemic defectiveness* (Boghossian 1996, 2003; Ross 2002) in the construction of epistemic objects through the use of concepts of "concept" and of having a "grasp of meaning". This defect exists, because the party being told that they are having this tumor shown in the image have an assumption about the inferential distance in the production of that image (and the diagnosis) and a certain belief about how accurate, correct, or true it is based on an assumed shortness of the inferential distance (i.e. the mistaken idea that medical imaging works like photography and produces transparency; in fact, it does not [Roskies 2008]) and the subsequent quality of the expert's judgment. In regard to the idea of "truth" a patient expects a certain "truth-value", which is actually often a bit further from "any truth" – if I may use this term here, for a moment – than they think. It is important that we accept here, too, that this doesn't mean that any false claims are made. Doctors simply make judgments with less certainty than patients think, aka patients assume that doctors using medical images are more (even absolutely) certain than they have any right to be. The question I was pursuing in my previous works on this issue was that this bestows upon doctors a particular kind of responsibility, which included, that they should take seriously the patient's testimony about their own on the one hand, especially when these stand in contrast to the medical image's presentation; on the other hand, doctors should be (response-)able to explicate and justify the actual inferential steps, i.e. that patients even have "a right to be justified to".[40]

It's important for me to emphasize that I think that the example of medical images is exactly that: an example for a more general case in knowledge practices. For example, in today's situation that we find, first, our political discourse and journalism in, and second, with regard to my research on the so-called global transition to (knowledge-based) Bioeconomy (Aguilar et al. 2009; Dabbert et al. 2017; Lewandowski 2017), where regimes of capitalist capture and extraction produce a form of neo-colonialism and precarization of living things – particularly in the Global South – through biotechnological knowledge practices. The case of introducing genetically modified organisms into the German market and public resistance is an exemplary case here (Motta 2014; Davison/Amann 2017): Among researchers and biotechnological industry representatives, GMO resistance is seen as based on myths about the technology, likened to "climate change denial", and the technology portrayed as perfectly safe,[41] simultaneously most lay people have too little understanding of the biotechnologies involved to make an informed and educated judgments but with numerous scandals and questionable behaviors between science, industry, and governments (such as in regard to glyphosate[42]) most people feel skepticism is indeed warranted; meanwhile, the use of GMOs is not entirely without remaining biological or ecological risks, as well as remaining socio-economic questions linger, which rarely enter the portrayal of GMO popularizes (many of them not themselves experts in the field such as psychologist Steven Pinker or historian Mark Lynas[43]).

The point is that even if there is no "absolute Truth", an orientation toward the ideal of truth with recognition and respect for facts is still in the cards. However, in terms of the epistemic defects in play in the earlier examples, we should distinguish between someone making claims to having a "grasp of the truth" and, what I think is more useful having a "grasp of meaning". In some ways, the structure of social interactions where experts or even so-called experts overrule lay people can be described as *persuasion without coercion with epistemic deficiency of its constituents due to an information asymmetry and an insufficient "grasp of meaning" between the interlocutors.*[44]

"Grasp of meaning" comes in three distinct flavors, as Paul Boghossian (2003: 16) shows:

> In one sense, a thinker T grasps S's meaning just in case T means some determinate thing or other by his use of S. Call this notion of grasp 'mere grasp of meaning.' In a second sense, T grasps S's meaning when he can correctly and knowledgeably state what S means. In this sense, it is not enough that T mean something by S – he must also have second-order knowledge of what S means. Call this notion of grasp 'knowledge of meaning.' In the third and most demanding sense, T's grasping S's meaning implies not only that T is able to state S's meaning knowledgeably but also that he understands that meaning well enough to know whether or not it means the same as some other sentence. Call this 'understanding of meaning.'

An actual expert must have at least a "(working) knowledge of meaning" to be able to put information to use. Preferably, though, an expert should have an "understanding of its meaning" – particularly, if one uses a technology or data/information *intentionally* as a tool *and* a medium that can affect the lives of others. A patient, for example, must at least have a "mere grasp" of meaning. If a patient does not have a "mere grasp of the meaning" of what it means to be sick in a clinical sense, for example, they could not comprehend a grasp of the meaning of "this is your tumour". For any technology to function as a persuasive technology and to fulfill the role of a social actor, such as news websites and social media for example, they must rest on conditions that connect with the quasi *a priori* conceptualizations that it helps produce, for example, a particular model of the "patient body" or the "society" they are engaging in knowledge practices about. I will get to an important distinction already used earlier here in detail: the distinction between *fact meaning* and *entitlement*, which I derive from the work of philosopher Tyler Burge (1993) on *content preservation*. With Burge we can distinguish *entitlement* opposed to *justification* as a type of positive *epistemic warrant:* Justification is the internalist form of warrant, and entitlement is epistemically externalist (see Casullo 2007; Graham 2010, 2012). The role of technologies and (social) media as persuasive technology lies in the (social) construction – not of truth or facts – but of *entitlements*. Warrant comes into play in social epistemology notion when there is disagreement between "peers" (Goldman 2009, 2010). There often exist asymmetries between peers, even if strictly speaking experts and lay people or, in the earlier example, patient and doctor may not immediately be thought of as "peers". But even non-experts often have a kind of special access to the matter, thing, or object in question: A patient, for example, has a unique access to their own body that a doctor has not; a lay person occupies a subject position and an (intersectional) positionality that an expert may not. In many cases, an expert has access to an abstracted and generalized (third person) where a lay person may have lived and experiential access themselves or be (inferentially) closer to someone who does. But neither type of access (lived experience expertise nor expertise) is per se epistemologically privileged! Instead, both are riddled with epistemic deficiencies and therefore interlocutors (experts and lay people) end up interacting *as if* they were peers in disagreement. It is important that this means that their disagreement *can be made reasonable*. Following a suggestion by Michael P. Lynch (2013), I consider that by giving reasons (regardless of their warrant)[45] someone is revealing their epistemic commitments through which they exercise epistemic agency.

Tyler Burge offers the following description of his distinction between entitlement and justification:

> Although both have positive force in rationally supporting a propositional attitude or cognitive practice, and in constituting an epistemic right to it entitlements are epistemic rights or warrants that need not be understood or even accessible to the subject. We are entitled to rely, other things equal, on perception memory, deductive and inductive

> reasoning, and on – I will claim – the word of others. The unsophisticated are entitled to rely on their perceptual beliefs. [... B]eing entitled does not require being able to justify reliance on these resources, or even to conceive such a justification. Justifications, in the narrow sense, involve reasons that people have and have access to.
>
> (Burge 1993: 458–459 [see additionally Burge 2003])

Admittedly, Burge's proposal was made in connection with epistemology and perception when talking about *warrant* and *entitlement*, and not about social epistemology, in responding to critiques that asked about short and long demonstrations of reasoning, judgmental force and the role of memory therein, he begged the question why memory should have that role therein, when he said explicitly that "[o]ne does not justify the demonstration by appeals to memory. One justifies it by appeals to the steps and the inferential transitions of the demonstration" (Burge 1993: 462).

Burge's "steps and inferential transitions" and Roskies's "inferential distance" do work together rather nicely when we want to understand why *epistemic expert systems* tend to gloss over the actual length of the inferential distances when demonstrating/explaining something to another epistemic agent who resides lower in the presumed epistemic hierarchy, i.e. is presumed to be of a status of *epistemic deference*, or, respectively, why experts do not correct – as they responsibly should – *epistemic viciousness* in their lay interlocutors. This is explained in the way that, according to Burge, Rhoderick Chisholm accounts for the role of memory – and even if memory itself does not play a role for us here, the general idea counts, I think, for justification all the same: "Long demonstrations are more fallible, and fallible in different ways, than short ones are" (Burge 1993: 462). Therefore, epistemic experts/expert systems assume positions of epistemic authority to shorten demonstrations for others. But, as they do so, they should be aware that they enter the realm of *epistemic defects* and *epistemic relativism* as well as the fact that technologies, tools, models, (social) media, and the like have themselves agentic properties (including epistemic agency) as well as functional properties, which bestows upon them *epistemic responsibilities*.

Fogg (2002) suggests that one of the functional properties is persuasive functions: These are, first of all not sinister but necessary, because in interactions we often must bridge epistemic gaps over large inferential distances. This requires in-built persuasion. However, ethical complications arise: People do make "belief investments" and are subject to what Gilligan Russel has quite insightfully identified as *epistemic viciousness*:

> Though the word 'viciousness' normally suggests deliberate cruelty and violence, I will be using it here with the more old-fashioned meaning, possessing of vices. Vices (such as avarice, alcoholism and nail-biting) are common, and most of us struggle with a few, but

> 'epistemic' means 'having to do with knowledge and the justification of belief' and so epistemic viciousness is the possession of vices that make one bad at acquiring true beliefs, or give one a tendency to form false ones.
>
> (...) Yet whatever the moral virtues of the well-trained budoka, it seems to me that the culture of training in many martial arts actually encourages epistemic vices, including both close-mindedness and gullibility, but also unwarranted epistemic deference to seniors and historical sources, lack of curiosity about important related disciplines and lack of intellectual independence.
>
> (Russell 2010: 130)

In any public interaction, for example on political matters or on science, we encounter reasons given, communications between experts, gatekeepers, regulators, and concerned and directly affected parties such as concerned stakeholders and/or side-affected publics. But my, admittedly, normative thesis is that what needs to be established at this point is not, as many would have it, a matter simply of trust: *It's not about how can scientists, doctors, journalists, or politicians regain our trust, it's about how we* ***mutually enable****(!) trust by understanding our respective epistemic responsibilities, and engage in our practices with others accordingly.*

I think a big problem here is how to mutually enable this, when we also need epistemic authority for interactive knowledge systems to function – i.e. experts truly cannot always explain everything (therefore, nor is the exercise of epistemic authority per se corrupt).

What seems important to me is that the exercise of epistemic authority should not become *epistemic paternalism* or *epistemic imperialism*. The deployment of social kinds and narratives that are *epistemically paternalistic* are, I claim, forms of *epistemic injustice*, such as various forms of *implicit bias*[46] (Haslanger 2015): Injustice, as Fricker, Haslanger, and others have long since shown us, is not merely something that emerges through explicitly observable action or expressed attitudes. Injustice obtains on the social structural level. I tend to come back to two aspects that Haslanger emphasizes:

> The *normative core* of what is wrong with racism/sexism lies not in the "bad attitudes" of individuals but in the asymmetrical burdens and benefits and inegalitarian relationships that societies impose on such groups.
>
> [...] The structural injustices may persist even when attitudes change.
>
> (Haslanger 2015:,2)

As for asymmetrical burdens and normative cores, I sense a problem that even when a political culture and individuals seems overtly dedicated to "free speech" or "democracy", an underlying antagonistic structure of the public/political discourse can persist – especially through constructions of the Other by means of the coloniality of power. Indeed, a large number of people, including especially on the political Left, may indeed not have racist attitudes, but that does not mean that they (myself and perhaps you, my dear reader, also included) do not enact, more or less often, white privilege or color-blind racism. In asking about the normative core, Haslanger basically asks of us that we unravel the epistemic and normative commitments that lie, first of all, in entitlements as a type of warrant; second and more importantly, though, her line of inquiry requires that we recognize the following: entitlements constitute and/or translate into another kind of entitlement, namely entitlements of the kind of *social privilege*. In social structures and, above all, knowledge practices we encounter configurations of power relations that reflect tacit calculations of what is possible (or not) for an agent within a given social order. These possibilities and the limits they are co-located with, expressed in knowledge practices, are important, because differentiation of knowledge and labor into expert systems is unavoidable – you cannot be your own doctor, plumber, and President at the same time. But in terms of the fact that most people are unaware of the actual inferential distances in the production and application of knowledge that affects them, experts in their field do have an epistemic responsibility to be able to justify their account: If they don't accept that responsibility (see also Origgi 2008), the culture of discourse turns at best into a deliberative culture based on entitlement warrants (an *elitocracy* like a *technocracy* or *noocracy*, for example, depending on what elite/knowledge is considered "relevant") and at worst into a pervasive antagonistic discourse of an aristocratic, despotic, or *anocratic* nature (see also Stingl 2015), where those in power enact of epistemic paternalism, acting as though *entitled to* be right (aka in possession of the truth). In an agonistic version, expertise can become in the extreme identical to hegemony, but a hegemony that is not absolute and that establishes in expert systems the right to *being justified to.*

Against Decadence: Creating New Possibilities through Epistemic Disobedience

What doesn't work at all – because it allows to arbitrarily either ignore (i.e. unreasonably) disagreeing views or cherry-pick one's entitlements, i.e. thus, to go "shopping" for arguments, facts, laws, as one pleases, to choose them (or deny them) as they fit one's own interests and ideology, in short, one cannot be arbitrary but must be systematic in the kinds of warrant one works with – I think this is true for both deliberative and agonistic models. All else would lead to nihilism and/or annihilative (and ultimately self-destructive) antagonism. I think that this can be reconciled well with decolonial options, even with regard to Western science, if science is willing to allow itself to conduct its discourse in a more agonistic fashion[47] – rather than deliberative or antagonistic. I think, Colin Scott's classic paper "Science for the West, Myth for the Rest" (orig. 1996, reprinted in: Harding 2011) – viewed as an inner critique in the Western academy – illustrates this quite adequately. What Scott shows is that the stance that Western scientific discourse has toward traditional/indigenous knowledges is not tenable: This stance, somewhat simplified, states that any knowledge not produced by science is not knowledge but myth; therefore any (proven to be "instrumentally useful") knowledge produced by indigenous (aka non-Western) people is not science but myth, unless it can be somehow translated into the system/language of science, but then it is science (and the property of science) and not indigenous anymore. This has a number of consequences – not the least of which has been the dispossession of indigenous peoples by means of bio-prospecting and property rights regimes (Merson 2000; Greene 2004; Shiva 2007; Blohm 2010; Torri 2011): For example, (epistemic) things do get lost in translation,[48] such as, say, the isolation of active ingredients from traditional remedies may lose us crucial aspects of the therapeutic regimen that attune to individual aspects of an illness (see here, also, an obvious link with the earlier discussion of Natasha Myers's research, regarding enactive, extended, and embodied practice components), among others, or an obviously effectual[49] regimen – even though its effects are visible, perhaps even (reliably) measurable – is ignored and discarded, because it cannot be translated into the scientific system. Scott, in his paper, clearly shows on the example of Cree language and knowledge that they have effectual knowledge that – even while not forming a (universalizable) "system" in the Western sense – is obtained as *systematicity*.[50] The point here is that in systematicity, warrant of the type of justification is possible and necessary. In a system, the system itself (autologic) or some of its institutions are often cited as entitlement-type warrant; both unsystematic ideologies and ideological systems, in my view, are pretty much characterized by the fact that they cannot provide warrants of the justification kind but only of the entitlement kind.

Science as a system allows, thus, for both kinds of warrants, but scientism (as an ideology) only allows for entitlement types of warrant and, subsequently, becomes antagonistic, similarly the problem with any extrem***ist*** forms of religion, national*ism*, or the forms of capitalist capture we know as neoliberal*ism* – this is, certainly, nothing shockingly new.

Therefore, it is wrong to declare systematicity, such as Cree knowledge, as untrue or as "mere myth" simply because it does not lend itself to be pressed into the system of science. So, in other words, Cree knowledge, since it is based on facts, can, indeed, be true. It is no less *the Truth* than Western science is. But perhaps both are aspects of truth, which allows us to demarcate clearly things that are *untrue*. Perhaps Western scientific and Cree knowledge cannot say much about or to the other, but that doesn't mean that they have to declare the other "untrue" if they find that each can give warrant by justification and citing of facts; for both, a position can, however, be *untrue* if there are no facts and justifications that can ever be given in support. Perhaps, Cree knowledge and science can both be true, if we accept that there is more than one truth: There are possibilities of multiple, perspectival ontologies and of equivocation[51] (Viveiros de Castro 2004), of univocity (Deleuze/Guattari 1987), and of "true" as a functional concept that allows for truth to be "one and many" (Lynch 2009).

However, this also does not mean that anything that claims to be knowledge/truth in such an account can automatically claim to be recognized as such. The point is, still, that this cannot be arbitrary. The issue for me is, here, that "anything goes" does precisely not mean "everything goes". For science to entertain the possibility of other kinds of knowledges and, subsequently, the possibility of productive (and, perhaps, "democratic"?) conversations with other kinds of knowledges does not mean the same as everything can be knowledge, everything can be fact. The same can hold for political discourses, its media ecology, and the respective publics.

For science as well as for political discourse and political news, I think it is quite adequate to channel Lewis R. Gordon (2006, 2014a, 2016) and his argument against *disciplinary decadence*:

> Failure to appreciate reality sometimes takes the form of recoiling from it. An inward path of disciplinary solitude eventually leads to what I call disciplinary decadence. This is the phenomenon of turning away from living thought, which engages reality and recognises its own limitations, to a deontologised or absolute conception of disciplinary life. The discipline becomes, in solipsistic fashion, the world. And in that world, the main concern is the proper administering of its rules, regulations, or, as Fanon argued, (self-devouring) methods. Becoming 'right' is simply a matter of applying, as fetish, the method correctly. This is a form of decadence because of the set of considerations that fall to the wayside as the discipline turns into itself and eventually implodes. Decay, although a natural process over the course of time for living things, takes on a paradoxical quality in disciplinary formation. A discipline, e.g., could be in decay through a failure to realise that decay is possible. Like empires, the presumption.
>
> (2014: 86)

Notes

1 www.welt.de/politik/deutschland/article167287260/In-diesem-Bundesland-hat-sich-die-Zahl-der-Migranten-mehr-als-verdoppelt.html.

2 The notion of histories or long durations of double-binded process (Bateson 1972; Elias 1983) comes to mind.

3 In the book, I add discussions of Lyotard's *Le Differend* (1989) and of certain ideas of Sara Ahmed (2004, 2012) to illustrate operative forms of annihilation in discourse.

4 Both these forms of annihilation and deliberation currently exist in contemporary societies side by side with interdependent effects of producing what I identified in the book as "inclusive exclusion" (Ophir et al. 2009; Decoteau 2013), the creation of inclusion-and-difference-engines (adopting, somewhat creatively, an analytic device by Steven Epstein [2007]), and (radical) "silent Others/Othered silents" (https://alexstingl.wordpress.com/2013/03/04/silent-othering/).

5 It is important to note here, too, that there exists an important and explicit distinction between "postcolonial" and "decoloniality"; however, this distinction is not one of the usual (intra-)academic rivalries between different theories, schools, or luminaries (about prestige or resources). In fact, it is not itself even a "rivalry", because postcolonial is primarily an academic orientation whereas decoloniality is explicitly not located in the "academy" but exists outside *as well as* across it (in for example seeing the [Western] university as is precisely as a colonial operation [Grosfoguel 2013] which is implicated and complicit in the destruction wrought not only [as postcolonialism has as its subjects] by colonialism [in the past] or by the neocolonialism of capitalist capture [in the present] *but* by coloniality). For instance, postcolonialism is not an enemy, but perhaps, at times an uneasy ally for decolonial thinkers and practitioners (see for a good summary the interviews between Mignolo and Vazquez [2013]). Within the academy and in scholarship, therefore, there can be nothing like, for example, a decolonial sociology, *but* decolonial options can certainly improve the sociological imagination; at the same time there is a place and also a dire need for postcolonial sociology (Go 2016; Boatcă 2016; Bhambra 2016) as a step toward provincializing sociological knowledge and opening the sociological imagination for decolonial options.

6 Along a similar line, non-human agents have not received much serious attention in the history of Marxist thought either. Swyngedouw (2006: 109f.) asserts quite correctly, in my view: "In other words, while mainstream economics equally forgot the natural foundations of economic life (only to rediscover this recently under the guise of environmental economics), much of Marxist theory became an exclusive 'social' theory rather than a socio-ecological one."

Economics, too, is still hampered by its restricted definitions of value and utility along the axis of a logic of exploitation and extraction. Few Alternative Economic principles have so far been offered; notable exceptions are Kaul (2007), JK Gibson-Graham (aka Katherine Gibson and the late Julie Graham: Gibson/Graham 2006a,b, 2008, 2011a,b; Roelvink et al. 2015), and feminist economics (Ferber/Nelson 1993; Agrawal 1995; 1997, 2010; Barker 2003;).

7 See also supra Fn. 7 and Fn. 8.

8 A term made popular through the work of Dipesh Chakrabarty (2008 [orig. 2000]), which can (and must!), however, take on slightly different meanings in different geopolitical locations and contexts.

9 See here, too, the seminal work of Michel-Rolph Trouillot (2002), which explicates the kinds of (colonial) work that categories and concepts do.

10 See in the question of the archival practices and the ontologies of sex, gender, and sexualities (see e.g. Stoler 2002; Oyewumi 1997).

11 For example: www.zeit.de/politik/ausland/2016-11/donald-trump-president-minority-liberality/komplettansicht; https://areomagazine.com/2017/03/27/how-french-intellectuals-ruined-the-west-postmodernism-and-its-impact-explained/; https://theoutline.com/post/2190/why-are-we-blaming-postmodernism-for-trump?zd=1&zi=437uix5c (all last accessed on June 20, 2018).

12 www.theguardian.com/us-news/2016/nov/30/political-correctness-how-the-right-invented-phantom-enemy-donald-trump.

13 "to interrogate" is another such "feminist theorist" term, which critics of anything smacking of a "postmodern flavor" like to hate, such as for example *The Washington Post* conservative columnist George Will, who has been waging this war against "interrogating" and "postmoderns" for over 20 years (see here from 1996 www.washingtonpost.com/archive/opinions/1996/05/30/smitten-with-gibberish/16b0bb62-3e2c-4bf0-a7ea-4b29d01a6a0a/?utm_term=.19219c155de2; and here on the Trump election: www.washingtonpost.com/opinions/higher-education-is-awash-with-hysteria-that-might-have-helped-elect-trump/2016/11/18/a589b14e-ace6-11e6-977a-1030f822fc35_story.html?utm_term=.2b225103c4e1 [either last accessed on June 20, 2018]). However, I think we should not only interrogate, but go even further, as Donna Haraway often suggests, that we "interrogate and trouble" texts, ideas, ideologies, concepts, practices, matters.

14 See for example: www.handelsblatt.com/impressum/nutzungshinweise/blocker/?callback=%2Fpolitik%2Fdeutschland%2Fzerreissprobe-fuer-cdu-und-csu-zerbricht-die-koalition-am-asylstreit-es-wird-wahrscheinlich-dramatisch%2F22686298.html (last accessed, June 20, 2018),

15 www.washingtonpost.com/gdpr-consent/?destination=%2fnews%2fworldviews%2fwp%2f2018%2f06%2f18%2ftrump-says-crime-in-germany-is-way-up-german-statistics-show-the-opposite%2f%3f&utm_term=.619f5cd343e4.

16 This has been subject of a PBS documentary titled "Crucible of Empire: The Spanish-American" (Miller 1999) www.pbs.org/crucible/frames/_journalism.html, last accessed June 20, 2018). Of course, this "(history of) media-reflexive" thesis as purported by "the media" themselves (such as said PBS documentarians) as well as being an account on the side of media-historians, has been equally scrutinized and critiqued, for example by Joseph W. Campbell (2003).

17 A longer footnote on the historical context had to be scrapped for space.

18 One is left to wonder what would have happened if this phrase included the option that women may define a situation as real – and here I am, obviously, doing more than playing with the ambiguities of the word "man" in the English language.

19 Despite the book's shortcomings in light of it being certainly quite dated, the methodological introduction (3–47) of Talcott Parsons's *The Structure of Social Action* offers to this day one of the most insightful treatments of these respective problems and a subsequent development of a concept of "fact".

20 Graff addressed, as do many who write on vagueness, the quite illustrative Sorites paradox, that – in a nutshell – states the question "when does a heap of sand of which individual grains are removed cease to be a heap".

21 https://tierneylab.blogs.nytimes.com/2008/02/14/dancing-dissertations/ (last accessed June 20, 2018).

22 I am indebted here to a discussion with Gilles Lhuilier, for reminding me of the productive ambiguities of this word.

23 The whole point about faith, which is why one must make that "leap of faith", is that one cannot *know* God. One can know facts, but one cannot know God; thus God is not a fact and has no place in science – specifically not as a causal agent.

24 See, however, this fun conversation for further points: https://gizmodo.com/i-watched-neil-degrasse-tyson-take-on-a-science-skeptic-1791380377.

25 But believes deriving from (or in) facts and believes deriving from faith are two different things.

26 www.msn.com/en-us/news/us/a-wealthy-televangelist-explains-his-fleet-of-private-jets-its-a-biblical-thing/ar-AAClpZV.

27 Originally, in this place, there was a section on how *white privilege* works based on a form of entitlement warrant as well, and how this ties back to the false meritocracy, social inequality, and why non-white lives matter less in the United States. But since the essay is already quite long (too long in fact), this section had to be dropped.

28 Habermas's validity claims being targeted at claims for truth, rightness, and truthfulness (see for a short introduction Niemi 2005).

29 www.huffingtonpost.com/entry/sean-hannity-mccain-fake-news_us_58af9677e4b0a8a9b780d36a, last accessed June 20, 2018.

30 Ironically, when one does consult the etymology (let's do Wikipedia for once, especially because I tell my students generally to avoid it if possible), one finds that "pundit" is itself a "migrant" word for "intellectual": Hindi पण्डित (*paṇḍit*), from Sanskrit पण्डित (*paṇḍitá*, "scholar, learned man, teacher, philosopher" Thus the word has three contemporary meanings, supposedly:

- A self-professed expert in a particular field, especially as called upon to provide comment or opinion in the media; a commentator, a critic. [from 19th c.]
- A learned person in India; someone with knowledge of Sanskrit, philosophy, religion, and law; a Hindu scholar. [from 17th c.]
- (historical) A native surveyor in British India, trained to carry out clandestine surveillance beyond British borders.

(from: https://en.wiktionary.org/wiki/pundit, last accessed: June 20, 2018).

31 Of course, it is a long-standing running gag among the political Left in the United States to call *Fox News*, the TV-channel this Mr. Hannity is working for, *faux news*.

32 Of course, Mr. Hannity is calling on to "truth" here.

33 Or did I?... Another interesting rhetoric point, whenever people start with a sentence with "believe me", we should be careful. Donald Trump, it has been noted, says it very often, and what come after is usually factually not true.

34 We stand, here, between Heidegger and Habermas.

35 There is an entire paper waiting to be written on plausibility-checking.

36 www.washingtonpost.com/news/fact-checker/?utm_term=.985079a1de10; www1.wdr.de/daserste/hartaberfair/faktencheck/index.html (last accessed June 20, 2018).

37 See on a general account of the so-called "destruction of Hilary Clinton" the book of this title by Susan Bordo (2017), but also reporting on several studies on the media coverage: www.washingtonpost.com/blogs/erik-wemple/wp/2017/08/25/studies-agree-media-gorged-on-hillary-clinton-email-coverage/?utm_term=.45b487e641da (last accessed June 20, 2018).

38 It has been done on isolated occasions.

39 Strangely current: www.zeit.de/2018/29/suendenbock-mesut-oezil-fluechtlinge-schuldzuweisung (accessed July 11, 2018).

40 This formulation clearly owes a lot to Rainer Forst's *right to justification*. Obviously, a detailed genealogy of my creative adoption of Forst's approach with particular emphasis on differences cannot be made here, but is subject of a chapter in a book manuscript I am currently working on.

41 This is also clear from my ethnographic field work at international conferences such as the Global Bioeconomy Summit 2018 or the INRA European Workshop on Bioeconomy 2017.

42 www.politico.eu/article/christian-schmidt-germany-admits-taking-glyphosate-decision-alone/ (last accessed: June 20, 2018.

43 https://geneticliteracyproject.org/2015/08/11/steven-pinker-crispr-bioethical-constraints-critics-get-way/; www.nytimes.com/2015/04/25/opinion/sunday/how-i-got-converted-to-gmo-food.html?smid=tw-share; https://ipscell.com/2015/08/stevenpinker/ (all last accessed on June 20, 2018).

44 In the following few paragraphs there might be some noticeable similarities to Stingl (2017), simply because my argument and the necessary reference, especially quotations from Burge (1993), are naturally the same. While the text here was written largely originally, any basic depiction of Burge's argument in the context I am making them will naturally look somewhat alike. This may be frustrating to some of those who enjoy policing any so-called "self-plagiarisms", but there's only so many ways one can cook noodles in water, after all.

45 It doesn't matter for now, but it becomes clear how it would matter in the following, that (while disagreement as simply discarding someone's knowledge for no reason at all is different than a reasonable disagreement) both statements are a reason given, though each of a different kind: (a) I disagree with your idea of basic social income because the President says that is a socialist idea and socialists are bad; or (b) I disagree with your idea of basic social income, because a consequence could be that people have no more incentive to work or innovate, the experimental study XYZ has shown that to be the case, and given the current financial reserves and the GDP of this countries we do not have the resources to create a basic income for all; we would need at least x amount of reserves and a GDP the size of y.

46 This refers explicitly not to any of over-simplified and debunked "implicit bias tests". Implicit bias is a complex matter, but has been itself used too broadly in media discourses.

47 I also would not think it wrong to believe that Alexander von Humboldt was already onto something here (see Wulf 2015), and perhaps inspiration, if taken with a decolonial grain of salt can come even from within the Western discourse on methodology, such as Goethe's "*zarte Empirie*" (Wahl 2005), especially since there exists a postcolonial engagement with Goethe by J.P.S. Uberoi (1984, see also Prasad 2016).

48 Also, again, it is important here, too, to keep in mind that one must explicitly pay attention to both objects of study and objects of observations in play.

49 I prefer to use "effectual" here to distinguish it from a Western concept of (instrumental) usefulness, (economic) utility, or efficiency.

50 Which is, precisely, "general" not "universal" (even and especially for Kant, cf. Stingl 2010).

51 Among German sociological theorists in particular, I have noticed there exists a genuine "fear of equivocation". Hardly any other theoretical criticism has been launched at me more often than the question whether this or that aspect I presented did not run the danger of equivocation. My response has always been that as a matter of fact, equivocation is an empirical fact of all social life (Stingl 2011).

9 Making Trouble in, for, and with "The Canon" – Toward a Sociology of Possibilities

In this chapter, using the book *Sociological Theory beyond the Canon* by Syed Farid Alatas and Vineeta Sinha as anchor point and grouping a series of other books around it in what is considered a "lateral movement" toward a sociology of possibilities, the current state of sociology as a discipline and its epistemological and ontological commitments are criticized. In taking issue with "the canon", the essay/review makes a case for engaging in *sociologies otherwise*, which is not a "sacrifice" of robustness, resilience, or validity, but for sociology, by looking at its own roots and genealogy, getting beyond them and back to actually "doing something". Sociology, in other words, can be descriptive or interpretative, but it must also be generative.

I am guided here by an unsettling of sociology, which considered itself the "science of modernity" – and unsettling that is enabled by Michel-Rolph Trouillot: "Then, I suggest that as a moment of a geography of imagination, modernity is necessarily plural" ("The Otherwise Modern": 222).

On Un-Settling the Context

Infamously, social researcher Jon Elster has railed on about and against the first winners of the Holberg Prize, as cited here:

> On the contrary, the winners of 2004 Julia Kristeva and 2006 Shmuel Eisenstadt are harming the prestige of the *Holberg Prize* according to Elsters *Aftenposten* article: "Shmuel Eisenstadt belongs to the B-team of international social science. It is not easy to prove this allegation, however, the following numbers give an indication: If you go to Google.scholar.com, you see that Eisenstadt's most cited work is cited 84 times. If you go to two social scientists that are clearly members of the A-team, John Ferejohn and Adam Przeworski, their most cited works are cited 455 and 534 times accordingly." "The award for Kristeva was particularly scandalous. It is an open secret that much of the French-inspired research in the humanities is close to being categorised as rubbish," Elster continues in his *Aftenposten* article.
>
> (www.nordforsk.org/en/news/holberg-prize-faces-criticism-550; last accessed May 2, 2018)

One can, of course, argue that Elster's criteria are arbitrarily, even capriciously, applied by him, for if he counts Eisenstadt's citation figures as indicative of anything, then Kristeva would, certainly, rule supreme with citations of individual works in the several thousands (at least that's what I get from Google Scholar, where *Pouvoir de l'horreur* is supposedly cited 13,788 times). What an exemplar of double standards Elster is, no?

My issue here is, however, that Eisenstadt – whose merits exceed the sole number of digitally counted citations (which today's "merchants of *relevance* and measurement" somewhat ludicrously call and worship as "impact [factor]") – proposed the idea that modernity was to be considered not as a monolith but a plurality, that is a small but not insignificant approach to suggest the possibility of taking the capitalization of the "M" in "Modernity" out of the equation. As a critique, as such, Eisenstadt is not itself particularly *radical* (in the true sense of this word), but for anyone who believes in Modernity as "the one and only", and, subsequently, in the Eurocentric holy grail of "the *one* Rationality", Eisenstadt's proposal was already too much of a provocation. However, Modernity-capital-"M" is nothing but a "useful" fiction – useful for a specific creed that is, because it was and, as a matter of "mentality/paradigm/pensée", still is the analytic category used to perform and legitimize North Atlantic/Western/Global Northern imperialism and hegemony, including the "we have come to improve your lives"-kind. The eminent and having too early have to be called the *late* "anthropologist of the West". Michel-Rolph Trouillot summarizes this accordingly, in a manner we find equally well and in detail analyzed by scholars such as Enrique Dussel or Sylvia Wynter:

> The world became global in the sixteenth century. Europe became Europe in part by severing Latin Christendom from what lay south of the Mediterranean, but also through a westward move that made the Atlantic the center of the first planetary empires. As such empires overlapped or succeeded one another within the modern world system, they brought populations from all continents closer in time and space. The rise of the West, the conquest of the Americas, plantation slavery, the Industrial Revolution, and the population flows of the nineteenth century can be summarized as "a first moment of globality," an Atlantic moment culminating in U.S. hegemony after World War II.
>
> [...]
>
> Modernity is a murky term that belongs to a family of words we may label "North Atlantic universals" and that are inherently ambiguous for reasons I shall try to expose. By North Atlantic universals, I mean words that project the North Atlantic experience on a universal scale that they themselves helped to create. North Atlantic universals are particulars that have gained a degree of universality chunks of human

> history that have become historical standards. They do not describe the world; they offer visions of the world. They appear to refer to things as they exist, but because they are rooted in a particular history, they evoke multiple layers of sensibilities, persuasions, cultural assumptions, and ideological choices tied to that localized history."
>
> (Trouillot 2002)

For sociology, this is particularly problematic, if not embarrassing, because its very origin story and, some would argue, its very mission statement are deeply intertwined with capital-M Modernity or, rather, *Euromodernity*. For a certain while, in the 20th century, the noun was happily used, both descriptively and prescriptively by social scientists as a verb: Modernization. This "verbalization" has, however, come under some critical scrutiny, because of the production of what one would only euphemistically call "discontents"; on these, the literature is, of course, already legion. It behooves us to "interrogate and trouble" sociology – to deploy a useful "methodological calling card" that has itself riled reactionary commentators who obviously do not like themselves or Western academic disciplines to be "interrogated", let alone "troubled" by anything – of course, which white, male privileged commentator or sociologist would (see here on "white fragility" in general DiAngelo [2011] and on sociology's color-blind racism Bonilla-Silva [2018]).[1] But in staring the ongoing and future problems in the face, "discontents" that are so "properly noun-ed" as *Anthropocene*, *Overpopulation*, *Social Inequality*, etc., we cannot but face the fact that our compensatory strategies that we claimed to spread under the names of Capitalism, Democracy, Free Market Society, and so on were ultimately merely "simplifications" (Dussel) that worked for a part of the Earth's population at the expense of Others – as in both: human and non-human Others.

Of course, we can – as do the Steven Pinkers of this world[2] – continue to tell the story of the world in the *proper* (aka simplified and ideology-wise usefully fictional) categories and criteria that exert an intellectual hygiene that requires to ever more often scrub the body more raw than just healthily clean and, thus, susceptible for skin-infections: After all, the etymology and troubled genealogy of "proper" should not be forgotten here (Sarasin 2001). But as we should know by now, the world is complex, and simplification makes it often only more complicated: This is why social scientists, and others, would do well to heed the warning that we must distinguish complexity and complication, as well as not fall for the trap of universalism, which tends to unveil itself, at the end, as a *Thomas-theorematic* reification. In this regard, it is fitting that as of the writing of this book meets with the 60th anniversary of Mills's *Sociological Imagination* (1959), which Thomas Kemple and Renisa Mawani (2009) re-reviewed for its 50th, with an eye toward even its "imperial shadows" graciously to show what a provincialization of such a classic can continue to instill and contribute to create a better sociological imagination. But ultimately, the necessary beginning lies before we can consider the diversity of lines of flight from the *disciplinary decadence* (Gordon 2014) that sociology "proper" is mired and implicated in today, prying out the ontological commitments that sociologists have made early on in the discipline's infancy, especially in its US-American version: And this beginning is, as Zine Magubane (2016a,b) has shown in her meticulously, diligently, and powerfully conducted[3] study of "American Sociology's Racial Ontology", deeply implicated in and imbricated with colonialism and slavery, as *monumentalized* (Foucault) in two 1854 publications of Henry Hughes's *Treatise on Sociology: Theoretical and Practical* and George Fitzhugh's *Sociology for the South*. Unlike (Whiggish) denialist accounts that settle on the discipline's collective amnesia (to riff on Herbert Gans's terminology), Magubane shows that the deployment of the term "sociology" by both authors (and their contemporaries in discourse) was no accident or disambiguation. Instead, it was as intentional and purposeful as it was fateful *for* and a structuring *of* the disciplinary history that would follow in its wake. Following my own recent witnessing of a panel at the American Sociological Association's Annual Meeting in 2018, which was supposed to deal with the necessity for and possibilities of a dialogue between the "history of sociology" and "sociology of science", the panelists were largely in agreement that while, of course, such a dialogue would be a nice thing, it was currently empirically not happening and it was doubtful that it would. It was further argued that a main point was because the object of study for history of sociology meant was to be open to intervention, whereas sociologists of science did not mean to intervene in their object of study – a (normative) evaluation to which one of the panelists, Joan Fujimura, vocally, argumentatively, and quite rightfully objected. My own intervention as a member of the audience pointed directly to the co-productive link between race/racism, slavery, science, and sociology in both history *and* present, as an important issue that illustrates the need for said dialog, an intervention that was largely ignored by most panelists (once again with Joan Fujimura being, for obvious reasons in light of her work, the exception) in favor of re-emphasizing in response that empirically no such conversation was currently happening. What such an event in its evident failure – which by my own past experiences and listening to many a colleague's account I feel qualified to deem *typical*, as well as Zine Magubane's eminent work – illustrates is precisely that we need to interrogate and trouble sociology as a discipline in order to come together and create a new and genuinely different and diverse sociological imagination, as Kemple and Mawani already proposed a decade ago.

A Global and Yet Provincialized Sociological Imagination?

Euromodern sociology is historically implicated in the extractivist *death projects* of Settler Colonialism and Modernization under capitalist/corporate/neoliberal capture, that is, those sociologists whose own "fragility-slash-privilege" (aka what I call elsewhere *White Collar academia* status) leads to their denial of this still continuing history have themselves become merchants of doubt(-slash-death) (Oreskes/Conway 2010).

It is, indeed, important in this context, as Kemple and Mawani stress, to provincialize sociology and do so in various ways. Kemple/Mawani (238) evoke Dipesh Chakrabarty's now seminal approach:

> In his seminal book, Provincializing Europe, historian Dipesh Chakrabarty (2000) argues that European thought is both indispensable to and inadequate for thinking through modernity in the non-West. Although Chakrabarty focuses on Europe, his insights can be extended to address the growth and hegemony of American sociology which preoccupied Mills. To provincialize sociology is not simply to bring sociological insights to bear on the study of imperial formations, or only to decenter the universal claims of American history and social life by 'adding' other non-western geographical and historical contexts to the sociological agenda (Burawoy 2005b). Instead, to provincialize is to critically explore, examine, and scrutinize how sociology's master categories all 'bear the burden' of European (and American) thought and history (Chakrabarty, 2000: 4). It is to consider how colonial and imperial expansion has figured centrally in the production of sociological knowledge and the formation of its epistemic boundaries. In the spirit of Mills (1959: 84), we can say that provincializing sociology means, in part, rendering the traces of its bureaucratic routines, ideological codes, and political meanings more explicit, legible, audible, and visible.

This gesture of provincialization today includes the question of what is to be done with the noun, adjective, and qualifier "*global*". Sociology addresses that which it deems "the Global", as well as matters of "global import", but therein it also "globalized" both itself and these matters and became (and continuous to serve as) a force of "the Globalization", which is a process of *capture* in finishing that "unfinished project of Modernity", as Jürgen Habermas calls it. But between capitalist capture of sociology on the one hand, and on the other hand those scores of Marxist sociologists who have never seen any reason for feminist, postcolonial, indigenous, non-ableist, non-human, or other perspectives[4] to "be given a voice" [*sic!*], for they claim to already "say what needs to be said in their stead", we need to have a sociology instead that listens to those who have been displaced[5] by the "global and globalizing discourse of (Euro)Modernity" in which sociology has been implicated and complicit while (and by) simultaneously studying these discourses as its objects. We can no longer "move along" with this Euromodern project, if we seek to genuinely reimagine – rather than just reboot and end up with the same – sociology. We must move laterally and think in different intensities, rather than merely extend sociology's grasp.

Coloniality remains, even after colonialisms have ended in geographically located places: Coloniality or the "colonial matrix of power" refers to how people think, see, perceive, articulate, construct, and so on, it's about aesthetics, ethics, epistemologies, politics, etc. I would never match the clarity of Anibal Quijano, Maria Lugones, Nelson Maldonado Torres, and others in describing this important concept, but I will not simply cite one of them here either, but instead invite readers who are not yet familiar with these ideas to go and explore. What I will do is to encourage the double-gesture of reading and stepping sideways into intensity. Or, perhaps, I will suggest to go into exemplary questions such as that of the (Euro)social construction of "intimacy", which Lisa Lowe pursued in her seminal study *Intimacies of Four Continents*. In her work on the colonial relationships between four continents, she views "intimacy" as

> the affective medium for republican citizenship and the subject's felt sense of individual belonging in liberal society; fantasy, sentiment, and desire in literature and popular culture produce the contours of intimacy that mediate the individual's inhabiting of everyday life in social relations.
>
> (Lowe 2015: 21)

Projects such as hers redefine precisely how we can study the "global". This is why a journey in theory such as Julian Go (2016) does in his study on postcolonial thought and social theory or the adventure as well as tragedy of discovering connectedness in and through sociology as Gurminder Bhambra (2017) does, is necessary, timely, and should be read as widely as possible. Their works are not the comforting, answer-providing, explanatory frameworks, but they leave us with questions. The point is, however, that not all of the questions are new, while many are, but that we begin to understand which formerly asked questions have remained unanswered and we understand better why not and why they must be. And sometimes, we find that these questions have different articulations and very diverse answers when asked from and for a different place. Sabelo Ndlovu-Gatsheni (2015) in exemplary fashion shows us what an intensification of inquiries looks like, when they are not only conducted about a place but also from that place. His work is not about Africa; it is from Africa and, moreover, from particular places in Africa: this means that his scholarly *intensity* is about not being constrained in viewing everyday lives from within Western social scientific categories, but instead uncovering what these lives are without these categories *and* what (damage) these categories have done. Especially in terms of "inequality", we encounter a key sociology line of inquiry and set of criteria, which have been economically and political influential. But inequality is not universal, and the idea of a universal sociological understanding of what inequality is and how it is experienced has itself been extremely problematic, as we learn from Manuela Boatcă (2016), who is teaching us to think differently on what is, perhaps, sociology's self-described mission, to make a difference on inequality, i.e. "the social question" (*Die soziale Frage*), which has always been a Euro-modern problem formulation.[6] But here we meet, also, precisely the (very Anglo-American) "cult of the individual", that also plagues sociological theory and methodology. The answer to "global" social problems is not found in ideologies of "emancipating" individuals economically, politically, ethically, as both neoliberalism and some Frankfurt School style thinkers propose the "finishing of Modernity". Enrique Dussel has long since become a primary and inspiring thinker who with unparalleled erudition un-Settles these categories and concepts. A re-imagining of the sociological imagination needs precisely these kinds of un-Settling. But that does not mean that we simply wipe away previous scholarship and declare it useless. Kemple and Mawani (2009: 240) remind us, in turn, that Mills *sociological imagination* was interested in a plurality of social worlds, in "human variety":

> Challenging the universal narratives of 'capitalism' and 'modernity' while addressing a plurality of 'ways of being human' places social scientific categories and modes of thinking into question, and opens up new possibilities for a sociological imagination that takes the epistemic presuppositions and political effects of empire seriously.

But they do go further and demand (ibid.: 241) that a renewal of the sociological imagination must expand Mills to include "non-human forces". I think this is correct, because the view of and from coloniality is not only Eurocentric and Androcentric; so any sociology that truly wants to be *otherwise* must also eschew (epistemically, ontologically, and politically) Anthropocentrism.

Catherine Walsh (Mignolo/Walsh 2018: 60–69) describes on the example of the Ecuadorian Constitution of 2008 how an epistemic-political *otherwise* can be created and what – of many – possible forms it could take; but she also shows how – in the process, and decoloniality is, first and foremost, a process, a doing, a becoming, not a (teleological, ultimate) end-state – this *otherwise* was never truly realized, but undermined from the start by neoliberalism and *coloniality*, installing, instead, what the

Ecuadorian government called a "*modern* capitalism", which follows the same old extractive logic.

Canon, Different Canon, No Canon, or What? Monologues Don't Create Possibilities

Therefore, Catherine Walsh cautions that, "[a]nother danger is the commodification of decoloniality as the property of a group of individuals (i.e. the modernity/[de]coloniality project) and as a new *canon* of sorts" (Mignolo/Walsh 2018: 82).

In this vein, the actual entry point for this essay, an entry point that has yet to make its actual appearance, was the publication of a book by Alatas and Sinha. The books title asks us to consider *Sociological Theory beyond the Canon* – an avenue, which the book travels along in an interesting and very productive fashion, in my view. It does beg the question, however, of the "canon" above all. Do we need it? Should we drop it? Should we replace it with another? What does "going beyond" – although I prefer the idea of going "before *and* beyond" in one move – mean, when we talk about "a canon"?

Actually, I have had many conversations about the mere issue of "canon" in the academy in general, not only in sociology (let alone sociological theory) in particular.

Canons and canonic authors are put to use in various ways. In a concrete discipline, such as sociology, the canon is a reflection of the discipline's history, and with that, the canon and its use(fulnes)s can be interrogated variously, as Chrisitan Dayé (o.f.) has recently demonstrated. But knowledge of, re-f/v-erence to, and conducting (gate-keeping and policing) operations of maintenance on and with a canon – as a general cultural canon or particular disciplinary canon – also can and does turn canon into a cultural capital and reservoir for the exercise of hegemony through memory politics (Kidman/O'Malley [o.f.]). Any canon represents, in many ways, an archive – among other things, an archive is, however, also a technology of selecting what is to be memorized and what is to be silenced qua forgetting. And the sociological canon, implicated in the techniques of Empire (see Steinmetz 2013), is also an art of governance aka a colonial archive (Stoler 2002).

But to simply say that, therefore, the canon in its current selection must be abolished and either reassembled or simply forgotten is not really an improvement. Just archiving different items instead of the current ones or abolishing the archive is not creating "archive practices *otherwise*" nor changing the meaning of what an archive is. Just making a new canon with more "inclusivity" does not solve how the canon operates as a technology of power. Meanwhile, just forgetting the canon means also to forget how its power worked and what world it used to construct. This is an ignorance, which we also cannot afford. The question is, thus, how to (re)claim, (re)occupy, and transform the canon/archive?

The "old" canon is still interesting to study, because it still helps understand how we ended up in the current situation. The way selections were made, who was forgotten and why, is highly interesting and very important. And it is also a fact: The canonic authors often still offer a lot of insight. To understand how, for example, Max Weber and the "Weberians" were and are mired in coloniality and complicit in the workings of Empire also does not negate that there is still potential and usefulness in Weber's work (see also Boatcă 2017, and Alatas/Sinha). On the contrary, the three Western "foundational classics" Weber, Durkheim, and Simmel remain full of creative potential and analytical robustness and methodological resilience, but: Theirs is a contribution that must be understood as situated, provincial, and as epistemologically, ethically, and politically ambiguous in terms of the effects. Thus, to conduct a "decolonial critique", if you like, of the canon would mean to put the canon in its geopolitical, geographic, and temporal-historic place. This canon is, thus, always the *canon-that-was*, and should be understood as such; in teaching, today, we should at the very minimum be empowered and enabled to open up the canon, to render it fluid; the canon as a permanent and never finished *canon-becoming*, or alternatively, should lead us to the acceptance of *variegated canons*.

I am myself inspired by the genealogical, interpretative analytical work of Michel Foucault, by the work on decoloniality by Gloria Anzaldúa, Sabelo Ndlovu-Gatsheni, Sylvia Wynter, Walter Mignolo, Oyeronke Oyewumi, Maria Lugones, and others, as well as the work on transdisciplinarity and teleological suspension by Lewis Gordon. These shape my attempts to deal with reoccupying and transforming concepts and methodologies, while making my positionality (with both its geopolitical and epistemic privileges and blind spots, as well as my lower class constraints) explicit, in what I call attempts to "fail spectacularly". These are attempts at what I have called the B'n'B approach, attempts at going "before and beyond" in one and the same movement, attempts at gestures that take an object, such as a specific canon, and try to go to a genealogical point where a decision becomes visible that established that object and silenced others, making that point and its *Before* visible and in that same act go *Beyond* that point and the exclusions it facilitated. I like the idea of the B'n'B as shorthand for this methodological attitude and gesture and style, because it is precisely a *Bed and Breakfast*: It is a place of hospitality, where travelers meet on a journey, where they can sleep safely and share a hearty breakfast together before traveling on or go exploring, but it is not the kind of "home" the Westernized, commercial *hotel* with its many controlling, surveillance, and exclusion techniques that are supposed to mimic foundational aspects of the Western "home", with its logic of *oikos*[7] and being a walled castle; the B'n'B, as an idea, seems to me of a more transitional nature, less policed, less walled, perfect for *nomadic traveling*. Perhaps we do need a bit more *nomadic scholarship*.

Now, Syed Farid Alatas and Vineeta Sinha, in creating *Sociological Theory beyond the Canon*, seem to me as having produced an effective version of one possibility of *variegated canon*. It's less *nomadic* in nature, and thus less radical than I would like personally, but it certainly will already prove to be too radical for many. But it is brilliant, yes I say: "brilliant", for it makes possible new ways and different travels. It's an offer and a gift, I think, to serve us recovering addicts from the sociological imagination of Empire as a halfway-house.

Alatas and Sinha – who split chapter duties, with each chapter devoted to a sociological author – begin their book, after an introduction that focuses on the role of Eurocentrism and Androcentrism in the establishment of "the canon", with a chapter on Ibn Khaldun. Having made it clear that why and how we must *also* look at non-Western and non-male scholars, in particular in sociological theory, the choice of a (in European chronology) 14th-century figure from the Muslim world as the starting point is refreshing, even if not unexpected for anyone familiar with Alatas's work.[8] But the chapter sets the tone for the book well, because it really draws out Ibn Khaldun's theory; as theory is something to *think with*, as I often say inspired by decolonial and feminist science studies authors such as Catherine B. Walsh or Donna Haraway, then Alatas gives us a concise and well-sized lesson in how *think with* Ibn Khaldun. This goes for the other chapters as well. Intermittently, we meet some old friends from the "traditional canon": Marx, Weber, and Durkheim[9] (although and despite the book already approaching 400 pages, I would have included either Tarde[10] or Simmel[11] as well, precisely because they offered each an alternative within the canon but one without transcending the colonial confines of Euromodernity). Marx is vividly reconstructed with special attention to his, ultimately problematic, reconstruction of the so-called "Asiatic mode of production". In conclusion, Alatas, who signs as author for the chapter, finds that

> [a]lthough there are problems with his understanding of the nature of Asiatic societies, informed by his Orientalist standpoint, our attitude is that rather than exclude Marx, he needs to be rescued from himself, from his Orientalist assumptions, in order that what remains useful in Marx and be salvaged. (78)

The inclusion of Harriet Martineau, while not that surprising to anyone familiar with the history of sociology, is still a wonderful (in all the senses of the word) boon for this book, because it pays a chapter's worth of attention to her, which her work rarely receives within the discipline. The chapter does, however, not shy away from also pointing out her Eurocentric *and* Androcentric shortcomings.

If I am allowed one critical remark[12] regarding Alatas and Sinha's book, it is that despite its important criticism of the sociological canon's Eurocentrism and Androcentrism, they remain with an ontological commitment to Anthropocentrism. Adding a chapter or two focusing on sociological scholarship that includes the theorization of ecosystems and non-human agents would have been, in my view, very important.

The other chapters in this book are dedicated to José Rizal, Said Nursi, Pandita Ramabai Saraswati, Florence Nightingale, and Benoy Kumar Sarkar. I would be surprised if there are many Western sociologists who are familiar with any one of these thinkers, with the exception of Florence Nightingale. And most would be surprised at her inclusion, given that she seems to be more a figure of legend. But the point is, precisely, that the history of Western sociology in the 19th century is deeply intertwined with social activism, nursing, and environmental education in which women were not only leading figures but also truly the leading minds – theoretical and methodological. Many other names of female scholars could and, perhaps, should be added in this discussion, who contributed effectively to the development of sociology.

Along those lines, Pandita Ramabai Saraswati is a particularly good point of comparison to many early women sociologists in the Global North, having her roots and focus of her work in India, but also having traveled the Global North and undertaken a substantial, we would today say ethnographic study of late 19th-century US-American society.

Finally, I want to point out the chapter on Filipino sociologist José Rizal in particular, because his work provides sociological and historical theories of colonialism and coloniality that holds back no punch. But it is precisely in terms of their bridging empirical and theoretical insights that they obtain their comparative dimension. Given that Julian Go in a more recent paper (2013) has outlined Pierre Bourdieu's theory of colonialism/coloniality, which is rooted in his experiences in Algiers and simultaneously foundational for his later theoretical apparatus (an apparatus which is as often used as it is misused in contemporary Western sociology), I think reading Go's early Bourdieu through the eyes of Rizal would be an invaluable correction to any Western sociological theory that tries to understand hegemonic struggle through a Bourdieusian lens.

For theorists and methodologists alike, Alatas/Sinha offer an entry into a number of necessary corrections we need to make as a discipline in order to make sense of as well as have a role to play in the world today. They offer an introduction we can use to embark on this journey not alone in our study chambers, but together with our students. It is a genuine in-*vita*-tion – a bringing to life (by bringing to light) – to sociology, because it allows the creation of a conversation: a conversation about other options of doing sociology, about *decolonial options*.

Let's use Symbolic Interactionism (SI) as a "good" methodological example from sociology – and, to be honest, I view SI primarily as a methodological alternative, rather than an ontological or epistemological one – that should in principle be particularly well-suited to enter a genuine conversation with decolonial options in social sciences. SI is, after all, the understanding that social life is (a) filled with dialog, and that (b) dialog does not simply happen, but can and must include differences, diversity, and, thus, often misunderstanding, which can lead to either conflict/destruction or creativity (Stingl/Weiss 2016). SI, as a methodology, must seek out the conversation and begin the long and difficult task *to think together with* other methodologies, methodologies that are decolonial and decolonizing (Smith 1999; Connell 2014). Instead of following in the (hegemonic, patriarchal, imperialistic) top-down "monologue of modern-Western reason" (Rafael Bautista quoted in Mignolo/Walsh 2018: 69), SI can, it should, yes, *it must* enter into the dialogue "from below" by understanding (that it has) its own positionality, situatedness, and provinciality as one,[13] however also truly fertile and creative methodological alternative among *other*, often equally interesting, valuable, and creative options. Having options means having and creating possibilities, and, as Eduardo Bonilla-Silva puts it – in reference to the relationship between sociological mentors and mentees in the academy: We shouldn't produce sociological clones of ourselves, because that doesn't make for good sociology. In other words, there should not be one sociology but sociologies in the plural.[14]

Going Before and Beyond Euromodernity's Sociology to a Sociology of Possibilities and Sociologies Otherwise

Asad Haider, in response to a dismissive review of his work that depicted him and his readership derogatively as "zombies", wrote that

> [t]he role of theory is not to comfort us, nor to provide us with the great book of answers. Theory unsettles; it strips away our certainties, and invites us to examine our world in a new light. This means theory does not always provide a clear and direct guide to action. It is nevertheless a necessity for meaningful action, as it provides a site for reflection and reinterpretation and gives us new terms and new concepts with which we can grasp our reality and attempt to intervene in it.
> (www.versobooks.com/blogs/4002-zombie-manifesto, last accessed Sept. 4, 2018)

A sociology, including *un-Settling* and *becoming meaningful* practices that are sociological theory, that becomes a sociology of possibilities and genuinely becomes *sociologies otherwise*, not only goes *before and beyond* – but moves laterally from – the confines of Euromodernity with its constrictive methodological nationalism and methodological individualism. Such a sociological theory becomes refocused on notions of "we" as well as concrete projects that are lines of inquiry into and lines of light toward what a "we" can mean – projects that Catherine Walsh analyzes as exemplars such as *Pueblos en Camino*. A *sociology otherwise*, as found in such a project, generates

> [s]paces of encounter, word, and action [...] spreads embodied thought-analysis as well as shared reflection and action. It weaves a praxis of

> political-epistemic-existence-based prospect as a way, this *we* argues, to sow life against the growing project of death of these times.
>
> (Walsh/Mignolo 2018: 39)

This, I argue in conclusion, presents for "thinking and doing *sociologies otherwise*" one very possible way to go.

Notes

1 To give account of my own positionality and make it possible for readers to have at least some idea of what privileges of my own I may or may not be aware of: I was born in Germany in the late 1970s, where I grew up as a white, male, into a lower class, i.e. non-academic family, working in bars and restaurants most of my adult life since my undergraduate years and way into my postdoc phase to survive financially. I have and continue to struggle with writing about decoloniality in ways that are not, in the end, appropriating or hegemonizing, as well as I have received criticism on why I do so – criticism not only from post- and decolonial writers (which I take very seriously, unless expressed in the form of personal threats) but also from white Leftist social scientists who do not see a need for decolonial or postcolonial scholarship (whose arguments are too often qualified by a lack of even being such, i.e. arguments).

2 Pinker's understanding of referenced authors such as for example and quite indicatively Norbert Elias is (to put it very, very euphemistically) questionable.

3 Why would I qualify her work with these three adjectives? Precisely, because I am quite aware how this kind of scholarship has been dismissed, because it is uncomfortable as well as its author's testimony on US sociology as that of a Black woman's born in South Africa is often subject to epistemic injustice, i.e. not taken as seriously as that of a white, male born in the United States. In using the three adjective qualifications here, I am making this fact explicit, fully aware that it will elicit a multitude of reactions itself.

4 Stuart Hall has made short process with these kinds of scholarship over 30 years ago. Bhandar (2018: 10) summarizes:

> He [*Stuart Hall*] rejected the economic reductivism of orthodox approaches to Marxist theory (which reduce the causes of race and racism to economic determinants or to the functional demands of social domination) and instead grasped the complex relationship between race and class through examining how they are articulated together as historically specific social forms of identity and domination.

5 I choose this term, taking inspiration from Zine Magubane's (2016a: 13) conclusion: "Modernity as a frame is inadequate for addressing the global and colonial in sociology because modernity has operated as cover for a set of displacements".

6 In her important study on the racial regime of property, Brenna Bhandar (2018) identifies an "ideology of improvement" that was installed in justifying landownership, an ideology which is similarly at work in Euromodern social thought on decreasing/erasing (global) social inequality:

> Those communities who lived as rational, productive economic actors, evidenced by particular forms of cultivation, were deemed to be proper subjects of law and history; those who did not were deemed to be in need of improvement as much as their waste lands were.
>
> (8)

That doesn't mean that inequalities don't exist nor that we shouldn't work to change them; it means, however, that the identification of inequalities and the means to work on them require critical reflection and provincialization themselves: So-called "developmental programs", including sociological work involved in them, were and are in themselves not colonially innocent and more often than not reiterate the fault lines of injustice they claim to overcome, as I have argued on the example case of *digital divide(s)*.

7 See here, Mohammad Nassifi's study on Weber, Polanyi, and Finley in the context of the so-called *oikos*-debate (2005).

8 While Alatas is not alone in having paid attention to Ibn Khaldun's sociology, as evidenced, for example, by Fuad Baali (1988), still Ibn Khaldun is neither regularly made reference to nor given substantial space in introductions to sociology.

9 George E. McCarthy (2002), in a very interesting study, identifies in precisely these three, for the course of sociology decisive figures, the discipline's main discourse's ancient roots in classic Athenian philosophy. I think that the same cannot be said for all influential figures within the history of Western sociology (whether we look at somebody like Georg Simmel or Dorothy Swaine Thomas).

10 Marc Schuilenburg (2015) has delivered a very interesting study on the issue of securitization, which delivers a good introduction to Foucault, Deleuze, Tarde each, and deploys them as an empirically, critical approach outlining the possibilities these three scholars present for socio-critical analysis.

11 Talcott Parsons, whose reverence for Durkheim and Weber is well known, was an admirer of Simmel's works in themselves but cautioned against any attempt at systematic deployment of Simmel as a sociological methodology, such as he saw attempted and leading to failure by Leopold von Wiese.

12 Perhaps one should also mention here that as a book on the material-technical level, it sadly suffers the increasingly common lack of editorial oversight in the production that would help prevent the occasional typos, grammatical quirks, and sometime reference errors, etc. which a decade or two ago was the task of in-house editing, copy-editing, and proof-reading to spot and help authors avoid. It's not particularly bad, but it reflects today's academic publishing culture and particularly postcolonial and decolonial authors being affected and makes it more questionable. It is not at a level that it would detract from the importance of this book nor the impact it hopefully has, but it is an issue that for any future editions someone should comb through it and make the necessary corrections here and there.

13 The project put together by Melissa Weiner and Antonio Carmona Báez (2018) with their edited book *Smash the Pillars!* is an exemplar of what sociological projects can look like that take such an approach seriously (full disclosure: the book appeared in a book series which I serve for as managing editor).

14 And I do not mean, here, the endless production of sub-/special sociologies (as inclusion-and-difference engines [Stingl 2015]), such as in "sociology of health, sociology of illness, sociology of medicine", which become internally also ever further differentiated, but a methodological as well as an ontoepistemological point of view.

10 Concluding *Matters*

Commoning Infrastructural Power in the Pluriversity – Going Before-and-Beyond *Oikos* with BluesCollarship

In his seminal *Ethics of Liberation*, Enrique Dussel has described a similarity of simplifications that felt familiar to me, a similarity that, however, I did not think a mere similarity, but that I deemed exercising mythical force in the way that, for example, we can discern from the lessons gifted us by Hans Blumenberg and Chiara Bottici:

> In fact, the modern simplifications (the dualism of an *ego-soul* without a body, teleological instrumental reason, the racism of the superiority of one's own culture, etc.) have many similarities with the *simplification* that Greek slavery produced in the second interregional system. The Greek *Weltanschauung* was advantageous to the modern man—not without complicity does the modern subject resuscitate the Greeks, as did the German Romantics.
>
> (Dussel 2013: 37)

I will conclude my writerly and mycelial text with musings on simultaneously nomadic and mycelial views of infrastructure as an alternative to *oikos*. With Dussel and others who give us decolonial options, we are given alternatives. We are invited *to think with* provincialization. These invitations do not simply mean "clarifying our position" but they mean to clarify our positionality and, thus, they invite us and help us to better (though it's never possible to fully and completely accomnplish this feat) understand from where, when, and how our positions come about; we are shown which the accompanying privileges and entitlements inherent in these positionalities are, which of them we should best suspend. Meanwhile, we should also understand that there are alternate views and ways even within the (geopolitical) Global North and (epistemological) West, except that these, too, need to acknowledge their own positionalities of articulation, which may have privileges of their own – Nothing is ever easy.

Lyotard's diagnosis is that *oikos* is a tragedy, and he installs a notion of *oikeion* to talk about the secluded and Othered. *Polis*, at first, does not seem to be in his purview, but it appears clothed in the German word *Öffentlichkeit*:

> Today, economics belongs to *Öffentlichkeit*, what we call *Wirtschaft* [the economy] is precisely part of the public sphere. I would even say that

> the economy is the very substance of *Öffentlichkeit*, because it is simply the regulation, the nomos, of goods, values, and services. When *oikos* gives rise to *oikonomikos* or *oikonomikon*, a complex transformation of the word *oikos* occurs.
>
> (Lyotard et al. 1993: 105)

This *Öffentlichkeit* and the *Wirtschaft*, we are often today told, are molded as *Digital Culture* – Henri Lefebvre, it is no accident, begins his 1990 essay on the "new contract" with such foresight: "The undeniable social power of information science has as its converse and inverse the ideological power of the media" (in 2017: 238).

If "the world" is subject to global digitalization – in the way it was said to be subject to global industrialization, to global mobility, and to global communication, then we are led to believe in a process that supposedly rests on structural foundations – material and cultural – that can be found in the – no less ambiguous – term *infrastructure*.

Infrastructure can be all the rage:

> Infrastructures produce collectivity on the material level. The pipes, channels, roads, and wires that organize life in modernity create a durable "in-between" (…) and provide a "connective tissue.
>
> For better or for worse … infrastructure unites distant strangers into new communities" (…). Infrastructures are therefore political matters (…). While the social study of technology has always drawn attention to the ways "artefacts have politics" (…), current debates in social and political theory go one step further: they take infrastructures as the vantage point for rethinking politics (…). The assemblages of pipes and wires become the ground on which a different understanding of material collectivity is sought (…). What are the implications of taking infrastructure as a key term for reenvisioning the common? What is the "we" of infrastructure that is implicitly presumed in these attempts to rethink material politics?
>
> (Optiz/Tellmann 2015: 121)

For Opitz/Tellmann, in their study of European *infrastructuralism*, the issue is about the interrelations between techno-material connectivity and political collectivity that are endemic of and/or constitutive to the post-war European project – in the end, their question is about the relation between (a largely techno-material) infrastructure and a kind of "commons" (starting their account with reference to Bruno Latour).

With their take on infrastructure, our "friend" Carl Schmitt makes a reappearance:

> If we take the nation-state and the empire as two ideal-typical versions of how infrastructures serve the production of political space with and

> without collectivity, Europe becomes a particularly interesting case: it articulates connectivity with collectivity through a politics of infrastructure without fitting neatly into this dual structure. Curiously and disturbingly, the main political theorist who has addressed this form of political space has been Carl Schmitt (…). He envisioned a political articulation of an extended and distinctly European infrastructural collectivity beyond the nation-state. However, Schmitt's European vision has an unacceptable political valence: the Großraum conceptualized as "living space" with an ethnic foundation. Despite this political allegiance, Schmitt's book Völkerrechtliche Großraumordnung (1941) is important for the current argument since it explicitly links technological infrastructures, spatial order, and collectivity in a political form that is neither state nor empire.
>
> (Ibid.: 176)

Various authors have contributed insightful ideas on infrastructure recently, including Isabelle Stengers and Brain Massumi ("infrastructure as ecology of practices", Stengers 2013), Brian Larkin ("ontology of multidimensional forms of infrastructure"; 2013), Hannah Appel ("infrastructural violence" 2012; Anand et al. 2018) Tim Ingold ("infrastructure disappearing under the surface or by *becoming surface*" 2017; Ansuas/Ingold 2015), and of course Michael Mann ("infrastructural power", 1983), to cite a few. When I speak of infrastructure, it is in terms of a line of flight of going "before and beyond *oikos*" that proposes to think of *infrastructure* as "matters of care and commoning" (Puig de la Bellacasa 2017; Papadopoulos 2018). My critical line of inquiry here, at the end of this book, leads me to trace infrastructure into the university, into academia, into education. I talk about infrastructure in terms of knowledge and culture work, of its precarization, and that White Collaredness is both a product and an intensification of precarization: While certain forms of elitarianism, privilege, and imperialism of White Collar academia have become implicated, White Collaredness has also intensified. Subsequently, there are many "perversities" in the current (university) system of (higher) education, which comprehensive decolonial critique must and is on the road to identify. In this respect, there are of late several publications emerging that feature already in their titles this very effort. I will not now pick one aspect randomly to give an example, but just hark back to how my previous book's notion of identifying a *digital coloniality of power and Being* at work is – sadly – finding confirmation in the university. Showing that instead of a genuine, caring, care-ful, and response-able *commoning*, coloniality of Digital Culture can occupy, precarize, and pervert a notion such as that of the commons: Pat Lockley's (2018) contribution to Bhambra/Gebrial/Nişancıoğlu's *Decolonising the University* not only illustrates this on the example of the intrusion of the metaphor of "open" into the university context, for example in terms of open access, open source but also drawing on other sources further shows how so-called Radical Left intellectuals either naïvely or – for their own purposes – willfully are complicit in the *digital coloniality of power*, reiterating the point that *digital coloniality* is, as I have claimed (Stingl 2015), just an extension of the already existing coloniality and colonial (cultural) history, and furthermore it is revealed that how the discursive connection to land and "homestead" – aka *oikos* – permeates this discourse:

> Copyright, and therefore, copyleft are both centered on the laws of the United States of America. [...] Openness in copyright (when done through these licenses [*e.g. Creative Commons; A.S.*] is therefore the colonial expansion of US law into other countries as much as the WTO (World Trade Organization) trade deal was and is.
>
> [...]
>
> Hemmungs Wirtén continues to discuss how Lange's work on the commons as analogical to the American West (and the associated acts of colonialism) ties into Locke's notion of unworked land as unowned and how this can be applied to copyright. Locke's thoughts are taken up by Eric Raymond (as mentioned, key to the founding of open source) in his work 'Homesteading the Noosphere'. Raymond asserts that individual ownership over a commons is key to running an open source project, and his use of the phrase 'homesteading' is close to Lange's linking of the commons to the expansion of the united States Westward into Native American and Mexican lands.
>
> The romanticism of the commons is not limited to openness. Hardt and Negri place open source and Creative Commons as a new form of peer production leading tot a revolutionary commons. Moe sees these licensed commons as an example of Lyotard's rejection of grand narratives.
>
> (Lockley in Bhambra et al. 2018: 154)

In this concluding chapter I offer *BluesCollarship* not as a solution, school of thought, method, theory, political ideology, or the likes, but as a writerly gesture and one among other options that is an in-*vita*-tion to think differently together with Others about knowledge work – an in*vita*tion to (re)build *infrastructures*.

Deploying a Dangerous Mode of Being (a Teacher) against the Precarization of Everything: Why We Need to Allow for Vulnerability, Our Own and that of Others ... Especially in and from within Academia

A little while ago, I attended a discussion on perspectives of teaching in higher education, which focused on the theme of the transformative roles that certain texts could and should play in learning in higher education, and what the goals of transformations through text might be. The two discussants were both friends of mine, faculty at an internationally oriented German university where I have taught courses for international and undergraduate students every winter for the past five years as one of the members of the (also in Germany, contrary to rumors of the opposite) growing academic *precariate* – it should be mentioned here that I do operate on a sociological diagnosis that argues that we are currently witnessing what can be roughly described as the *precarization of everything,* a process that is an extension of what Peruvian sociologist Anibal Quijano once identified as the *coloniality of power* and which Maria Lugones extended to critique as the *coloniality of gender,* however now turning in on its own geography of origin in the Global North. My intervention in the discussion between my friends and colleagues was born from a line of thought I only had recently begun to follow based on the American and British campus debates regarding trigger warnings, safe spaces, and the ambiguities of freedom of speech, specifically in academic and other public spheres: the question of *vulnerability.*

My intervention began with an idea about why and how lecturers choose certain texts. The discussants had originally chosen to speak about texts that they themselves found specifically transformative and were, naturally, somewhat fond of; in this respect, they both were hinting at experiencing a certain unease when having students have a go at these texts. But here lay exactly the point of the idea that was forming in my mind, as I was listening to them: What if it is this very feeling of unease and uncanny-ness that should be the explicit substance of the teaching moment? And, if so, what is it that we must make explicit and in doing so allow for?

I proposed that the key aspect around which the exercise of a "transformatory" learning experience by text should be constructed was *vulnerability.* In choosing a text, specifically choosing a text one has a connection with – such as a transformatory experience of one's own – and in perhaps even making this connection known to the students, does reveal, no, it actually produces a teacher's vulnerability. Communicating that vulnerability and, more importantly, allowing it, I argued, is the essential lesson, precisely because it has the potential to create a space that teachers (I consider professors here in their role as teachers) and students share in. And the best part of it – it is something that already happens in many class-rooms quite naturally. So my intervention, really, was more about making the argument that we need to think about this experience that many teachers and students already make, a little bit harder and really flesh out that aspect of vulnerability and its essential qualities, precisely because I think that vulnerability is (a) key to academic learning and practice in general, (b) the important (poetics) dimension at stake in the debates on speech politics on campus and beyond, and (c) the very stake in the regime of the *precarization of everything.* The importance of connecting (a) to (b) to (c) dawned upon me beginning with the aforementioned intervention, but kept progressing through a series of conversations with (junior) scholars, such as my learning form from Hannah Gibson (from New Zealand) who is thinking hard about vulnerability and forms of therapy, and finally at a conversation session with Tamura Lomax which the brilliant Vilna Bashi Treitler had organized and which C. Shawn McGuffey moderated with no less brilliance at the annual meeting of the Eastern Sociological Society in Boston in March 2016 (where I had myself organized a mini-conference on decolonial options for social sciences).

So, before going on about *infrastructure*, I will attempt to initiate a conversation on several zones of importance of *vulnerability*, focusing for good measure on academia's problems with *vulnerability* and being utterly convinced that allowing for *vulnerability* is so important.

The academy today, in its incarnations of the *university* and *college*, is a place that is, to some degree, *not conducive of vulnerability*. Whether it is, for example, the more recent "attack on tenure" or the opposite question whether tenure continues to protect individuals who have been harassing and hurting others, the academy can be quite a political mine-field. At the same time, "academic social capital" is very unevenly and unequally distributed: Gender, sex, race, class, and other issues do play a significant role in its distribution, specifically and most importantly in intersectional terms.

We are, increasingly dealing with this: The *University Implified*.

The *implification* of the university, the moment in the second half of the 20th century, that it enters into cognitive structure of cognitive capitalism, to paraphrase Rosi Baidotti (https://vimeo.com/254360134; ca. 21:30 ff), "the moment the production of knowledge is lost to the university and goes private" (ibid.) which "intensifies" with the development of the new technologies that we consider "digital". It is an *implification* (implication and intensification [Stingl 2015]), because these technologies were – if not already developed, then at least – popularized and proliferated by the very academics in a discourse that was implicating neoliberalism, capitalism, and so on, and the new technologies were actually viewed as a democratizing force against neoliberalism and (its) grand variety of corruptional and corruptive forces of and in the production of knowledge – they were viewed as such, even if, today, of course, nearly everybody claims to have from the start been a critic, a skeptic, and wise to the downsides from the start; they were viewed as a force that would go and implicate the corruption, and while indeed the implications, the indictments, followed, they were met with an intensification of what was implied, not with a remedy, an intensification in the very place, the very location, the very institution from where the implication was enunciated.

Matthew Showering: The Burning of Academic Social Capital and Economies of Relevance of Funding Institutions

In recent years, colleagues and myself have often talked about our experiences with funding institutions in higher education and research, which I have recently begun to re-evaluate these experiences under the conceptual-empirical lenses I call *economies of relevance* and *White Collar academia*. Using these concepts, I have come to the conclusion that while supposedly "helping" researchers with funding them, these organizations often hamper critical research as well as the careers of scholars with the promise to have critical impetus and clout.

> *Two short clarifications regarding this paragraph: By 'critical', I evoke quite intentionally simultaneously the multiple dimensions of the word. And secondly, before anyone starts the whole 'how is it a legitimate use of tax-payer money, anyway', 'how is any research funding warranted', 'academics feel entitled', etc., etc., that discussion is another discussion entirely and I am more than happy to have it – respectively, I have had it, I have said my bits and pieces here and there (including on my blog as well as in the book that I co-wrote with Weiss and Restivo), and I will have it for years to come. But I am after something else here and now.*

My concern is, here, about the issue of accountability and transparency, about a "culture of suspicion", and about the deployment of social capital, which has become a lopsided affair for those academics who are not already under the "golden *Matthew shower*". I am, of course, referring here to the (in)famous *Matthew effect*, as coined by Robert Merton in the 1960s (though hardly the first to notice this), which refers to a quote from the King James version of the gospel of Matthew and basically says that *those who already have will be given even more*. In academia, as Merton and others noticed, those who are already blessed with credentials of some kind will receive more and easier so, whereas those who do not have these credentials will not have a chance of obtaining any. Hence, one can hear people saying that the "first research grant is the hardest, but afterwards…". In the so-called 3P academic job economy of pedigree, (department) politics, and publications this quote can be seen in terms of pedigree (where have you graduated and who are your reference writers or where are they from). As one of many consequences, I have argued, several critical research areas and questions receive little or no attention (let alone funding), because they require "dirty work", which does not match up with any of the 3Ps and is a hard sell. For example, (a) social research questions that are basic research, (b) research questions that are uncomfortable, or (c) research questions that would require developing at the same time a (often genuinely interdisciplinary) line of inquiry and simultaneously new methodologies that reveal how insufficient some of the existing methods are in order to ask some of the harder questions or, worse, that there cannot even be any seriously sufficient methods, because some things just can't be measured. The toughest and yet "shooting your career in the foot" thing one can say is: "Look, there is this major issue, and we really know jack-shit about it". It's an unspoken "no-no" to propose fundamentally new and unproven research like that – and not only for the quite illustrious use of expletive, but it's actually the research we'd need to do. In this respect, I have come to call this kind of "getting your hands dirty research", where the *blue collar* comes into BluesCollarship, and continue to critique and criticize the prevailing culture of entitled, privileged *White Collar* scholarship. There are, of course, also (intersectional) issues of class, ethnicity, gender, race, (dis)abelism, included in this notion, which all deserve to be lifted and acknowledged, but for the moment, I can only work with a focus on the ones I understand best (from the positionality of a male, white, German, transatlantic scholar with a lower class background, i.e. first in my family to even be eligible for higher education, let alone obtaining a PhD), which are the global mechanisms of White Collar scholarship and the forms in which they work the "class"-differences.

Let me say one thing about the "schizophrenic" issue of social capital and academic prestige: Without profiting from it myself, I do represent a form of "prestige coin" for the German and European Research Area domestically and abroad, including the schools I am affiliated with, however loosely. Some time ago, a representative of an institution I enjoyed only a "contingent affiliation" with introduced me – and later explained to me, that this was being said in all seriousness – to an audience as "one *of* the school's luminaries". But as is the case anywhere, being *contingently affiliated* (i.e. not having regular, full-term employment with health and social insurance benefits) makes it hard to receive research funding: Being an adjunct at a minor league college versus being an (associate/full) professor at a major research university means that the former has little chance of receiving funding, often enough they cannot even apply, whereas the latter's chances are comparatively fairly good. Similarly, think about something as simple and supposedly applying for a small-sum such as for travel funding, when going to a conference: Supposedly we are talking about "small-sum" here, but for someone who actually has a tenured, full-time academic job this small-sum (is still a good chunk of money, but it) doesn't register as measured in "months of health insurance payments" as it would for an adjunct. Some conferences, kindly, offer registrations fee rates for un-/underemployed participants, although some don't and, interestingly, they don't like to be questioned about reasons – as I have tried to find out what reasons could be, while citing that as a member of the organization hosting the conference, I should expect some transparency about the cost structure. But when it comes to applying for travel funding – and travel we must, because presenting your work and networking are the only ways of "un-contingency-fying" your academic career – with funders, one can again count on the fact that a funding institution is likely a *Matthew shower*. Most likely, funding is given to people who are at high maintenance universities, i.e. those who already allot some travel funds to their faculty, not to applicants who actually need financial help, for: "How dare they?"– yes, funding is a reward for being (already) "and part of the "awesomeness of a prestigious elite" – by "awesome" I mean, of course, "relevant": But really, think about it, it makes sense on the part of the funders, they need to show to politicians, tax-payers, wealthy alumni, and other stakeholders, that the money they are in charge of is given to people who visit places. And how can you guarantee that people visit places? By choosing people who already are at places. So, they don't have that much to give to people who are in "no-places". Of course, there have to be a few poster boys and poster girls to deflect criticism, but even these are, often enough, already somehow and strangely "in the game", as well as some programs that are "charitable" and simultaneously maintain the illusion and ideology of "meritocracy" while making the "recipients" continuously feel that they are just that: "recipients".

Now, in many ways, we could just say, "why even give a shit?" Maybe, because those who are not *Matthew Showered*, still try for funding, and have to – even knowing how small the chances are – continue to believe in the research they are doing, continue to believe in the important difference their research could make, and because "we (all) want to believe" that at least every now and then a worthwhile project can get a win, and that it is not only risk-free "career-trajectory observant projects". And there is this one issue for BluesCollars that concerns me here, and that is that these funders require to burn a lot of (academic) social capital in the process of applying for funding. I use the word "social capital" a bit freely and unspecifically here for a sociologist, but it will become apparent what I mean.

"Only" an anecdote of course (but how many would raise their hands with similar experiences[1]): Not too long ago, I applied for travel funding to a large conference where I was presenting a paper as well as having organized an entire mini-conference with nine panels, several of which I am also supposed to preside over; of course, the mini-conference was on Decoloniality and involved aspects of the Global South, and my own paper was critical of the Global Northern "economies of relevance and precarity". Did I get funding? No. And normally, I'd be a bit disappointed and that'd be it. But there was something else in the process, and that's something I find increasingly typical in these affairs, which I find worthy of some hefty criticism, precisely because it displays one of the sources of the inequalities that make the *Matthew shower* so unfair and perpetually so: I was asked to deploy some of my social capital in order to even be considered eligible for the travel funding, and I had to spend it needlessly. After being accepted into the conference program, I sent the funding organization a copy of the program since, by their own rules, all that is required is an official receipt of acceptance (what's more official than being on the program?) and only in certain cases a personal letter of invitation (Mind that if you are receiving that funding, you will also be reimbursed only months after the event itself, no one cares if you can afford to travel otherwise in the first place, duh; and you have to give proof afterward that you actually attended the meeting – had I received it, I would have been tempted to include pictures). Now, attending a large academic association's conference, even as organizer of a small sub-conference, does not normally mean that someone will just write you a personal invitation for laughs – that should be common sense, even for someone working for an official semi-*governmentous* funding organization. However, that personal invite was what I was told is required of me. So, I headed out (electronically) to ask for some personal favors of people I knew only by mediation through other friends (and that already puts me in a privileged position of sorts, I have worked hard to have some friends who know people who could help here). And, by kindness of strangers, I got my personal letter of invitation – at this point, I should also mention, one of the speakers for the mini-conference already bowed out after he did not receive funding from, I was made to understand, the same organization; subsequently, I had to ask for program changes (not the last, for similar reasons each time: lack of funding) to be made from the same people who kindly provided me the letter. You see where this is going and how not receiving the funding is then increasing the value of the academic social capital that I burned with these very kind people. And that is one of the many aspects we often do not talk about in the "economy of third party funding", that there are several forms of social capital at stake, and that the academic social and cultural capital of BluesCollar knowledge and cultural workers gets burned more often and with more severe consequences than for the already privileged White Collar academicians. Let alone that, to begin with, BluesCollars entangled with the academy often have to work harder and in far more fields and contexts to make a living and keep an open profile (to apply for a variety of positions), only to be told that it makes them a harder "sell" on the job and funding market because either they seem "unfocused"/"undedicated" or "scare" reviewers/committee members because they can do so many, many things – well, they are, of course, (contingent) "luminaries", after all.

Some research funders ask very specific details about "research populations and where to find them". While they will keep the information certainly confidential, for most of them ask you to have elaborate "data management plans" appended to grant proposals. But here's the issue: you have to find, groom, and maintain a research population and its gatekeepers. How is that supposed to work, if the decision on a grant takes up half a year or more (longer if you have to revise and resubmit, which for some funders is now the usual procedure as everyone knows that with some "you never get funding before the second round and if, only in the third, really")? If your chances of getting the grant are small? If you have to try different funders (and aren't even allowed to submit simultaneously to several)? Then the point is you can only seriously groom a research population if you don't have to depend on the funding to do the research in the first place. So, anyone who already has a well-paying job at a good research university, anyone who is already "in the game", is already enjoying an advantage over anyone who hasn't, because they can just count on the fact that they will do their research, no matter what – funding for them is rather a matter of scale not of "if". It's also easier to work on a White Collar issue with a White Collar access: So an Ivy Leaguer (Not to be misunderstood: not all Ivy Leaguers are White Collar academicians, not all White Collar academicians work at an Ivy League school; it is actually quite surprising that one can find the White Collar mindset in so many odd places, and often precisely with people who by any right shouldn't support it) who studies social inequality, using a social welfare program run by a church affiliated with a charity funded by an alumni of another Ivy League school or a program supported by the government and represented by an official with academic credentials, will have an easier time gaining and maintaining access, than someone currently working at a community college, who is trying to recruit random people in a poor neighborhood to "tell it like it is". Of course, these examples are a bit simplified, but they flesh out the contrast well enough, I think, and they are not too far from the legion of conversations I had with scholars and researchers over several years. Not too long ago, I approached a person who does an interesting and effective (but very Blue Collared) project combining social developmental work with martial arts training. His response to my approach was that he would love to have research done around the project, but would only consider allowing me to mention his name once the research funding was already granted, since he already gave his name several times to other researchers for the funding process (none of them White Collar academics, from what I understood between the lines), and no one ever got the funding. This is precisely my point, since with this very involved, energetic project, a lot of social capital has already been burned: In order for a researcher to apply for funding, his organization would need to be named as a research site with any major funder, but something he is no longer willing to do. In the same way, I could ask: How many times will people do people favors in writing letters of invitation, recommendation, and so on? With regard to letters of recommendation, I have repeatedly stated that these are a means of maintaining the current inequalities in higher education: If you are not already at a highly recognized university, and thus endowed university (many of which often aren't as good as they pretend to be, by the way), you will have to be on the job market very aggressively, meaning a lot of applications with a very low chance of success and a lot of letters of reference/recommendation (by writers the more prestigious – and thus busy and disinclined – the better) that need to be written and send off. Yup, social capital at work.

Regardless of whether it is about the decision who will be hired for a job (when often the short list is already clear before the actual process starts) or how a teensy bit of travel funding is decided, applicants are always told that "decisions can't be explained" to them for xyz reasons. However, applicants are asked to be transparent and to help make sure that the funders and their officers can do their jobs in ways so that they can be considered accountable; only, it is the applicants who have to burn academic cultural capital for it, and if they are BluesCollar knowledge and cultural workers, the value of social and cultural capital they have to burn in order to do so weighs heavier than it does on privileged White Collar academicians who already have higher chances to score the jobs and grants, because they are already under the *Matthew shower.*

As Lyotard said: "I am interested in remaining a child"

Universities and related institutions dedicated to higher education and/or research have been and continue to be for the foreseeable future subject to a social and political regime bent on demoting, demolishing, and de-intellectualizing them. This is conducted both from the so-called Rights, Lefts, and Neoliberals. This regime of demolition has explicit and implicit support both from within these institutions – let's go on to call these *neoliberal academia* – and from populations without an apparent stake in academia – let's call these, broadly, *anti-intellectuals.* It's done both in and under various names such as the names of progress, freedom, equality, profit, and austerity. It's both conducted in practice and given warrant in various ways, too, which, however, both go against many of the pillars of academic discourse itself and its particular sub-fields (largely those so-called [mistakenly] "soft", "not-STEM nor business" fields). Reactions by "intellectual" academics have varied, but more often than not, those already with a membership in academic institutions with a job or with a serious stake in one seem to resort to choose strategies that are of a collaborative nature with the regime of destruction – regardless whether they chose the passive strategy to merely duck and hope not to be noticed if only they keep their mouths shut, a passive-aggressive strategy that means saying a few snarky things but overall don't resist actively (and make it clear you won't), or even contribute actively to the destruction (in some schizophrenic cases, academic actors have literally talked an intellectual talk while walking the anti-intellectual walk). However, this destruction of higher education and research is undermining the future of our societies, specifically the future of and in the Global North, and, as a consequence, for some – though not all – dependent institutions, social groups, and countries in the Global South.

Before going deeper into the issue I am trying to elaborate, a few things should be said about the *geopolitical* aspect of my intervention. While I do understand that some critical voices from or concerned with the Global South, including the decolonial discourse of which I am a supporter, will have good reason to say "Why should we care?" and may even claim that such self-induced demise of the North (by means of causing its own "brain-death", so to speak) might be well deserved, I would say that one could argue that this will not change the global situation and thus the situation for the Global South for the better but may make it worse; it will also not create any kind of justice in the Global North itself, since there is a "global South" within the Global North, i.e. people who will suffer from this, who are in many ways except geographically closer to the South than the North, and it is not just in terms of unfairly punishing future generations. My own approach here is, precisely, about the need to *provincialize* the Global North (as a geopolitical location) and the West (as an epistemic system), which means to strip away the North's privilege, entitlement, and the (epistemic-)imperialist *claim to* universality. Tentatively and somewhat simplifying but for the present purpose adequate, let's say that doesn't mean, for example, that knowledge produced by the North in the Western form of knowledge practice cannot be used in a generalized fashion, but that it cannot consider itself as an unprerequisited foundation. Also, and most importantly, I am not defending a mere return to any of the classic *University* ideas – from Humboldt to the Neoliberal University, or else. However, I will certainly claim that we can always learn something from these various ideas and the reasons that were deployed from their inception. But in understanding these reasons, in considering which of their goals and ideas were and still may be worthwhile as partial connections and occasional ideals to try out, I would still say that we must recognize and understand their failures and acts of violence, but recognition also means owning them and owning the responsibility for them, and to not be silencing them (it is here, where I do think many contemporary more aggressive debates on "safe spaces" run into dangers – but this will have to be another discussion for another day). That said, I am, after all, a proponent of new ideas for *re-creating* the University as a *pluriversity*, but the current and recent destruction of these institutions, I am afraid, not only does not help facilitate such a re-creation, it also is counter-productive to it. The destruction of these institutions will result in a remaining type of a University and higher education and research system, which is with regard to the general mind-set that both governs it and that it propagates and instills in its students a much worse universal *monoculture*.

A dozen or so years ago, I first articulated a basic *infrastructural* tenet, that I have not abandoned since, namely that any larger territorial state, its society, and national or trans-/supranational organized collective have to continually and seriously invest into and own (in terms of guaranteeing freedom from external control) on all its members behalf its infrastructure, and by *infrastructure* I mean, roughly but in that order of importance (knowledge enables the others), the following four (partially interlocked) areas:

- ***Knowledge&Information***, including the (a) child development, school, higher education, research, science broadly defined (including not only technology/engineering, but also all "natural sciences", social/cultural sciences, arts, and humanities); (b) knowledge and cultural(!) labor (structures), and (c) information communication channels, which refer to guaranteeing (c.1) that there is no political (including no economic) monopoly/hegemony over knowledge and information distribution and (c.2) the technological pathways of information communication (phone lines, "wave spaces", etc.).

- ***Health Care and Aging***, to guarantee equal access to all its members to all sectors of the organization/society.
- ***Traffic***, to guarantee that passengers and cargo can be transported quickly, safely, and in ecologically sustainable ways from point A to point B.
- ***Security***, by which I mean police, clandestine services, penal and rehabilitation system, forensic psychiatry, and military.

Something like "*the* Economy" is precisely not part of the infrastructure for a number of reasons.[2] Including among those reasons, which I cannot, of course, go into detail here, are that the economy – precisely because some parts of it consider themselves(!) structurally global, some local, some constantly transition between both – cannot be an infrastructure, e.g. that the subsidy structure for businesses established is often hampering, negatively affecting or in some extreme cases close to destroying other elements of the infrastructure (which precisely is conceptually and empirically prohibited for an element of a structure to be part of its infrastructure); most importantly that infrastructure should be able (encapable, empower) to *enable* and *constrain* actors to participate in the economy on fair and equal terms. Here, one could, for example, also use some philosophical proposals about distinctions between different orders/types of goods, and remind people that there are types of (social) goods that exist to enable the possibility and availability of other goods. In addition, unlike some overzealous, pseudo-scientific Wikipedia-style definitions, we should understand that, "No!", infrastructure is not what is needed for "economy to function", but it is the term for the explicit structures and institutions that sustain an organized collective; an/the "Economy" can be such a collective itself or considered a sub-system of a higher order collective/system, e.g. society, but it is precisely not (in either case) part of the infrastructure. In this respect, I do think that, for the foreseeable future, investment in a military is important, but I also believe in the idea that a soldier is a "citizen in uniform" (German: "Bürger in Uniform") aka in service of people, and I do believe in a very demanding concept of citizenship as a form of belonging (although and yet precisely, citizenship is a contested, very particular, Western form of belonging, and far from the only one, but both an interesting one and for the Global North, a constitutive one). These latter aspects (security and restricted citizenship), which I also will not unfold here further, do not sit well with some Left-oriented colleagues and interlocutors of mine. But mine should not be mistaken for the naive views many self-styled (structural) conservatives hold. It is precisely through the provincializing of a concept such as "citizenship" that its (zones of) usefulness and areas of in-applicability, as well as its restrictions can be made to show. By restriction, I do mean, precisely, that being born into a territory or to parents of a specific nationality should not make one automatically a citizen, because with citizen rights come citizen responsibility, which a citizen must be able to fully understand and willingly accept to fulfill. Citizenship, if viewed as a relation of belonging, participation, and provisioning, exists in the relational framework of *care power information*, and, as such, should be characterized by *response-ability* and not be constituted in national sovereignty and Euromodern subjectivity/personhood. Again, this will need to be further explored in future work, but it should be clear that I view it as a state's duty to enable all the people, either living on its territory or claiming another form of (justifiable) belonging to it, to become a citizen; and this is, clearly, an aspect covered in the "knowledge&information" infrastructure. Furthermore, in the "hierarchy" of infrastructural premises, it should be said that different temporal priorities mean that some individual demands render the hierarchy "not fully transitive": It should be clear that, while generally an investment in research should always be considered more important than an investment in the army, in times of a genuine external threat on the horizon, an "informed decision" would lead to a military investment; but clearly, an "informed decision" requires people who are knowledge-able and who have access to knowledge to make these kinds of judgments in the first place. And, of course, the elements are interlocked: Security knowledge occupies a place between or in both sectors. But this also already explains why, if we take infrastructure seriously, blindly investing in weapons and surveillance technologies cannot generate more security: The issues and threats security deals with are very complex and so must any solutions be that aim to be more than a temporary, quick fix – here, such as in other areas, I argue that concepts such as *sustainability* are interesting intra-system demands, for which the knowledge sector has the function to elaborate what the system-specific *criteria* would have to be that define what "sustainability" would mean in each case. This makes, however, also clear that there is no universal idea of "Sustainability" (Capital-S), and "sustainability" is dependent on infrastructure. In this respect, I also argue that concern for "the environment", i.e. what environmentalism practically means (theoretically, also, because of what we have learned about the fact that there is no mere nature nor mere culture, but rather naturecultures), also depends on infrastructure. A final matter that should be mentioned, I think, is that the common idea of "the Economy" is also deeply thwarted, for example since the theoretical premises of its science, economics, in many of its contemporary schools ignore at their own peril many historical premises it was built on (for example, it still upholds some of Jean-Baptiste Say's ideas but neglects the constraints these rested upon, for example on the very restrictive functions of money); at the same time, there are different players/actors in the economy who are not comparable nor even commensurable, when the idea of a "free market" is applied – in part, the problem lies in what is understood as "free", because the often found assumption that "freedom" (or "autonomy") could be something that is unconditional is, as I think should be tragically obvious, both conceptually and empirically a *contradiction in itself.* For many so-called multinational corporations, a genuine *free market* would actually be in opposition to their goals and functions. This is also why infrastructures help make "free markets" possible, whereas "*the* Economy" is not co-extensive with "the (idea of) Free Market"; quite on the contrary, there are sectors of "*the* Economy" that operate either despite of or antagonistically toward the Free Markets, including a disposition against the infrastructures it is pre-requisited by. Therefore, in a "weak" account if my argument has merit to at least be taken into consideration as a possibility, then we should be careful whenever considering whether any aspect of "the Economy" should be put in charge of infrastructure in any way, and look for justifications in each individual case. In a "strong" account, it is clear that "the Economy" (nor its "actors") should have no place whatsoever in infrastructure considerations of any kind. Here, I would argue that any critical, reflective reader, even if politically disinclined toward me, should be able to at least be willing to allow for the weak account. Not to do so would mean to take an extremely ideological (and fully destructive, antagonistic) view. What I will have to say in the further discussion here should be acceptable to consider from the position of the weak account, although I will freely admit that I, personally, lean more toward the strong account, although I do think that an intermittent position is, at least in select cases, pragmatically possible, where – within limits – convergence/contact zones can be established between infrastructural and Economic sectors.

So to situate and make one long story short, knowledge and cultural work and the institutions that enable and carry knowledge work are part of the infrastructure, which *enables* economic sectors (and not the other way round). Introducing epistemic monocultures, such as the Economic paradigm, into the infrastructure has, thus, a potentially negative ("weak account") or self-destructive ("strong account") consequence, which may/will subsequently have negative/destructive effects on the higher level of the organized collective.

One of the many problematic consequences of mono-culturalization is an increasing lack of variability: We end up increasingly with the same types of knowledge, and even what could be considered "out-of-the-box"-thinking or alternative thinking is already predetermined in terms of how it must be expressed; in other words, it is already pre-established what is allowed to count as an alternative. Everything that is neither official knowledge nor officially recognized as alternative knowledge is not considered knowledge and effectively silenced. This state of affairs is, of course, nothing particularly new. However, even this type of critique has, in the meantime, become very formalized in the way it is allowed to be expressed. Establishing genuinely alternative discourses is extremely problematic. Our problem with constructs such as any so-called subaltern is that these are often part of a hegemonic discourse themselves. In other words, this picture that there are two connected processes at work – whether in society or in a knowledge economy –, namely inclusion and exclusion, is insufficient. I have argued this point elsewhere in more detail. But the gist of this argument is that in order to be excluded but to have a claim toward inclusion or toward expressing one's exclusion and the consequences of exclusion as one's official status hinge on a form of recognition that not all agents (in a collective) or all forms of knowledge (and cultural labor) enjoy. Besides being an included member or included by presenting official knowledge, "membership or knowledge that is recognized as excluded" is still formally a kind of knowledge, namely a knowledge that is excluded but may be allowed to make legitimate claims toward inclusion. My – unfortunately empirically quite pressing – example is the question of who has access to social and/or health services, when access is increasingly dependent on (digital/electronic) information and communication technologies (ICTs), with no more non-ICT access paths available. An important part here, as at least a few studies have shown, is that providing access to (and access to competencies to use) ICTs alone does not suffice to guarantee that all excluded people will be able to use these services. But here the problem really begins: We have people included in society, who do not need to use social services for the excluded, we have people who are considered excluded but who have a claim to these social services, and what about people who – for whatever reason – cannot express their claim, i.e. cannot express their state of exclusion, people who are effectively silenced? Second, how do we generate and express this third kind of knowledge about these silent and silenced people, when knowledge is considered knowledge about inclusion, alternative knowledge is considered knowledge about exclusion? What about knowledge about those who do not exist in the exclusion-inclusion system, those who have been silenced by that system? Making these people heard, giving them a voice, helping them become "being known" in the first place are difficult tasks, precisely because (a) they "do not exist" in the permitted way of being excluded, and (b) "giving them a voice" is no less a colonial, imperial, hegemonic act that denies those who are silenced any agency. For any researcher the second part (b) is an ethical issue that we can only in the future resolve by working together with one another; however, when only "working on" knowledges about these silenced people, it is (ad a) very difficult to be heard by other researchers, by policy-makers, or by the public, precisely because it is very hard to proof the existence of a problem (or of people), the expression of which cannot be made in system-permissible forms. In various types of conversations and communications, I have made the experience that there are numerous experts, specifically among social scientists, who work on health care, yet who completely deny the existence of "people who are silent", i.e. of people who are not counted by the system as excluded but that are effectively invisible to the system. For these social scientists (and many policy-makers), if people are not considered by the system as existing as excluded, they don't exist and, thus, there cannot be any knowledge about them. Whereas some people, whose job it is to actually collect official or commercial data for care systems, admitted to me that there are indeed people who actually exist but whose data – even data about exclusion – could not be collected, and who, thus, do not appear in the system either as included or excluded. It becomes a further issue that without acknowledging these people as well as knowledge about them, the problem will only proliferate and expand. But the fact is that people still are materially present in the world and will require care services, some of which are still provided in *(material) "vital" zones* that serve as *boundary/transitional/contact zones* (for example emergency rooms), but the services are then sometimes denied, sometimes delivered but not accounted for. This lack of accountability will, eventually, be countered, but the question is whether by more denial of service to any person without documentation or by effectively allowing for more comprehensive knowledge production, i.e. the permission to acknowledge other(ed) forms of knowledge and, subsequently, changing the system toward becoming more *pragmatic* (in the actual meaning of pragmatism). Otherwise, exclusion and silencing become, thus, increasingly excessive.

In the system of knowledge and cultural work, we see a similar effect of exclusion and silencing mechanisms, for example through the proliferation of certain, very specific styles of academic habitus and academia specific forms of symbolic/social/cultural capital. While I do not disagree with the proposition that science (and I include the social/cultural sciences, liberal arts, and humanities here into the general concept of "science") should be somewhat "useful", I disagree with the complacent and/or lazy and/or anti-intellectual attitude that many people have when it comes to the meaning of "usefulness", as well as the impoverished (and increasingly shallow and techno-reductionistic) understanding of – not only what science is, but more importantly – how science works. There are numerous infamous caricature versions about scientists and even more so about philosophers (as the archetypical arts and humanities scholars), which are all versions of the archetypical story of the Thracian servant who was laughing about Thales of Miletus, when he fell into well because he was thinking without paying attention to where he was walking. This is seen as illustration of the impracticality and, thus, uselessness of all things (too) intellectual. In this respect, the ideal of useful science has become the technological or engineering ideal. The problems with this set-up, however, are, first and foremost, that in the minds of most people, the engineering ideal of science and intellectual forms of scholarship is considered as antagonists in general, and, particularly, as rivals in a zero-sum game over scarce (and ever scarcer) resources – thus "usefulness" becomes equated with impact maximization of minimal resources. While this may fit with a "neo-darwinistic" worldview, we do know that "evolution" does not work in purely antagonistic ways, but that many (if not most) processes work agonistically and/or symbiogenetically. It is important that we understand that we need different types of knowledges that challenge each other, but not in antagonistic destructive ways, but in constructive ways. While also a cartoonish caricature, but the simplified idea that science can tell you how to clone a tyrannosaurus rex and humanities can tell you why that might be a stupid idea is an interesting illustration – even if you've never seen the cartoon, of course you can guess that the scientists get chased and eaten [and yes, of course, there are reasons why we cannot ever clone a dinosaur; although a lot of people think we can, and here, too, it is alternative types of scholarship that could teach us about how communication works between science and its publics]. Science and technology have never worked in straightforward ways. To begin with, there is this general idea of (universal) "Progress", which is still "publically" quite popular, which many "engineering"-types of scientists also still believe in, but which many critical scholars have long viewed as debunked. And yet, we could counter and say, there may be something like *progress*, but what counts as *progress* is always dependent on a variety of contexts, relations, criteria, and so on. The same is true for ideas of *usefulness* or *sustainability*. And these criteria and knowledge about relations and contexts cannot be developed from within the engineering-ideal itself, because these are the criteria by which the engineering ideal is judged. In a sense, you do not ask a group of professional basketball players what the ideal height for a kitchen table is; what you really want is – actually – a kitchen table with an adjustable height. While this is, of course, both simplifying as much as it is hyperbolic, it serves as a provocation worth thinking about.

But in terms of intellectual knowledge production, the problem that knowledge should be more than just useful but be actually effective, it is being required to take into account varying temporalities; and that is something that intellectual scholarly disciplines are actually good for. And, to uncover these variabilities and express their many potential effects and options requires a lot of work by a lot of people over long periods of time, not to mention many conversations between different disciplines and different ways of thinking and doing. So it's not about the "engineering"-ideal being bad (or good). It's about the "engineering-ideal" having almost already become the monoculture of how all science is done: Knowledge, to be considered knowledge, must be expressed within the terms and/or structures of this (linear teleology of) ideal of engineering, including any criticism and alternative having to be expressed in permissible form – which means critique of any kind must be either construed in engineering-ideal or antagonistic terms and/or structures. But this denies many effective and/or novel ideas to see the light of day. Fewer types of knowledge exist as a consequence of the current way that science is done. Think about the data collection problem and whether it is a problem at all. When would the fact that people are not considered (as included or excluded) in the system become a problem for the system (and not only for a few individuals)? It's a matter of the scale of the problem. But that's the very point; we don't know the scale of the problem, because we do not generate knowledge about it. But this is precisely the point: Science should tell us what it is that we don't know and help us to turn what we don't know into knowledge. But that's not how science is done anymore. Science deals more and more with what is already known: This is precisely the gist of the "engineering ideal" and how it has influenced the funding of research and scholarship. If more and more science requires to be funded by external and/or taxpayer funded institutions, and if this kind of funding seeks to allot its funds according to a blurry concept such as impact, then this way of "doing science" cannot deal with the unknown, precisely because "science of the unknown" cannot say in advance what it's impact will be. That is why anything unknown is an intellectual problem to begin with – the kind of problem that doesn't get funded in contemporary society. As a consequence, scholars from the humanities and social sciences either corrupt their scholarship to look like it fits with the engineering ideal or they opt to express criticism in the known forms of criticism and propose known alternatives – the recent resurgence of vulgar Marxisms (or should I rather say Marxist vulgarity?) being a good example.

I want to pin these last points down to an analytical point I have been making repeatedly, namely that one effective analytical tool to deploy is the notion of *economies of relevance*. I posit that these vague concepts, notions of "impact", "usefulness", and so on derive from what I call *economies of relevance*. Of course, "relevance" is itself rather broad, but it seems to me that this is the kind of umbrella concept that very neatly and comprehensively describes what's going on with these other vague concepts that are often deployed in actual situations (for example, when explaining your "impact" in a research grand proposal). So, the types of questions we need to ask are, for example "What does it take to seem relevant today?" There are interesting examples, for example, why do politicians both talk so much about "*the* Economy" and chum up with business-folks, and appear to look themselves more and more like either MBA-trained managers or corporate lawyers, rather than people who tackle by means of policy-making the social problems of the communities and societies they have been given responsibility for? That is, of course, because one of the main metaphorical vectors has become the economic angle, or maybe a kind of "business accountant"-speak or "business imaginary". Just think about how austerity politics is presented as "household discipline" on the one hand, while one of the severest moves – but no way the first, nor the most historically meaningful one – toward austerity politics was made following a financial crises but was made hand-in-hand with a politicians saving financial institutes and, subsequently, watering down legal consequences (both punitive ones as well as future-oriented regulative ones). Politicians who do not seem "business-talk"-savvy are hardly taken seriously by both their constituents and by "business and Economy" representatives. With voters, the problem is doubly complicated, since voters expect their preferred politicians to be able to both be business-savvy in order to "not waste tax-payer money", i.e. by keeping household discipline like a good common person or "housewife" would (maintaining the mythology of the *oikos*), in a sense, also "investing smartly", and do as much "useful stuff" with as little of *their* (tax-payer) money as possible, while at the same time being able to both sweet-talk and tough-talk to business leaders on an eye-to-eye level. So, in order to "appear relevant", political workers (be they politicians or other kinds of political workers) must seem to be able to "talk business". But the more that politicians buy into this *relevance economy*, the more we find a monoculture (of economy-oriented minds) take control in terms of how problems, their origins, and their solutions *can* be imagined. And the household-metaphor is one prime example within this monoculture (of *oikos*): Most people have become accustomed to think of "the state" as national government in terms of a household; this or similar metaphors are often dragged up, especially during campaign time: It is noteworthy that in some political geographies, such as the United States, there is now almost always "campaign time" (and "the state" is, thus, conceived and re-temporalized accordingly), there seems to be no more space for actually acting in the political: In the United States, the United Kingdom, and large parts of the European Union (EU), as a consequence, we encounter a Global North where we find effectively almost all political action replaced with *politics* – anyone who doesn't do *politics* or acknowledges that's how things are *actually* done is called out as "very naïve". In short, the only *realpolitik* in town is *politics* (in the negative definition of that term): This *realpolitik* only functions because so many believe in it, while many *actually real* problems remain unsolved and a status quo of the political structures, including insufficient infrastructure, is preserved – in *The Digital Coloniality of Power*, I mentioned a distinction between *structural conservatives* (to which both the Left and the conservative Right belong, regardless what name they give themselves Democrats, Social Democrat, Libertarians, Republicans, Brexit Parties, Labour, Tories, Tea Party, Christian Democrats, etc.,) and genuinely *value conservatives*.

Knowledge and cultural work are important and necessarily receive recognition (and, thus, fair resources and adequate *infrastructural power*) in a *value conservative* political framework, but not in a (in today's) *structural conservative* politics framework. This has many reasons, but one is that for politics (and the Economy) to sustain on a larger scale its hegemonic status, knowledge and cultural labor must be over-powered by politics and the Economy, for otherwise it would be revealed how they are only "efficient" in view of their internal problem constructions, but ineffective in helping people overcome or deal with their actual (political, social, etc.) problems. *Politics* and *the* Economy are about managing (*oikein*) problems, *the* Political is about overcoming problems, while knowledge and cultural workers stand in a mutually constitutive and interdependent relation to *the* Political. By forcing politics (and its inherent rhetoric, poetics, and economies of relevance) onto knowledge and cultural labor, this work becomes ineffective and, subsequently, infrastructure suffers. There are various aspects of this development that inhere in very long-running structural and metaphorical mechanisms that derive from our thinking in terms of *household-as-oikos* - this relates back to the Global North's ancient Greek intellectual-political origins, which sort of shaped the operating software that our (Western/Northern) societies run on.

If the *economy of relevance* had a currency, I think it would be "attention". Not for nothing is one of the main areas of medicalization, including the medicalization of childhood, constructed around the idea of *attention deficit* (aka ADHD). Attention medications are even deployed by the demand of parents wanting to help their children be successful (and "not losers") in life, in cases where there is no ADHD clinically indicated ["Good intentions..., and so on, no?"]. Not for nothing do we say "Pay attention!" While this "command" may seem like a mere accident of language, I do not necessarily think so, because there are some social and science historical aspects to the rise of "attention" as a concept of this social and clinical magnitude (which I explore[d] elsewhere), and there is this general treatment of attention in the education, cultural, and labor structures we build, that treat it like a scarce resource. Indeed, I think what we see, instead, is that "attention" is constructed like a kind of scarce currency – it is, in a way, the actual gold of the gold standard for and within the *relevance economy*. An *attention economy* seems to be the main type of mechanism of the economies of relevance (I will leave out here the certainly important discussion, in how far this notion of economy with regard to relevance and/or attention should be conceptualized in term of *political economy*). But doing knowledge and cultural work, I think I have come to understand something crucial here: We should – careful: irony! – be "paying more attention" to the *ecologies of attention* and not the *attention economy;* who knows, perhaps *the* Economy is really only *the Attention Economy* (*General*). I am not the first, of course, to speak of "attention ecology", but my point is actually about different, diverse, and variable *attention ecologies* existing simultaneously, while there is only one "type" being constantly co-produced and given hegemony by the *attention economy*. This is something of a shift of focus from *attention economy* to *attention ecologies* that I have been advocating for a while now. I will, however, have to trouble and complicate this distinction somewhat, because we really must understand that we inhabit always different *attention ecologies* simultaneously, and these *ecologies* and *economies of attention* shape each other mutually (largely through temporalizations and spatializations through practices; practices being the key issue in my theoretical and methodological works, which is why I am not going into much deeper detail here, since this is not the place for a long theoretical exposition). Think of it this way: The world around us, with all its lovely complexities, affords us a lot of opportunities that we could direct our attention to: a lot of things to perceive, manipulate, interact with, and so on. There may be an interesting plant, or a plant that needs watering, there might be a trail of ants, there is a car driving, which you hear or see, a cell-phone is ringing, a screen shows an advertisement, etc. But most of us are not really universally receptive to everything that is afforded to us in these manners simultaneously (or even ever at different points in time). Some people may never be able to register the ants marching and always only attend to moving macro-objects that might be dangerous to them and primarily the "objects of technological civilization", whereas other people might be constantly distracted by all the little creepy critters and greens around them. Of course, this description is a gross oversimplification, but you get the idea that different people are geared toward existing in different *attention ecologies* that exist around them. Furthermore, in societies and collectives, we do assign particular values to certain aspects in our ecologies, which are co-related with certain functions and certain practices (again, let's not do too much theory here). The point is that, as we continue to fill our world with screen-based technologies and interfaces, more and more people will attend more and more to screen-based affordances and practices and other attention ecological affordances will be part of what they would even be able to understand as a discernible part of their ecology affording perception or interactions with. So, the more certain actors in a society understand, create, and manipulate social practices to function according to the idea of an attention economy, the more the attention ecologies people can live in are transformed by this economy. So, while companies struggle for your attention with advertisements for you to buy their products, considering, qua economic paradigm, that your attention is a scarce resource, they will not only compete over existing advertisement spaces but also attempt to create or conquer new advertisement spaces. Think about a very simple example: 30 or 40 years ago, a restaurant customer would have looked at a menu and among the various drinks have looked for drink groups and ordered the drink group s/he liked, nowadays instead of drink group one regularly reads one brand name after the other, often presented by the brand logo which is also often replicated in various other places in the restaurant (napkins, shades, etc.). More recently, one can see this gone further at many airports, where restaurants have tablets on the tables which – besides menu information – run advertisements. Now ask yourself, have you ever observed insects that live in an airport? Or considered the – sometimes actually visible – fungi that live in the various cracks and nudges of the table, floor, and wall of that airport? And, be honest with yourself, if you were sitting in an airport restaurant, what would you really think of a person who would notice these things and point them out to you? Now, consider your responses to these questions and why you think these are your responses, and think further, what do you think would the responses of the next generation of people be and the generation after that, given the way the "world continually transforms our immediate environments"? The point I am trying to make lies, here, specifically in terms of *doing critical thinking* (and hasn't *critical thinking* become a wonderful slogan for education reforms of late??? "We need to teach people critical thinking skills, yeah!"): We need to shift our view from *attention economy* – from thinking about attention as a scarce resource – to the *attention ecologies* in order to understand (before we can think about economies at all) what is afforded to draw our attention and why and how these ecologies transform in the ways they do. Again, this matter could and should be further complicated on various theoretical and methodological levels (for example through French philosopher Gilbert Simondon's ideas about individuation and milieu, to name only one). But for the purpose of illustrating the most critical issue with the specific way "attention" works when it is constructed as the currency of relevance, this should suffice for now. What should also be clear is that, to a large degree, we cannot afford to be oblivious to what's going on around us. The simplest example is how people have accidents because they "don't pay attention" to the street but to their smart phone. Think about how we attempt to solve these problems: We urge people to "pay attention", yes, but in terms of technology, we look for the driverless car. However, while I think in the distant future, driverless cars will be a fabulous thing, this is not the solution to the problem. Instead of "paying attention to" we should "be(come) mindful of" *attention ecologies*. An important aspect here is, too, that if we want to build a driverless car, we need to make these cars function in ways that don't just pay attention to the street, but that are "mindful of" far more than the elements that we think that attention should be paid to. That's what creating artificial intelligences (that can run complex aspects and, thus, are becoming part of the collective organizations we call societies) means – don't kid yourself: a driverless car is an entity[3] exerting social force and causes social transformation, the same way that the introduction of trains did. Carving these complexities out and thinking them through, that's a lot of knowledge and cultural work that has to be done. But instead of recognizing the many areas where (massive) knowledge and cultural work has, is, and will have to be conducted, the current mood is to diminish the relative contribution of these forms of labor, to diminish their recognition, and to diminish their remuneration (including, above all, the financial remuneration). While there is, actually, an increasing need for knowledge (and cultural) work, the official demand for knowledge work is strictly regulated and diminished in terms of the economy of relevance, leading to a continuous *precarization* of knowledge workers.

The "impact" of the *economy of relevance*, remaking for example academia into one of the *economies of relevance*, on knowledge work can be grasped in many phenomena and illustrated in many examples and analogies. But whether expressed in fashionable vague terms of "being useful", "impact", "efficient", "sustainable" and so on, knowledge work is generally both criticized (in the negative) and (suspiciously and skeptically) commented on, which always runs back to the same kind of struggle, the *struggle for appearing relevant* (what is not relevant disappears, never appears, or is rendered as having never appeared.). Something appears more relevant if, for example, it has a "practical application", or an "immediate effect", or "creates a patent", or helps "maximize outcome/profit", or if it is "done by a member of a BIG NAME University", etc. Most people will, I guess, nod knowingly at this list and be able to add to it. Although it is also clear that while most readers will nod knowingly and agree that there can certainly be merits to be had by projects that don't fulfill this list, many readers when hearing about a project idea will also immediately evaluate in their heads if they think the project is worth doing – yes, we all do that, myself included. And we always look for criteria to make that evaluation. So, what do we do other than ask: But is it relevant (and, if, then to whom)? All knowledge and cultural work live under that regime of having to justify their existence in terms of relevance. Think about cultural work: How do we determine funding for it? It has to be successful, and to be successful – measured in viewers/readers/profits – it has to be entertaining. Think about it: Reputedly, £31.6 million (45.2 in US dollars) in British tax-payer money (in the form of tax reliefs) were contributed to the making of *Star Wars: The Force Awakens*, which cost £204 million to make and made about £1.4 billion (2 billion US dollars) in sales by March 2016. Did a mega-corporation like Disney, which stood to make a huge profit, need tax-payer money? The argument is often made that these subsidies help the local economy and creative industries. But is that so? The problem is that it is indeed very hard to calculate these matters *validly and(!) reliably*. But it goes without saying that with £31 million, a number of smaller, local cultural projects (and subsequent jobs) could have been financed as well, which, too, would have had net effects in the local economy – perhaps even more sustainable ones. So, why, overall, does a mega-corporation receive this funding and why are the reasons brought up in favor being heard and do trump the discussion while reasons for an opposing view are hardly ever being even discussed? Because of *relevance*. Again: The same money could be used to finance, say 31 smaller projects and these could benefit the local economy in no less positive ways than the mega-corporation would, but it still would not be seen as equally relevant. To some degree, this has been discussed under the terms of the *Matthew effect* of accumulated advantage (those that already have, will be given more, i.e. the rich get richer and so on). The *economies of relevance*, however, have complicated and worsened the problem, because we find several additional developments that come with it. To begin with, well, there is the issue of lobbying. *Relevance work*[4] on a grander scale often includes lobbying structures. They also include a buying up of channels and actors that could facilitate a more equal playing field: We see this, in analogy, in the US-American *patent/litigation economy* or in the "stacking the bench" with talented players that "rich soccer"[5] clubs have been accused of conducting, i.e. in buying far more talented players than they can use, just to make sure no one other club can become a rival. But the most important issue, I think, is that we must consider that relevance often works in a requirement of occupying the *attention ecology*. We can see this mechanism in the question of "why advertise at all". Advertisement of products is used to be thought of in common terms as either informative or manipulative: Advertisement in this simple version was considered as alerting customers to a new or specially priced product or a new company; alternatively, a (now lay-)psychological view would argue that ads were a clever way of manipulating customers to buy a specific product or prefer one company's product over the other. And while these accounts are not entirely wrong (especially in terms of pharmaceutical ads in the United States), they have been insufficient explanations in the early age of advertisement already and they are almost misleading today. Advertisement does many things, but most importantly, I claim advertising is about "being relevant" at all. Customers have for the longest time been overflowed with ads, many of them constructed cleverly and focused on their target group, and so on. But they are almost all done too well. There is no reason why a customer should prefer car A over car B, or car C, or car D (and I am not even entering the [in]transitivity problem of preferential orders here), based on advertisements alone – unless, say the ad for car D was really shitty or created a shitstorm because of some political issue. But here is the rub: What about car 5. Now you will say, "5? What?" Exactly. There is an economy of what a car company (as a product itself) needs to be in order to be considered relevant at all; it needs to be in line with certain norms and it needs to be able to receive attention. If a potential buyer has not seen an ad for car 5, s/he will not be very likely to buy it, even if the salesman shows it to him/her. The customer may be looking for a family car, and A, B, C, D, and 5 are – as products – roughly equal, 5 may even be slightly better. The customer will not be too interested to purchase car 5. Obviously, it still happens sometimes that car 5 will be bought, but the car 5 company will still not rise to be a major player, because it's not relevant. Car manufacturer A knows that advertisement is not usually making it more likely that their car will be preferred over B, C, or D, but they have a sense for the fact that we have built ourselves an *economy of relevance* where they must advertise or, like car company 5, they will become non-relevant. Advertisement is more and more conducted to *maintain* relevance (to remain part of, or even dominate, a market) and less and less to sell more of a product. Many chains (supermarket, restaurants) have opened branches in so many places that some analysts have asked whether some of the branches can even be profitable. But here, too, the decision is made in terms of relevance (which relates, of course, to the whole construction of the stock-market/shareholder idea). A chain has to have a certain presence in order to be considered relevant; sometimes, the idea is to be present in a place to prevent a competitor from being able to receive attention. This is not to say that this won't, sometimes, backfire. There is such a thing as over-saturation. Even if customers often do not have as much choice in a market as they think (many mega-corporations own a lot of smaller companies selling similar products, or control various productions pathways, and so on), it is important that the illusion of choice is maintained. It is the most intriguing phenomenon that some customers are fully aware that the degree of choice in a market is an illusion, but they still act as though there were a high degree of choice. But to reiterate and summarize: It's all about appearing relevant.

Now, I went through all this explanation, because in contemporary society, we have reconstructed (and devalued) knowledge work in terms of the same economy of relevance. In academia, the world of higher education and research, that's become all it's about. Ask in the United States or British academia (and to a large degree most continental European academia), why a person was hired for a job, and the answer you often get is "3P", which stands for *politics* in the hiring department/school, *pedigree* (prestige of their degree's institution), and *publications* (in prestigious [e.g. according to impact factor] journals). There are many career-building steps that are built around relevance. The most relevant factor has become third-party funding. There are many relevance markers, career-pathways, and so on that are built along this *academic economy of relevance.* For example, the whole issue of status maintenance of Ivy League universities is an interesting example. But to be frank, most Ivy League scholars don't do better research (let alone good teaching); they only appear more relevant while they really are producing just as much average and mediocre research as almost everybody else does[6] – and one could actually argue that these Ivy League universities, precisely, must produce *mediocrity itself,* i.e. that the label of "excellence" which is thrown about so much is merely another mechanism in terms of relevance, and that in truth "excellence = mediocrity". Why? Because it's nearly impossible to get funding for research that is truly inventive and new, for the results of this research are always uncertain by definition. Of course, an Ivy League University still is a place where some of that unique research yet happens on occasion, because they do offer better equipment than less endowed colleges can. But my concern is that these inventive results happen *despite of* and *not because of* the "excellence" status of these institutions. Furthermore, we look to so-called excellent universities to define what excellence is, and measure all universities by these standards of excellence, and the people we vote for (or hire) who have to decide on funding for research and education often come from universities who bought in these ideals of excellence: Tautology and reification, anyone? Who decides, for example, how and by what criteria research grants should be evaluated? Research grants are written in a particular writing style and structure, which are unsurprisingly geared toward making the research seem relevant. But this style was developed and is taught by a certain class of people, and it is difficult to learn from people not privy to this type of implicit knowledge. As one person, who used to work for a major funding organization once explained it to me, at Ivy League universities, unlike smaller or less prestigious colleges, students are included in the learning about the style of grant writing by professors who are already part of this grant writing culture and its implicit knowledge stock – meaning they do have a lot of implicit knowledge about the economy of grant writing, an economy that works, largely, by writing a proposal that has as few points it could be negatively criticized for as possible, since peer reviewers are looking for reasons to reject a proposal (as quickly as possible) not to the actual contribution of a project. In cloning these cultural techniques and themselves with each student generation, a cadre of researchers creates its own "replication crisis" that way. Even a category such as "merit" itself works through an economy of relevance, i.e. it has to be written in a certain way, rather than actually describing a genuine merit. The problem isn't that this is something that's not known. Everyone in academia knows it and sometimes, in certain dark corners, it is even sometimes talked about. But to talk about this publicly and in the open and to actually propose how to change it, and say why it really needs to change, is something most people shy away from, because they are afraid nobody will pay attention to them afterward anymore. Giving voice to an open secret one runs the risk of losing relevance, not because one can actually lose relevance, but because the question becomes why a person gives voice to this problem. If a person does that, the question becomes whether they are actually an irrelevant person to begin with. Who but "car 5" company would complain about how car companies A and B have fixed the market, and car companies C and D play along? So many would say (and some do): Only "a loser" will complain about that and reveal how the game is rigged, yes? So, a main reason why in academia we don't change the way grant writing, job applications, tenure promotion, publication review, peer review, and so on are working and work very badly and unfairly – why we don't even really talk about the reasons why they work the way they do – is that while we can agree, on a *collective level* (e.g. the Left, adjunct unions), that there are these issues (*implication*), on the other side, as a *collectivity* of many individuals, most of these individuals fear that if a dissenting voice was attributed to them individually, they would have no longer a chance of being identified as someone with relevance (*intensification*). Dissent, articulated in a non-permitted form, renders one's "future silent", i.e. irrelevant or "othered into silence".

We find the need to "stay relevant" in the world of advertisement and of branching corporations hand-in-hand with an inflation of sites for advertising and branching, that, paradoxically, intensifies this process of seemingly reducing the number of opportunities – something we find, similarly, in academia. While it is clear by now that a product is placed, often enough, not to sell more of it, but to maintain one's position in the market, we also seem to observe markets are growing, that places for opening a business or placing a product seem to grow in number, and yet, these places more often than not seem to be taken up by big brands rather than small businesses or "craft/artisan" products. In academia, we find similar processes happening. While the number of journals seems to constantly grow, acceptance rates (in peer review) seem to be ridiculously small – similarly, whenever research grant funding agencies proclaim a rise in budget, their acceptance rates also seem to decrease. Not to mention that those entries or projects accepted also deserve to critiqued in terms of why certain institutional types or researcher profiles seem to have an edge, why peer review is conducted in a certain way even if it has been shown that it must be reformed, and even though proposals on how to do so have been made. It is, clearly, tough to disentangle these processes and to conduct good and critical research on these matters about academia from within academia itself for a number of reasons (and, no, "privatizing" it is not the solution, because the structure behind "privatizing" is at the heart of the problem). But the complexities between publications, third-party funding through grants, prestigious (well-funded) institutions, and other biases with regard to peer-review and hiring in academia have been discussed across many discussion forums – just not "relevantly". Again, what I am concerned with is that the conjunction of what seems like an inflation of opportunities on the one side, and a decrease of non-precarious jobs in academia on the other side, seem to effect an overall decrease in real opportunities to participate in academia in both meaningful and livelihood sustaining ways, and that the mechanism through which all of this works in this way can be reconstructed in terms of what I call *economy of relevance*. In various ways, relevance is the opposite of or anti-thesis to knowledge and cultural work. I do not mean to say that knowledge/cultural work are not ever relevant, but the precise point is that we do not know when and if it will be relevant, because we can never really know what will become needed as relevant in the future. We can try to pre-determine some of future relevance, but that is, in a sense, only accomplished by rigging the game. And it never works as well as people think it will, because many parties try to rig the game to their advantage, once again causing additional uncertainty, contingency, and chaos. This is, however, also a basic argument, why we need redundancies in certain systems. Why do we need redundancy with health care (and other) insurance, with police, or many government offices? It is, of course, easy to see only the immediate costs, and, perhaps, also include the "probability" of certain events happening. However, when these events happen and the service is required but not sufficiently available, because redundancies were eliminated, the consequences can be extreme. Let's try a little thought-experiment: If the argument to reduce the number of full-time firefighters is that over the year only x-amount of fires is accepted in a county, what are the possible consequences – besides more unemployed people. Let's say that, indeed, a number of firefighters are let go, some older ones get an early retirement, some recent hires are laid off, and future hires are for part-timers with only short-term renewable contracts. For the firefighters of this county as an organization this will equal – first of all – a destruction of implicit knowledge (knowledge you only learn by working together with others for a while) by elimination of some of the carriers of this knowledge, and constrain the further communication as well as creation of implicit knowledge, because short-term contract hires often do not have the opportunity to create, learn, and then teach others any implicit knowledge – this is what happens, by the way, in many organizations today, including in higher education and research: *Precarization* is, among other things, the destruction of implicit/tacit knowledge. But for organizations to function not only efficiently but also effectively, implicit/tacit knowledge is crucial. Indeed, efficiency (for example cost-efficiency) can be increased to a point, meanwhile effectivity reaches zero levels. The pharmaceutical business is a good example: While it has become extremely efficient as a market and company's profit margins can be substantially exploited, the effectivity of medication, research, coverage, and so on appears to have decreased dramatically (and only some of the reasons are understood, while others are under-researched, such as, e.g., the increase in and diversification of placebo-effects). Now, our firefighters are in a process of being reduced dramatically. How do you retain your job or get one of the few new (part-time) ones, if these are at all attractive? We can be certain that some way it's gonna be about relevance in one way or the other. But, again what is relevant? What interests me more in the thought-experiment here, now, is that the elimination of redundancies erupts when the events why redundancies were built into the system in the first place actually occur. Yes, that's the very point, redundancies are part of systems for a reason, and while it may be true that some redundancies may emerge for political reasons and some may exist even though they will not be required, most redundancies do not fall under that category – regardless of what some austerity-fanatics who promise to lower taxes and bemoan the squandering of tax-payer money claim. While your now super-efficient fire department may be well-staffed to handle a house on fire, what happens if two or three are on fire at the same time? Who you gonna call? And has anyone thought about the fact that there might be an increasing number of bush and forest fires thanks to climate change? And so on and so on. The interplay between contingencies and redundancies is important to understand, but it cannot be understood in terms of relevance because if something is relevant it is precisely not thought of as contingent. But deal with contingencies we must – in real life, that is. And one of the ways to enable us to deal with contingencies individually, as communities, as organizations, as societies, is knowledge and cultural labor. Why? Because the interplay between contingency and redundancy is an infrastructural problem.

Another knowledge work area where "relevance" has taken over to dramatic effect is the teaching aspect in higher education. When I am teaching in Germany, one of the most-asked questions is "*Ist das relevant für die Prüfung?*" i.e. "Is this relevant for the exam?". There is a noun for this: *Prüfungsrelevanz* – "Exam-relevance". We have even come to the point that we must mark what is relevant in advance and cannot ask anything but what the students know to be relevant. Administrations even ask us to define more and more "relevancies" in advance, such as in terms of "teaching goals" in a syllabus *and* do so in predefined language. There is a whole market for "accrediting" universities around such practices. Anyone who defies the "ways and rhetorics this is done" is not fit to work in higher education today. And while many of the "old ways" of the university maintained the authority and privilege of the (mostly white, male) professors, the new ways have not changed the reality of exclusion, inherent coloniality, or the continuation of privilege. They are just done differently.

It has been remarked by many authors, myself included, that the neoliberal academic practices confuse training (for a predefined goal with a predefined temporality) with education (which enables people, among other things, to identify goals, create goals, set goals, or to acquire means to reach goals they cannot anticipate yet). Turning *higher education* into a place where people train practices and routines for the present job market is, yes, relevant – but this relevance itself is fleeting the present job market; the skills and routines it requires are temporary, and having a job, while certainly an important part of social life, is not the only and, perhaps, not even the most important part of social life or function in society. My argument is, however, that while the infrastructural aspects are more important, being equipped by (education) to help along with these infrastructural aspects will also enable you to acquire skills that may be required in a future job market. Being trained for the present job market will neither equip you to help along the infrastructure to function nor to be able to acquire future job skills. The Liberal Arts and Humanities, as well as the core social sciences (such as sociology), in particular, are about that infrastructure. The attempts to make them more "marketable", "relevant", and "quantifiable" only negate their true infrastructural power, which is, however, increasingly needed. This is, I think, the alternative we must choose from: exam-relevance or Liberal Arts. In choosing Liberal Arts and Humanities, students acquire meaningfulness, mindfulness (in the sense of not only working by rote and routine, i.e. not working "mindlessly", which is not effective and leads to fatal errors), implicit knowledge, and skills for creating, maintaining, evaluating, and transferring implicit knowledge. We must create Liberal Arts without and against "*economies of relevance*". The inherent precarity, precariousness, and the precariate are a major concern, because we are witnessing what I consider the *Precarization of*

Everything. I am concerned with the emergence of what I consider a Green Precarity/Precariate/Precariousness becoming the main status of all agents in the emerging Post-Risk Society. The monocultural programmatic that guides the idea of "*the* BioEconomy" as well as Digital Culture is the main driver behind this development. The imbrications of biotechnologies, digitalization, and various dimensions and structures of societies need to be understood in meaningful ways, and that refers precisely to concepts such as sustainability and responsibility that are often deployed as absolute but empty and impotent (meaningless) metaphors. The point is not to say: But who'd argue against sustainable production? The point is to ask: What is sustainable, for whom and why, in what time-frame, and so on? The use of phrases and terms like "sustainability" is a way to make something seem *relevant* but it doesn't make it necessarily meaningful. Instead, we should consider new ideas in terms of, for example, *generative justice* and *generative values*, and these are conversations I encourage readers (YOU!) to join in to and become active for. For scholars to intervene on this basis in public debates means considering those contemporary news cycles that are mediated through so-called social media; there are two lines of "reporting" in which I observe the following two generalized trends that I have coined "the precarization of everything" and "your doing your brain all wrong": Liberal Arts, therefore, means studying political cognitive embodied cultures; it means to understand the political *and* the biological, since the infrastructure of organization (political) is not without its messy, living bodily organisms (biological).

This begs the question, of course, what is the *political project* of Liberal Arts and Humanities? Some Liberal Arts programs and even the university that host them and often host them for that particular reason (more and more in an act of "corporate social responsibility"), enthusiastically, optimistically, and even with a hint of utopianism – and all that is in and of itself not the result of a bad attitude or wrongful intentions – sketch their approach in the formation of citizenship ideas, a citizen's discourse, and participation in or proliferation of civil society. But there is also a problem, because citizenship is historically a Western and in some ways deeply colonial political mode of belonging, which has undergone some transformation but at its root contains many of presuppositions that are problematic in terms of enabling political participation, of allowing for alternative modes of political action and different ways of belonging. Being a citizen is not the same as participating politically, but often one's political agency is taken seriously only if one is a citizen or is considered at least eligible for citizenship. But that hinges on a paradox: On the one hand, the ideas that constitute what makes one a citizen are very narrow ideas, but on the other hand, there is also room for arbitrariness, which allows for mechanisms of exclusion that are more a form of – once national, once social psychological – protectionism: As some recent research has shown, for both the United States and Europe, immigrants and foreigners are the main topic that concerns those social groups responsible for the rise of extremist parties and politicians and for the attacks on welfare state principles and institutions, as well as the rampant anti-intellectualism. It becomes quite dangerous, when even those people who qualify themselves as "tolerant" people (and there'd be a lot to say about the problems with and violence of "toleration") say things, such as "But one has to, at least, understand that these people have fears…". The problem is not the fears, but, again the way that these "fears" are about *relevance*, namely the relevance of eligibility for the type of belonging that is citizenship. Furthermore, that citizenship has become even narrower and more arbitrary at the same time in the past decades has to do with its re-casting itself as a mode of belonging which some have called *neoliberal citizenship*, characterized by nouns such as "authenticity", "self", "structured competition", and so on. But specifically competition means here a competition for relevance and efficiency, on a person-level defined via wealth and/or job, i.e. the model of the *rentier* on the one hand or, on the other hand, the pro-/con-sumer-hybrid. But this kind of citizenship is a mode of belonging without *civics* (civics is a process, not a defined set of traits for civility and/or citizenship) and *solidarity* (which I understand in terms of integrative practices of being together with Others). Not remaining stuck in the (neo)coloniality of both the neoliberal transformation of academia and White Collaredness, what Liberal Arts, as their political project, therefore must do and ask is "What other forms of belonging, for example in "*the* political" (post-foundational) sense, i.e. in an infrastructural relation, are there, could be there, and how can people become sensitive to them?" Liberal Arts must be provincialized and be opened to decolonial options. Liberal Arts and Humanities as a political project means learning what to "think with" for how to "think through" civics as the never-ending process (becoming/individuation) of figuring out what the transition from *political* to *civic* can mean and would look like infrastructurally, and what it means locally (for the local organization of [political] agencies and their translocal ties) – perhaps, therein, this political project is also a project to develop an empirical, concrete, and integrative ethics of care. In order to follow through with a genuinely political project in general, and in particular a project that is about different and perhaps even new modes of belonging, the Liberal Arts and Humanities must cease their ties to what I have called (in various talks, forums, but also publications, including throughout *Care, Power, Information*) White Collar academia, which is certainly characterized by what others, following Lewis Gordon, have called *disciplinary decadence*, *urban normativity*, and so on. What I contrast these with is the knowledge and cultural work of *BluesCollarship*. *BluesCollars* are political, empirical, nomadic, and in the thick of things, as well as they are doing theory, but for them theory means – evoking the wisdom of Donna Haraway – "thinking, nomadically travelling, and *staying with the trouble* together with Others": Nomadic scholarship, such as is *BluesCollarship*, is not traveling alone. Theory should not be a monologue or a monoculture of mind, and Others is a very open notion that means to take all kinds of Others, and their knowledges and agency seriously. And that is a lot of work. And a lot if happens in, with, and through infrastructure, and it always significantly maintains, transforms, and creates infrastructure. And infrastructure is the basis of everything else, such as the economy or the job market – and precisely not the other way round.

This brings me to the final and important point of this exercise. I encourage you to think: Why and how we need higher education and research, why and how we need scholarship and scholars to thrive, to be funded, to be respected, but also to be committed, why and how we need academia, aka institutions of higher education and research to become? But also why we need more pluriversity than university, BluesCollars instead of White Collar academicians, and more intellectuals than anti-intellectualism – wherein, however, being an intellectual does not mean that one is ignorant, privileged, or elevated with regard to non-academic knowledges or material concerns. I could expand these lists, but you get the idea, and, yes, it is, as it should be, an open and writerly list to begin with. And Liberal Arts should, perhaps, mean exactly that "arts" (perhaps in the many meanings of the Latin "ars" and it precursors) should be "liberal" in the sense that they are "open" and "opening": they should be open and opening knowledges and techniques – aka Liberal Arts as the open and opening techniques and knowledges of infrastructure. Therein, they become important for the future. Liberal Arts and Humanities should, indeed, be there for people and made by people to make people political and make people matter politically: "To be political is to emerge, to appear, to exist" (Gordon 2014a: 88).

Lewis Gordon also gives us a hint about how this could be done, gives us a hint of how, what I call *BluesCollarship* could work in "doing knowledge and cultural labor differently":

> A few years ago I was to address a community of academics and activists in Grahamstown, South Africa. I had shown up very early for the talk and noticed a piano in the meeting hall. As few people were there, I decided to give it a go. One of my hosts immediately walked up and informed me it was out of tune.
>
> "It doesn't matter," I replied, and then sat down and played some blues and jazz until it was time to speak.
>
> My hosts came over after the audience's applause and voiced, perplexed, that they thought the piano was out of tune. "It is," I answered.
>
> "It didn't sound so," one answered.
>
> "Yes, it's out of tune. I just played it differently."
>
> What many musical artists from the blues to jazz to R & B to hip- hop know is that the ultimate goal of musical performance is to make music. All "out of tune" means is "tuned to a different

> frequency." A similar observation could be made about time, rhythm, and other aspects of musical performance. The philosophical message here is that the ability to express music in different ways is a manifestation of freedom and also a metacritical reflection on what constitutes music. Its implications for other forms of human practice — such as building and governing social institutions — are similar. What ultimately is the point of calling for democracy if we cling to models that don't groove — that is, don't exemplify the people's living engagement, participation, and responsibility for institutions of power and their creative transformation?
>
> (Gordon 2018: 24/25)

After the rise of neoextractivism, neo-r/f-acism, and neocolonialism, we must invest in the social sciences, humanities and liberal arts furnished in BluesCollarship, as an investment in the future, an investment in the true infrastructure. STEAM, not STEM. We must learn that this "investment" is not only one of "capital" but that it requires us to provincialize the Western knowledge basis and to create new spaces of learning – investment means "to clothe", after all: What clothes are we wearing how were they produced, etc.? Within the academy, we need the move away from university and monoculture to pluriversity and new, open forms of scholarship and conversation, transdisciplinarity not disciplinary decadence; instead of White Collar Academia we need something like *BluesCollarship*.

Many "culprits" can be identified for the current political state. It is not wrong but also too easy to say: It is white, wealthy, and(!), above all, *wanna-be wealthy* voters[7]; it is people who enjoy privileges and, thus, mindlessly execute racism as if it is a privilege; people who voted for Trump and Brexit and right-wing parties, it is an anti-intellectual movement, but it is also many well-meaning Left-leaning voters, activists; it is, as well, a number of people who didn't vote, who is not participating in the political process, who is not feeling as though they belong at all. All of that and more consumes and destroys the true infrastructure by precarizing knowledge and cultural labor and which now turns to precarize everything. We see, however, an increasing lack of political imagination; it is the lack of political society, when all that seems left are either options for belonging that are turning belonging into the form of participation rooted in the colonial idea of Enlightened civil society on the one hand or to subscribe to the imperialistic, neocolonial privileged anti-intellectual camp of property-owners, relevance-worshippers, and wanna-be-rich-individualists.

Instead, we need to and we can rebuild an actually real *infrastructure* – knowledge and culture work – but rebuild in using decolonial options and provincialize in the same movement and gesture. The time is now, the place is here, as we are faced with mounting political and ecological problems, but as I wrote in *The Digital Coloniality of Power*, it is not anymore about doing problem-oriented scholarship, it is about creating opportunities from and within the problem. Let's think together, learn together, have conversations, and let's move together, not just occupy but inhabit together: Let us build together a new *infrastructure*, in which it will be possible to inhabit the future together with Others.

Notes

1 And yes, I have listened to many accounts from colleagues.
2 See also Grossberg et al. (2014).
3 Once driverless cars were to become a reality, there would also be patterns of malfunctions of them, some of which, I can predict with ease, will be likened to the "machines experiencing road rage".
4 And, yes, we need a theory of and empirical studies into *relevance work*.
5 Given that many readers may be from the United States, I say soccer, but it's, of course, the real football I am talking about.
6 We have come to think of mediocrity as a bad thing, but in truth, it may be the most important research: In clinical research, in psychological research, and so on we have a so-called replication crisis. But the point is that we might not have that, if we accepted that not everything needs to be the "most relevant" but that mediocrity in research can also sometimes be a virtue.
7 (www.theguardian.com/us-news/2016/nov/09/white-voters-victory-donald-trump-exit-polls).

Bibliography

Abraham, M. (2015). Conceptualizing Academic Freedom after the Salaita Affair. *First Amendment Studies, 49*(1), 8–12.

Adams, V., Murphy, M., & Clarke, A. E. (2009). Anticipation: Technoscience, Life, Affect, Temporality. *Subjectivity, 28*(1), 246–265. doi:10.1057/sub.2009.18.

Adock, R. & Bevir, M. (2005). The History of Political Science. *Political Studies Review, 3*, 1–16.

Agamben, G. & Radisch, I. (2015). Europa muss kollabieren. Retrieved from www.Zeit.de/2015/35/Giorgio-Agamben-Philosoph-Europa-Oekonomie-Kapitalismus-Ausstieg (last accessed April 21, 2019).

Agrawal, A. (2005). Environmentality: Community, Intimate Government, and the Making of Environmental Subjects in Kumaon, India. *Current Anthropology, 46*(2), 161–190. doi:10.1086/427122.

Agrawal, B. (1994). *A Field of One's Own: Gender and Land Rights in South Asia.* Cambridge: Cambridge University Press.

Agrawal, B. (1997). "Bargaining" and Gender Relations: Within and Beyond the Household. *Feminist Economics, 3*(1), 1–51.

Alfredo, A. et al. (2009). Biotechnology as the engine for the Knowledge-Based Bio-Economy, *Biotechnology and Genetic Engineering Reviews, 26*(1), 371–388,

Agyeman, J. & McEntee, J. (2014). Moving the Field of Food Justice Forward through the Lens of Urban Political Ecology. *Geography Compass, 8*(3), 211–220. doi:10.1111/gec3.12122.

Ahmed, S. (2004). Affective Economies. *Social Text, 22*(2), 117–139.

Ahmed, S. (2012). *On Being Included: Racism and Diversity in Institutional Life.* Durham, NC: Duke University Press.

Ahmed, S. (2014). *Willful Subjects.* Durham, NC: Duke University Press.

Alatas, S. F. & Sinha, V. (2017). *Sociological Theory beyond the Canon.* London: Palgrave MacMillan.

Albrow, M. (1996). *The Global Age.* San Francisco, CA: Stanford University Press.

Alexander, J. C. (2001). Theorizing the "Modes of Incorporation": Assimilation, Hyphenation, and Multiculturalism as Varieties of Civil Participation. *Sociological Theory, 19*(3), 237–249. doi:10.1111/0735-2751.00139.

Alexander, J. C. (2006). *The Civil Sphere.* Oxford and New York, NY: Oxford University Press.

al-Farabi. (1998). *On the Perfect State* (R. Walzer trans.). Chicago, IL: Kazi Publications.

Amoore, L. (2013). *The Politics of Possibility: Risk and Security beyond Probability.* Durham, NC: Duke University Press.

Anand, N., Gupta, A., & Appel, H. (eds) (2018). *The Promise of Infrastructure.* Durham, NC: Duke University Press.

Anderson, B. (2010) Morale and the Affective Geographies of the "War on Terror". *Cultural Geographies, 17*(2), 219–236.

Anderson, B. (2016). *"We Will Win Again. We Will Win a Lot": The Affective Styles of Donald Trump.* Retrieved from http://Societyandspace.Org/2017/02/28/We-Will-Win-Again-We-Will-Win-a-Lot-the-Affective-Styles-of-Donald-Trump/ (last accessed June 20, 2018).

Anusas, M. & Ingold, T. (2015). The Charge against Electricity. *Cultural Anthropology, 30*(4), 540–554. doi:10.14506/ca30.4.03.

Appel, H. C. (2012). Walls and White Elephants: Oil Extraction, Responsibility, and Infrastructural Violence in Equatorial Guinea. *Ethnography, 13*(4), 439–465. doi:10.1177/1466138111435741.

Arendt, H. (1965). *On Revolution.* New York, NY: Viking Press.

Arendt, H. (1970). *On Violence.* New York, NY: Harcourt, Brace & World.

Arendt, H. (1981). *The Life of the Mind.* Chicago, IL: The University of Chicago Press.

Arendt, H. (1990). *Lectures on Kant's Political Philosophy* (3. Dr.). Chicago, IL: The University of Chicago Press.

Arfi, B. (2012). Reconfiguring the (Lacanian) Real: Saying the Real (as *Khôra — χώρα*) qua the Impossible–Possible Event. *Philosophy & Social Criticism, 38*(8), 793–819. doi:10.1177/0191453712461097.

Asdal, K. & Moser, I. (2012). Experiments in Context and Contexting. *Science, Technology, & Human Values, 37*(4), 291–306. doi:10.1177/0162243912449749.

Asher, N. L. (2011). *The Technician.* London: Tor.

Ashwood, L., Harden, N., Bell, M. M., & Bland, W. (2014). Linked and Situated: Grounded Knowledge. *Rural Sociology, 79*(4), 427–452. doi:10.1111/ruso.12042.

Atkins, P. W. (2010). *Liquid Materialities: A History of Milk, Science and the Law. Farnham*: Ashgate. London: Routledge.

Ault, B. A. (2007). Oikos and Oikonomia: Greek Houses, Households and the Domestic Economy. In *Building Communities: House, Settlement and Society in the Aegean and Beyond* (pp. 259–265). British School at Athens Studies, vol. 15. London: British School at Athens.

Baali, F. (1988). *Society, State, and Urbanism: Ibn Khaldun's Sociological Thought.* Albany, NY: SUNY Press.

Baehr, P. (2001). The "Iron Cage" and the "Shell as Hard as Steel": Parsons, Weber, and the Stahlhartes Gehäuse Metaphor in the Protestant Ethic and the Spirit of Capitalism. *History and Theory, 40*(2), 153–169. doi:10.1111/0018-2656.00160.

Bakker, J. I. (ed) (2015a). *The Methodology of Political Economy: Studying the Global Rural-Urban Matrix.* Lanham, MD: Lexington Books.

Bakker, J. I. (Hans) (2015b). Thick Description, Nomological Laws and Ideal Types: Which Methodology Helps Most with Praxis? In H. F. Dahms (ed), *Current Perspectives in Social Theory* (Vol. 33, pp. 267–292). doi:10.1108/S0278-120420150000033011.

Barad, K. (2008). *Meeting the Universe Halfway.* Durham, NC: Duke University Press.

Barad, K. (2017). Troubling Time/s and Ecologies of Nothingness: Re-turning, Re-membering, and Facing the Incalculable. *New Formations, 92*(92), 56–86. doi:10.3898/NEWF:92.05.2017.

Barker, D. K. (ed) (2003). *Toward a Feminist Philosophy of Economics*. London: Routledge.

Barker, J. (2015). The Corporation and the Tribe. *American Indian Quarterly*, *39*(3), 243. doi:10.5250/amerindiquar.39.3.0243.

Barker, J. (2018). Territory as Analytic. *Social Text*, *36*(2), 19–39. doi:10.1215/01642472-4362337.

Barkley, R. A. & Peters, H. (2012). The Earliest Reference to ADHD in the Medical Literature? Melchior Adam Weikard's Description in 1775 of "Attention Deficit" (Mangel der Aufmerksamkeit, Attentio Volubilis). *Journal of Attention Disorders*, *16*(8), 623–630. doi:10.1177/1087054711432309.

Barnes, E. (2010). Ontic Vagueness: A Guide for the Perplexed. *Noûs*, *44*(4), 601–627.

Bataille, G. (1995). *Consumption* ([Nachdr.]). New York, NY: Zone.

Bateson, G. (2000). *Steps to an Ecology of Mind*. Chicago, IL: The University of Chicago Press.

Battersby, C. (2011). Singularity and the Female Self: Encountering the Other. *Women: A Cultural Review*, *22*(2–3), 125–142. doi:10.1080/09574042.2011.561113.

Bauman, Z. (2008). *Modernity and the Holocaust*. Cambridge: Polity Press.

Baumeister, A. A., Henderson, K., Pow, J. L., & Advokat, C. (2012). The Early History of the Neuroscience of Attention-Deficit/Hyperactivity Disorder. *Journal of the History of the Neurosciences*, *21*(3), 263–279. doi:10.1080/0964704X.2011.595649.

Beck, S. (n.d.). Von Praxistheorie 1.0 Zu 3.0—Oder: Wie Analoge Und Digitale Praxen Relationiert Werden Sollten. In *Digitale Praxen–GoeU FFM, 19.–21.1.2015*. Retrieved from www.academia.edu/10952692/Von_Praxistheorie_1.0_zu_3.0_oder_wie_analoge_und_digitale_Praxen_relationiert_werden_sollten (last accessed April 17, 2015).

Beck, U. (1986). *Risikogesellschaft: auf dem Weg in eine andere Moderne* (Einmalige Sonderausg). Frankfurt am Main: Suhrkamp.

Beck, U. (2002). *Macht und Gegenmacht im globalen Zeitalter. Neue weltpolitische Ökonomie*. Frankfurt am Main: Suhrkamp.

Bellacasa, M. (2011). Matters of care in technoscience: Assembling neglected things. *Social Studies of Science 41*(1), 85–106.

Benbow, H. M. (2003). Ways in, Ways Out: Theorizing the Kantian Body. *Body & Society*, *9*(1), 57–72. doi:10.1177/1357034X030091004.

Benhabib, S. (2003). *The Reluctant Modernism of Hannah Arendt*. Lanham, MD: Rowman & Littlefield.

Benjamin, W. (2002). *The Arcades Project*. Cambridge, MA: Harvard University Press.

Bennett, J. (2010). *Vibrant Mater*. Durham, NC: Duke University Press.

Benson, S. M. (2017). Democracy and Unfreedom: Revisiting Tocqueville and Beaumont in America. *Political Theory*, *45*(4), 466–494. doi:10.1177/0090591715616851.

Bernal, M. (1987). *Black Athena: The Afroasiatic Roots of Classical Civilization*. New Brunswick, NJ: Rutgers University Press.

Bernanke, B. S. (2015a). Germany's Trade Surplus Is a Problem. Retrieved from www.Brookings.Edu/Blog/Ben-Bernanke/2015/04/03/Germanys-Trade-Surplus-Is-a-Problem/ (last accessed April 21, 2019).

Bernanke, B. S. (2015b). Greece and Europe: Is Europe Holding Up Its End of the Bargain? Retrieved from www.Brookings.Edu/Blog/Ben-Bernanke/2015/07/17/Greece-and-Europe-Is-Europe-Holding-up-Its-End-of-the-Bargain/ (last accessed April 21, 2019).

Bevir, M. (2010). Rethinking Governmentality: Towards Genealogies of Governance. *European Journal of Social Theory, 13*, 423–441.

Bevir, M. & Phillips, R. (2017). Genealogies of European Governance. *Comparative European Politics, 15*(5), 685–704. doi:10.1057/s41295-016-0080-8.

Bhambra, G. K. (2015). Citizens and Others: The Constitution of Citizenship through Exclusion. *Alternatives: Global, Local, Political, 40*(2), 102–114.

Bhambra, G. K. (2016). Postcolonial Reflections on Sociology. *Sociology, 50*(5), 960–966.

Bhambra, G. K. (2017). *Connected Sociologies*. New York, NY: Bloomsbury.

Bhambra, G. K., Gebrial, D., & Nişancıoğlu, K. (eds) (2018). *Decolonising the University*. Chicago, IL: Pluto Press.

Bhambra, G. K. & Narayan, J. (2017). European Cosmopolitanism: Colonial Histories and Postcolonial Societies. Retrieved from http://public.ebookcentral.proquest.com/choice/publicfullrecord.aspx?p=4729744 (last accessed April 21, 2019).

Bhandar, B. (2018). *Colonial Lives of Property. Law, Land, and Racial Regimes of Ownership*. Durham, NC: Duke University Press.

Biagioli, M. (2006). *Galileo, Courtier: Practice of Science in the Culture of Absolutism* [Nachdr.]. Chicago, IL: University of Chicago Press.

Binder, D. K., Schaller, K., & Clusmann, H. (2007). The Seminal Contributions of Johann-Christian Reil to Anatomy, Physiology, and Psychiatry. *Neurosurgery, 61*(5), 1091–1096. doi:10.1227/01.neu.0000303205.15489.23.

Birch, K. (2016). Emergent Imaginaries and Fragmented Policy Frameworks in the Canadian Bio-Economy. *Sustainability, 8*(10), 1007. doi:10.3390/su8101007.

Birch, K. (2017a). Rethinking *Value* in the Bio-economy: Finance, Assetization, and the Management of Value. *Science, Technology, & Human Values, 42*(3), 460–490. doi:10.1177/0162243916661633.

Birch, K. (2017b). The Problem of Bio-Concepts: Biopolitics, Bio-Economy and the Political Economy of Nothing. *Cultural Studies of Science Education, 12*(4), 915–927. doi:10.1007/s11422-017-9842-0.

Birch, K. (2019). Technoscience Rent: Toward a Theory of *Rentiership* for Technoscientific Capitalism. *Science, Technology, & Human Values*, 016224391982956. doi:10.1177/0162243919829567.

Bischof, M. (2010). *Salutogenese – Unterwegs zur Gesundheit: neue Gesundheitsmodelle und die Entstehung einer integrierten Medizin*. Pulow: Drachen Verl.

Blaser, M. (2009a). Political Ontology: Cultural Studies Without "Cultures"? *Cultural Studies, 23*(5–6), 873–896. doi:10.1080/09502380903208023.

Blaser, M. (2009b). The Threat of the Yrmo: The Political Ontology of a Sustainable Hunting Program. *American Anthropologist, 111*(1), 10–20. doi:10.1111/j.1548-1433.2009.01073.x.

Blaser, M. (2014). Ontology and Indigeneity: On the Political Ontology of Heterogeneous Assemblages. *Cultural Geographies, 21*(1), 49–58. doi:10.1177/1474474012462534.

Blinder, A. & Lewin, T. (2015). Clerk in Kentucky Chooses Jail Over Deal on Same-Sex Marriage. Retrieved from www.nytimes.com/2015/09/04/Us/Kim-Davis-Same-Sex-Marriage.Html (last accessed November 15, 2018).

Blohm, C. (2010). *Zugang zu humangenetischen Ressourcen indigener Völker Lateinamerikas: eine Stakeholderanalyse (1. Aufl).* Wiesbaden: Gabler.

Blumenberg, B. (2015). *Eine Frage der Belichtung.* Retrieved from www.nzz.ch/feuilleton/buecher/eine-frage-der-belichtung-1.18578697 (last accessed November 15, 2018).

Blumenberg, H. (1987a). *Das Lachen der Thrakerin: eine Urgeschichte der Theorie* (1. Aufl). Frankfurt am Main: Suhrkamp.

Blumenberg, H. (1987b). *Die Sorge geht über den Fluss* (1. Aufl). Frankfurt am Main: Suhrkamp.

Blumenberg, H. (1988). *Work on Myth.* Boston, MA: MIT Press.

Blumenberg, H. (1989). *Höhlenausgänge* (1. Aufl). Frankfurt am Main: Suhrkamp.

Blumenberg, H. (1996). *Arbeit am Mythos.* Frankfurt am Main: Suhrkamp.

Blumenberg, H. (1997). *Die Vollzähligkeit der Sterne* (2. Aufl). Frankfurt am Main: Suhrkamp.

Blumenberg, H. (2001). *Ästhetische und metaphorologische Schriften* (1. Aufl). Frankfurt am Main: Suhrkamp.

Blumenberg, H. (2007). *Theorie der Unbegrifflichkeit* (1. Aufl). Frankfurt am Main: Suhrkamp.

Blumenberg, H. (2012). *Die Legitimität der Neuzeit* (Erneuerte Ausg., 6. Aufl). Frankfurt am Main: Suhrkamp.

Blumenberg, H. (2014a). *Beschreibung des Menschen* (1. Aufl). Frankfurt am Main: Suhrkamp.

Blumenberg, H. (2014b). *Schiffbruch mit Zuschauer: Paradigma einer Daseinsmetapher* (6. Auflage). Frankfurt am Main: Suhrkamp.

Blumenberg, H. (2015). *Paradigmen zu einer Metaphorologie* (6. Aufl). Frankfurt am Main: Suhrkamp.

Blumenberg, H. (2018). *Die Lesbarkeit der Welt.* Frankfurt am Main: Suhrkamp.

Blumenthal, U.-R. (1988). *The Investiture Controversy: Church and Monarchy from the Ninth to the Twelfth Century.* Philadelphia: University of Pennsylvania Press.

Boatcă, M. (2016). *Global Inequalities Beyond Occidentalism.* Abingdon and New York, NY: Routledge.

Boatcă, M. (2017). "Modernity, Citizenship and Occidentalist Epistemology in Max Weber and Beyond". Retrieved from https://trafo.hypotheses.org/category/provincializing-epistemologies (last accessed August 22, 2018).

Boeckenfoerde, E.-W. (1991). *Recht, Staat, Freiheit* (pp. 92–114). Frankfurt am Main: Suhrkamp.

Boesch, E. E. (2002). The Myth of Lurking Chaos. In H. Keller, Y. H. Poortinga, & A. Schölmerich (eds), *Between Culture and Biology* (pp. 116–135). doi:10.1017/CBO9780511489853.007.

Boghossian, P. (1996). Analyticity Reconsidered. *Nous*, 30, 360–391.

Boghossian, P. (2003). Epistemic analyticity. *Grazer Philosophische Studien*, *66*, 15–35.

Boltanski, L., Chiapello, E., & Elliott, G. (eds) (2018). *The New Spirit of Capitalism.* London: Verso.

Bonilla-Silva, E. (2018). ASA Presidential Address "Feeling Race: Theorizing the Racial Economy of Emotions", given at the 113th Annual Meeting, Philadelphia. Retrieved from https://vimeo.com/284778934 (last accessed August 22, 2018).

Bordo, S. (2017). *The Destruction of Hillary Clinton.* Brooklyn, NY: Melville House.

Bourdieu, P. (1996). The Forms of Capital. In J. Richardson (ed), *Handbook of Theory and Research for the Sociology of Education* (pp. 241–258). New York, NY: Greenwood.

Boureau, A. (1998). *The Lord's First Night: The Myth of the Droit de Cuissage*. Chicago, IL: The University of Chicago Press.

Boyd, D. (2014). *It's Complicated: The Social Lives of Networked Teens*. New Haven, CT: Yale University Press.

Boyd, D. (n.d.). The Not-So-Hidden Politics of Class Online. Retrieved from www.academia.edu/3977846/The_not-so-hidden_politics_of_class_online (last accessed January 24, 2015).

Bozalek, V. (2017). Towards a Response-able Pedagogy across Higher Education Institutions in Post-Apartheid South Africa: An Ethico-Political Analysis. *Education as Change, 21*(2), 62–85. doi:10.17159/1947-9417/2017/2017.

Bradley, K. R. (1987). *Slaves and Masters in the Roman Empire: A Study in Social Control*. New York, NY: Oxford University Press.

Bradley, K. R. (1990). *Servus Onerosus*: Roman Law and the Troublesome Slave. *Slavery & Abolition, 11*(2), 135–158. doi:10.1080/01440399008575004.

Bradley, K. R. (1994). *Slavery and Society at Rome*. Cambridge and New York, NY: Cambridge University Press.

Bradley, K. R. & Cartledge, P. (eds) (2011). *The Cambridge World History of Slavery: The Ancient Mediterranean World*. doi:10.1017/CHOL9780521840668.

Braidotti, R. (2002). Between the No Longer and the Not Yet: On Bios/Zoe-Ethics. *Filosofski Vestnik, 23*(2), 9–26.

Brandt, E. J., Elliott, R. W., & Swank, R. T. (1975). Defective Lysosomal Enzyme Secretion in Kidneys of Chediak-Higashi (Beige) Mice. *The Journal of Cell Biology, 67*(3), 774–788. doi:10.1083/jcb.67.3.774.

Brannon, M. M. (2013). Standardized Spaces: Satellite Imagery in the Age of Big Data. *Configurations, 21*(3), 271–299. doi:10.1353/con.2013.0021.

Brannon, M. M. (2017). Datafied and Divided: Techno-Dimensions of Inequality in American Cities. *City & Community, 16*(1), 20–24. doi:10.1111/cico.12220.

Breithaupt, F. (2009). *Kulturen der Empathie* (1. Aufl). Frankfurt am Main: Suhrkamp.

Breithaupt, F. (2012). The Blocking of Empathy, Narrative Empathy, and a Three-Person Model of Empathy. *Emotion Review*, 4, 84–91.

Breithaupt, F. & Hamilton, A. B. B. (2019). *The Dark Sides of Empathy*. Ithaca, NY: Cornell University Press.

Brenner, E., Marshall, R. (2014). Interview. *3AM*. http://www.3ammagazine.com/3am/the-ethical-machiavelli/ (last accessed: March 15, 2018).

Brisson, L. (2016). The Intellect and the Cosmos: The Demiurge in the Timaeus. *Méthodos*, (16). doi:10.4000/methodos.4463.

Brock, R. (2013). *Greek Political Imagery from Homer to Aristotle*. London and New York, NY: Bloomsbury.

Brown, W., Forst, R., Di Blasi, L., & Holzhey, C. F. E. (2014). *The Power of Tolerance: A Debate*. New York, NY: Columbia University Press.

Buckland, W. W. (2010). The Roman Law of Slavery: The Condition of the Slave in Private Law from Augustus to Justinian. doi:10.1017/CBO9780511707360.

Buck-Morss, S. (1991). *The Dialectics of Seeing: Walter Benjamin and the Arcades Project*. Cambridge, MA: MIT Press.

Burawoy, M. (2005). "Provincializing the Social Sciences", in George Steinmetz (ed), *The Politics of Method in the Human Sciences: Positivism and its Epistemological Others*. Durham, NC: Duke University Press.

Burawoy, M. & Behbehanian, L. (2012). A Pedagogy for the Global, Contexts, *11*(1), 80–83.

Burge, T. (1993). Content Preservation. *Philosophical Review, 102*(4), 457–488.

Burge, T. (2003). Perceptual Entitlement. *Philosophy and Phenomenological Research, 67*(3), 503–548.

Butler, J. & Spivak, G. (2011). *Who Sings the Nation-State?* Calcutta: Seagull Books.

Callon, M. (2005). Why Virtualism Paves the Way to Political Impotence. Callon Replies to Miller. *Economic Sociology (European Electronic Newsletter), 6*, 3–20.

Campbell, W. J. (2003). *Yellow Journalism: Puncturing the Myths, Defining the Legacies.* Westport, CT and London: Praeger.

Carey, C. (2016). Breaking the News: Telegraphy and Yellow Journalism in the Spanish-American War. *American Periodicals: A Journal of History & Criticism, 26*(2), 130–148.

Carolan, M. S. (2005a). Realism without Reductionism: Toward an Ecologically Embedded Sociology. *Human Ecology Review, 12*(1), 1–20.

Carolan, M. S. (2005b). Society, Biology, and Ecology: Bringing Nature Back Into Sociology's Disciplinary Narrative Through Critical Realism. *Organization & Environment, 18*(4), 393–421. doi:10.1177/1086026605281697.

Carolan, M. S. (2010). The Mutability of Biotechnology Patents: From Unwieldy Products of Nature to Independent 'Object/s.' *Theory, Culture & Society, 27*(1), 110–129. doi:10.1177/0263276409350360.

Carolan, M. S. (2019). Automated Agrifood Futures: Robotics, Labor and the Distributive Politics of Digital Agriculture. *The Journal of Peasant Studies*, 1–24. doi:10.1080/03066150.2019.1584189.

Carroll, P. (1996). Science, Power, Bodies: The Mobilization of Nature as State Formation. *Journal of Historical Sociology, 9*, 139–167.

Casullo, A. (2007). What Is Entitlement?" *Acta Analytica, 22*, 267–279.

Cavalcanti, C. (2002). Economic Thinking, Traditional Ecological Knowledge and Ethnoeconomics. *Current Sociology, 50*(1), 39–55. doi:10.1177/0011392102 050001004.

Cepek, M. L. (2011). Foucault in the Forest: Questioning Environmentality in Amazonia. *American Ethnologist, 38*(3), 501–515. doi:10.1111/j.1548-1425.2011.01319.x.

Cerella, A. (2016). Encounters at the End of the World: Max Weber, Carl Schmitt and the Tyranny of Values. *Journal for Cultural Research, 20*(3), 266–285. doi:10. 1080/14797585.2016.1141833.

Chakrabarty, D. (2008). *Provincializing Europe: Postcolonial Thought and Historical Difference.* Princeton, NJ: Princeton University Press.

Chalcraft, D. J. (ed). (2008). *Max Weber Matters: Interweaving Past and Present.* Farnham and Burlington, VT: Ashgate.

Chandhoke, N. (n.d.). The "Civil" and the "Political" in Civil Society. *Democratization, 8*(2), 1–24.

Chatterjee, P. (2006). *The Politics of the Governed.* New York, NY: Columbia University Press.

Chatterjee, P. (2008). Economic and Democratic Transformation in India. *Economic and Political Weekly*, April 19, p. 57.

Chiapello, E. & Fairclough, N. (2002). Understanding the New Management Ideology: A Transdisciplinary Contribution from Critical Discourse Analysis and New Sociology of Capitalism. *Discourse & Society, 13*(2), 185–208. doi:10.1 177/0957926502013002406.

Christen, K. (2011). Opening Archives: Respectful Repatriation. *The American Archivist*, *74*(1), 185–210. doi:10.17723/aarc.74.1.4233nv6nv6428521.

Christen, K. (2015). Tribal Archives, Traditional Knowledge, and Local Contexts: Why the 's' Matters. *Journal of Western Archives*, *6*(1), Article 3.

Code, L. 2010. Testimony, advocacy, ignorance: thinking ecologically about social knowledge. In A. Haddock, A. Millar, & D. Pritchard (eds), *Social epistemology* (pp. 29–50). Oxford, New York: Oxford University Press.

Cohen, E. (2009). *A Body Worth Defending: Immunity, Biopolitics, and the Apotheosis of the Modern Body*. Durham, NC: Duke University Press.

Cohen, M. & Otomo, Y. (eds) (2017). *Making Milk: The Past, Present, and Future of Our Primary Food*. London: Bloomsbury.

Collins, R. & Collins, R. (1994). *Four Sociological Traditions*. New York, NY: Oxford University Press.

Collins, R. (2005) *Interaction Ritual Chains*. Princeton: Princeton UP.

Connell, R. (2014). Using Southern Theory: Decolonizing Social Thought in Theory, Research and Application. *Planning Theory*, *13*(2), 210–223.

Connolly, J. J., Glessner, J. T., Kao, C., Elia, J., & Hakonarson, H. (2015). Attention-Deficit Hyperactivity Disorder and Pharmacotherapy—Past, Present, and Future: A Review of the Changing Landscape of Drug Therapy. *Therapeutic Innovation & Regulatory Science*, *49*(5), 632–642. doi:10.1177/2168479015599811.

Conradie, E. M. (2007). The Whole Household of God (Oikos): Some Ecclesiological Perspectives (Part 2). *Scriptura : Journal for Contextual Hermeneutics in Southern Africa*, *94*, 10–28.

Coolsaet, B. (2016). Towards an Agroecology of Knowledges: Recognition, Cognitive Justice and Farmers' Autonomy in France. *Journal of Rural Studies*, *47*, 165–171. doi:10.1016/j.jrurstud.2016.07.012.

Crouch, C. (2000). *Coping with Post-Democracy*. London: Fabian Society.

Crouch, C. (2010). *Post-Democracy* (Repr). Cambridge: Polity Press.

Crouch, C. (2011). *The Strange Non-Death of Neoliberalism*. Cambridge: Polity Press.

Crowe, M. (1997). A History of the Extraterrestrial Life Debate. *Zygon*, *32*(2), 147–162.

Crowe, M. (2011). *A History of the Extraterrestrial Life Debate 1750–1900*. Mineola, NY: Dover Publications.

Crowston, C. H. (2013). *Credit, Fashion, Sex: Economies of Regard in Old Regime France*. Durham, NC: Duke University Press.

Dabbert, S., et al. (2017). *Knowledge-Driven Developments in the Bioeconomy*. New York, NY: Springer Berlin Heidelberg.

Daigle, M. (2019). Tracing the Terrain of Indigenous Food *Sovereignties*. *The Journal of Peasant Studies*, *46*(2), 297–315. doi:10.1080/03066150.2017.1324423.

Danks, D. (2014). *Unifying the Mind: Cognitive Representations as Graphical Models*. Cambridge, MA: MIT Press.

Davies, G. (2005). Researches into the Ancient Greek Countryside. – The Contributions of Paul Giraud, William Heitland and Alfred Zimmern. *Topoi*, *12*(1), 385–406. doi:10.3406/topoi.2005.2017.

Davison, J. & Ammann, K. (2017). New GMO Regulations for Old: Determining a New Future for EU Crop Biotechnology. *GM Crops & Food*, *8*(1), 13–34.

Dayé, C. (n.d.). A Systematic View on the Use of History for Current Debates in Sociology, and on the Potential and Problems of a Historical Epistemology of Sociology. *American Sociologist*, Online First.

Decoteau, C. (2013). Exclusionary Inclusion and the Normalization of Biomedical Culture. *American Journal of Cultural Sociology, 1*(3), 403–430.

De La Rosa, S. (2014). Hannah Arendt im Spannungsfeld zwischen Säkularisierung und Sakralisierung. *Leviathan*, 174–190. doi:10.5771/0340-0425-2014-2-174.

Deleuze, G. (1988). *Le pli: Leibniz et le Baroque*. Paris: Editions de Minuit.

Deleuze, G. (1993). *The Fold: Leibniz and the Baroque*. Minneapolis: University of Minnesota Press.

Deleuze, G. & Guattari, F. (1987). *A Thousand Plateaus*. Minneapolis: Minnesota University Press.

de Lorenzo, V. & Schmidt M. (2018). Biological Standards for the Knowledge-Based Bioeconomy: What Is at Stake. *New Biotechnology, 40*, 170–180.

Demailly, K.-E. & Darly, S. (2017). Urban Agriculture on the Move in Paris: The Routes of Temporary Gardening in the Neoliberal City. *ACME: An International E-Journal for Critical Geographies*, University of British Columbia, Okanagan.

Denkler. (2015). Griechenland-Krise. www.Sueddeutsche.de/Politik/Griechenland-Krise-Hoert-Auf-Europa-Kaputtzureden-1.2566774.

Depew, J. F. (2016). Foucault Among the Stoics: Oikeiosis and Counter-Conduct. *Foucault Studies*, 22. doi:10.22439/fs.v0i0.5012.

Deresiewicz, W. (2015). The Neoliberal Arts How College Sold Its Soul to the Market.Retrievedfromhttps://Harpers.Org/Archive/2015/09/the-Neoliberal-Arts/ (last accessed April 21, 2019).

Derrida, J. (1986). Point De Folie — Maintenant L'architecture. Bernard Tschumi: La Case Vide — La Villette, 1985. *AA Files, 12*, 65–75.

Derrida, Jacques. (2006). *Schurken*. Frankfurt am Main: Suhrkamp.

Derrida, J. & Bass, A. (2009). *Writing and Difference* (Repr). London: Routledge.

de Sousa Santos, B. (2001). Toward an Epistemology of Blindness Why the New Forms of 'Ceremonial Adequacy' Neither Regulate Nor Emancipate. *European Journal of Social Theory, 4*(3), 251–279. doi:10.1177/13684310122225109.

Detel, W. (2005). *Foucault and Classical Antiquity: Power, Ethics, and Knowledge*. Cambridge: Cambridge University Press.

DiAngelo, R. (2011). White Fragility. *International Journal of Critical Pedagogy, 3*(3), 54–70.

Dierksmeier, C. & Pirson, M. (2009). Oikonomia versus Chrematistike: Learning from Aristotle About the Future Orientation of Business Management. *Journal of Business Ethics, 88*(3), 417–430. doi:10.1007/s10551-009-0128-7.

Dietz, M. (2012). Between Polis and Empire: Aristotle's Politics. *The American Political Science Review, 106*, 257–293.

DiMaggin, P. & Hargittai, E. (n.d.). From the 'Digital Divide' to "Digital Inequality": Studying Internet Use as Penetration Increases. Center for Arts and Cultural Policy Studies, Working Paper #15, Summer.

Dmitriev, S. (2016). The Protection of Slaves in the Athenian Law against Hubris. *Phoenix, 70*(1/2), 64. doi:10.7834/phoenix.70.1-2.0064.

Dolphin, R. & van der Tuin, I. (2012). *New Materialism*. Ann Arbor, MI: Open Humanities Press.

Dosse, F. (2009). *Gilles Deleuze, Félix Guattari: biographie croisée*. Paris: La Découverte.

Driessen, C. & Heutinck, L. (2015) Cows desiring to be milked? Milking robots and the co-evolution of ethics and technology on Dutch dairy farms. *Agriculture and Human Values, 32*(1), 3–20.

Dupuy, J.-P. (2013). *The Mark of the Sacred*. Stanford, CA: Stanford University Press.

Dupuy, J.-P. & DeBevoise, M. B. (2013). *The Mark of the Sacred*. Stanford, CA: Stanford University Press.

Dussel, E. (2013). *Ethics of Liberation: In the Age of Globalization and Exclusion* (A. A. Vallega, Ed., E. Mendieta, C. P. Bustillo, Y. Angulo, & N. Maldonado-Torres, Trans.). Durham, NC: Duke University Press.

Easton, G. (2010). Critical Realism in Case Study Research. *Industrial Marketing Management*, *39*(1), 118–128. doi:10.1016/j.indmarman.2008.06.004.

Eaton, W. M., Gasteyer, S. P., & Busch, L. (2014). Bioenergy Futures: Framing Sociotechnical Imaginaries in Local Places: Bioenergy Futures. *Rural Sociology*, *79*(2), 227–256. doi:10.1111/ruso.12027.

Ecks, S. (2010). Near-Liberalism: Global Corporate Citizenship and Pharmaceutical Marketing in India. In Ai-hwa Ong & N. N. Chen (eds), *Asian Biotech: Ethics and Communities of Fate*. Durham, NC: Duke University Press.

Eglash, R. & Garvey, C. (2014). Basins of Attraction for Generative Justice. In S. Banerjee, Ş. Ş. Erçetin, & A. Tekin (eds), *Chaos Theory in Politics* (pp. 75–88). doi:10.1007/978-94-017-8691-1_5.

Eisenstein, E. (1979). *The Printing Press as an Agent of Change*. Cambridge: Cambridge University Press.

Elden, S. (2013). *The Birth of Territory*. Chicago and London: The University of Chicago Press.

Elden, S. & Mendieta, E. (eds) (2001). *Reading Kant's Geography*. Albany, NY: SUNY Press.

Elias, N. (1983). *Engagement und Distanzierung* (2. Aufl). Frankfurt aM: Suhrkamp.

Elias, N., Dunning, E., Goudsblom, J., & Mennell, S. (2000). *The Civilizing Process: Sociogenetic and Psychogenetic Investigations* (Rev. ed). Oxford and Malden, MA: Blackwell Publishers.

Ellenberger, H. F. (1970). *The Discovery of the Unconscious: The History and Evolution of Dynamic Psychiatry*. New York, NY: Basic Books.

Engesser, S. et al. (2017). Populist Online Communication: Introduction to the Special Issue. *Information, Communication & Society*, *20*(9), 1279–1292.

Eppler, E. (2005). *Auslaufmodell Staat?* (Originalausg., 1. Aufl). Frankfurt am Main: Suhrkamp.

Epstein, B. (2015). *The Ant Trap: Rebuilding the Foundations of the Social Sciences*. New York, NY: Oxford University Press.

Epstein, S. (2007). *Inclusion: The Politics of Difference in Medical Research*. Chicago, IL: University of Chicago Press.

Erdkamp, P. (2014). How Modern Was the Market Economy of the Roman World? *Œconomia*, *4-2*, 225–235.

Escobar, A. (2018). *Designs for the Pluriverse: Radical Interdependence, Autonomy, and the Making of Worlds*. Durham, NC: Duke University Press.

Esfeld, M. (2002). *Holismus in der Philosophie des Geistes und in der Philosophie der Physik* (Orig.-Ausg., 1. Aufl). Frankfurt am Main: Suhrkamp.

Esfeld, M. (2008). *Naturphilosophie als Metaphysik der Natur* (1. Aufl). Frankfurt am Main: Suhrkamp.

Esfeld, M. (2013). Ontic Structural Realism and the Interpretation of Quantum Mechanics. *European Journal for Philosophy of Science*, *3*(1), 19–32.

Esfeld, M., Lam, V. (2011). Ontic structural realism as a metaphysics of objects. In Alisa and Peter Bokulich (eds), *Scientific structuralism* (pp. 143–159). Dordrecht: Springer.

Esposito, E. (2002). *Soziales Vergessen: Formen und Medien des Gedächtnisses der Gesellschaft* (1. Aufl). Frankfurt am Main: Suhrkamp.

Esposito, E. (2011). *The Future of Futures: The Time of Money in Financing and Society*. Cheltenham and Northampton, MA: Edward Elgar.

Esposito, R. (2011). *Immunitas: The Protection and Negation of Life* (English ed.). Cambridge and Malden, MA: Polity.

Feingold, H. (1983). How Unique Is the Holocaust? In A. Grobman & D. Landes (eds), *Genocide: Critical Issues of the Holocaust*. Los Angeles, CA: The Simon Wiesenthal Centre.

Feingold, H. (1995). *Bearing Witness: How America and Its Jews Responded to the Holocaust* (1st ed.). Syracuse, NY: Syracuse University Press.

Ferber, M. A. & Nelson, J. A. (eds) (1993). *Beyond Economic Man: Feminist Theory and Economics*. Chicago: The University of Chicago Press.

Fischer, M. M. J. (2003). *Emergent Forms of Life and the Anthropological Voice*. Durham, NC: Duke UP. Retrieved from https://ezproxy.aub.edu.lb/login?url=. doi:10.1215/9780822384953.

Fischer-Lescano, A. & Teubner, G. (2006). *Regime-Kollisionen*. Frankfurt am Main: Suhrkamp.

Fishel, S. R. (2018). *The Microbial State: Global Thriving and the Body Politic*. Minneapolis: Minnesota UP.

Fishel, S. R. (2017). *The Microbial State: Global Thriving and the Body Politic*. Minneapolis: Minnesota UP.

Fitch, M. L. (2013). Native American Empowerment Through Digital Repatriation. Electronic Theses and Dissertations Paper 2291. Retrieved from http://dc.etsu.edu/etd/2291 (last accessed April 21, 2019).

Floridi, L. (2005). Information Ethics, Its Nature and Scope. *ACM SIGCAS Computers and Society*, *35*(2), 3–3. doi:10.1145/1111646.1111649.

Floridi, L. (2006) Information Ethics, Its Nature and Scope. *ACM SIGCAS Computers and Society, 36*(3). doi: 10.1145/1195716.1195719.

Floridi, L. (2012). *The Philosophy of Information*. Oxford: Oxford University Press.

Florini, A. (2005). *The Coming Democracy*. Washington, DC: Brookings Inst. Press.

Fogg, B. J. (2002). *Persuasive Technology*. San Francisco, CA: Morgan Kaufmann.

Forst, R. (2002). *Contexts of Justice: Political Philosophy Beyond Liberalism and Communitarianism*. Berkeley: University of California Press.

Forst, R. (2012). *The Right to Justification: Elements of a Constructivist Theory of Justice*. New York, NY: Columbia University Press.

Forst, R. (2013). *Toleration in Conflict: Past and Present*. doi:10.1017/CBO9781139051200.

Forst, R. (2015). Noumenal Power. *Journal of Political Philosophy*, *23*(2), 111–127.

Foucault, M. (1982). The Subject and Power. *Critical Inquiry*, *8*(4), 777–795.

Foucault, M. (1988). *The History of Sexuality* (1st ed.). New York, NY: Vintage Books.

Foucault, M. (1995). *Discipline and Punish: The Birth of the Prison* (2nd ed.). New York, NY: Vintage Books.

Foucault, M. (1996). Truth and Juridical Forms. *Social Identities*, *2*(3), 327–342.

Foucault, M. & Bischoff, M. (2003). *Die Wahrheit und die juristischen Formen* (1. Aufl., [Nachdr.]). Frankfurt am Main: Suhrkamp.

Foucault, M., Nigro, R., & Briggs, K. (2008). *Introduction to Kant's Anthropology*. Cambridge, MA: MIT Press.

Fox, W. (1995). *Toward a Transpersonal Ecology: Developing New Foundations for Environmentalism*. Albany: State University of New York Press.

Frank, A. G. (1998). *ReOrient*. Berkeley: California University Press.

Fraser, N. (2014). *Transnationalizing the Public Sphere* (K. Nash, Ed.). Cambridge and Malden, MA: Polity.

Freeland, C. (2011). The Tea Party vs. The "Freeloader." *New York Times*. Retrieved from www.nytimes.com/2011/07/08/us/08iht-letter08.html.

Freire, P. (2000). *Pedagogy of the oppressed* (30th anniversary ed.). New York, NY: Continuum.

Frickel, S., Gibbon, S., Howard, J., Kempner, J., Ottinger, G., & Hess, D. J. (2010). Undone Science: Charting Social Movement and Civil Society Challenges to Research Agenda Setting. *Science, Technology, & Human Values*, *35*(4), 444–473. doi:10.1177/0162243909345836.

Fujimura, J. (2003). Future Imaginaries: Genome Scientists as Socio-Cultural Entrepreneurs. In A. Goodman, D. Heath, & S. Lindee (eds), *Genetic Nature/Culture: Anthropology and Science Beyond the Two Culture Divide* (pp. 176–199). Berkeley: University of California Press.

Fukuyama, F. (2004). *Statebuilding*. Ithaca, NY: Cornell University Press.

Fukuyama, F. (2013). *What Is Governance?* CGD Working Paper 314. Washington, DC: Center for Global Development.

Fumagalli, A. & Mezzadra, S. (eds) (2010). *Crisis in the Global Economy: Financial Markets, Social Struggles, and New Political Scenarios*. Los Angeles and Cambridge, MA: Semiotext(e) and Distributed by the MIT Press.

Gad, C. & Ribes, D. (2014). The Conceptual and the Empirical in Science and Technology Studies. *Science, Technology, & Human Values*, *39*(2), 183–191. doi:10.1177/0162243914522304.

Gawoll, H. J. (1993). Impulse der Lebensphilosophie. Anmerkungen zum Verhältnis von Heidegger und Simmel. *Simmel Newsletter*, 3(2): 139–151.

Gerhardt, U. (2007). Much More Than a Mere Translation: Talcott Parsons's Translation into English of Max Weber's "Die protestantische Ethik und der Geist des Kapitalismus": An Essay in Intellectual History. *The Canadian Journal of Sociology/Cahiers Canadiens de Sociologie*, *32*(1), 41. doi:10.2307/20460615.

Gerhardt, U. (2016). The Crux of Authenticity Comparing Some Translations of Max Weber's Works into English an Essay in Intellectual History. *Max Weber Studies*, *16*(1), 11–37. doi:10.15543/MWS/2016/1/3.

Gibson-Graham, J. K. (2006a). *The End of Capitalism (As We Knew It): A Feminist Critique of Political Economy* (1st ed.). Minneapolis: University of Minnesota Press.

Gibson-Graham, J. K. (2006b). *A Postcapitalist Politics*. Minneapolis: University of Minnesota Press.

Gibson-Graham, J. K. (2011). A Feminist Project of Belonging for the Anthropocene. *Gender, Place & Culture*, *18*(1), 1–21.

Gibson-Graham, J. K. & Roelvink, R. (2011). The Nitty Gritty of Creating Alternative Economies. *Social Alternatives*, *30*(1), 29–33.

Giesecke, M. (1991). *Der Buchdruck in der fruehen Neuzeit*. Frankfurt am Main: Suhrkamp.

Go, J. (2013a). Decolonizing Bourdieu Colonial and Postcolonial Theory in Pierre Bourdieu's Early Work. *Sociological Theory*, *31*(1), 49–74.

Go, J. (2013b). For a Postcolonial Sociology. *Theory and Society*, *42*(1), 25–55. doi:10.1007/s11186-012-9184-6.

Go, J. (2016). *Postcolonial Thought and Social Theory*. New York, NY: Oxford University Press.

Godfrey-Smith, P. (2016). *Other Minds*. London: William Collins.

Goldberg, S. (2009). Experts, Semantic and Epistemic. *Noûs*, *43*(4), 581–598.

Goldberg, S. (2011). The Division of Epistemic Labor. *Episteme*, *8*(01), 112–125.

Goldman, A. I. (1991). Epistemic paternalism: communication control in law and society. *Journal of Philosophy*, 88, pp. 113–131.

Goldman, A. I. (2009). Systems-oriented social epistemology. *Oxford Studies in Epistemology*, *3*, 189–214.

Goldman, A. I. (2010). Epistemic relativism and reasonable disagreement. In R. Feldman and T. Warfield (eds), *Disagreement* (pp. 187–215). Oxford, UK: Oxford University Press.

Gómez-Barris, M. (2017). *The Extractive Zone*. Durham, NC: Duke University Press.

Gordon, L. R. (n.d.-a). Black Aesthetics, Black Value. *Public Culture*, *30*(1), 19–33.

Gordon, L. R. (n.d.-b). Is the Human a Teleological Suspension of Man? Phenomenological Exploration of Sylvia Wynter's Fanonian and Biodicean Reflections. In A. Bogues (ed), *After Man*. Basingstoke: Palgrave.

Gordon, L. R. (2006). *Disciplinary Decadence: Living Thought in Trying Times*. New York, NY: Routledge.

Gordon, L. R. (2008). *An Introduction to Africana Philosophy*. Retrieved from doi:10.1017/CBO9780511800726.

Gordon, L. R. (2011). Shifting the Geography of Reason in an Age of Disciplinary Decadence. *TRANSMODERNITY: Journal of Peripheral Cultural Production of the Luso-Hispanic World*, 1(2). Retrieved from https://escholarship.org/uc/item/218618vj (last accessed April 21, 2019).

Gordon, L. R. (2014a). Disciplinary Decadence and the Decolonisation of Knowledge. *Africa Development*, *XXXIX*(1), 81–92.

Gordon, L. R. (2014b). Is Philosophy Blue? *Johannesburg Salon*, 7, 15–20.

Gordon, L. R. (2016). Disciplining as a Human Science. *Quaderna: A Multilingual and Transdisciplinary Journal*, *3*.

Gordon, L. R. (2018). Black Aesthetics, Black Value. *Public Culture*, *30*(1), 19–33.

Gordon, S. (2011). *Horrible Bosses*. Burbank, CA: Warner Bros.

Gorski, P. (2013). What Is Critical Realism? And Why Should You Care? *Contemporary Sociology: A Journal of Reviews*, *42*(5), 658–670.

Gottlieb, R. & Joshi, A. (2010). *Food Justice*. Cambridge, MA: MIT Press.

Govindrajan, R. (2018). *Animal Intimacies: Interspecies Relatedness in India's Central Himalayas*. Chicago, IL: The University of Chicago Press.

Graeber, D. & Jeffries, S. (2015). David Graeber Interview: So Many People Spend Their Working Lives Doing Jobs They Think Are Unnecessary. *The Guardian*. Retrieved from www.theguardian.com/books/2015/mar/21/books-interview-david-graeber-the-utopia-of-rules (last accessed November 15, 2018).

Graff, D. (2000). Shifting Sands: An Interest-Relative Theory of Vagueness. *Philosophical Topics*, *28*(1), 45–81.

Graham, P. J. (2010). Testimonial Entitlement and the Function of Comprehension. In D. P., A. M., & A. Haddock (eds), *Social Epistemology* (pp. 148–174). Oxford: Oxford UP.

Graham, P. J. (2012). Epistemic Entitlement. *Nous*, *46*, 449–482.

Gratton, P., Harman, G. (2010) Interviews: Graham Harman, Jane Bennett, Tim Morton, Ian Bogost, Levi Bryant and Paul EnnisPeter Gratton, Graham Harman, Jane Bennett, Tim Morton, Levi Bryant & Paul Ennis. *Speculations* 1 (1), 84–134.

Greene, S. (2004). Indigenous People Incorporated?: Culture as Politics, Culture as Property in Pharmaceutical Bioprospecting. *Current Anthropology, 45*(2), 211–237.

Griffiths, T. F. W. (1998). "Ethnoeconomics and Native Amazonian Livelihood : Culture and Economy among the Nipode-Uitoto of the Middle Caqueta Basin in Colombia." Ph.D., University of Oxford. Retrieved from http://ethos.bl.uk/OrderDetails.do?uin=uk.bl.ethos.264879 (last accessed April 21, 2019).

Groh, R. (1998). *Arbeit an der Heillosigkeit der Welt: zur politisch-theologischen Mythologie und Anthropologie Carl Schmitts* (1. Aufl). Frankfurt: Suhrkamp.

Grosfoguel, R. (2013). The Structure of Knowledge in Westernized Universities: Epistemic Racism/Sexism and the Four Genocides/Epistemicides of the Long 16th Century. *Uman Architecture: Journal of the Sociology of Self-Knowledge, 11*(1), 1–19.

Grosfoguel, R., Boidin, C., & Cohen, J. (2013). Special Issue: Decolonizing the University: Practicing Pluriversity. *Human Architecture, 10*(1).

Grossberg, L., Hardin, C., & Palm, M. (2014). Contributions to a Conjunctural Theory of Valuation. *Rethinking Marxism, 26*(3), 306–335. doi:10.1080/08935696.2014.895543.

Grosz, E. (2011). *Becoming Undone.* Durham, NC: Duke University Press.

Guattari, F. (2005). *The Three Ecologies.* London and New York, NY: Continuum.

Gushing, K. (2005). *Reform and the Papacy in the Eleventh Century: Spirituality and Social Change.* Manchester: Manchester University Press.

Guyer, J. I. (2007). Prophecy and the Near Future: Thoughts on Macroeconomic, Evangelical, and Punctuated Time. *American Ethnologist, 34*(3), 409–421. doi:10.1525/ae.2007.34.3.409.

Habermas, J. (1973). *Legitimationsprobleme im Spätkapitalismus* (1. Aufl., [Nachdr.]). Frankfurt am Main: Suhrkamp.

Habermas, J. (1987). *Philosopisch-Politische Profile.* Frankfurt am Main: Suhrkamp.

Habermas, J. (1999a). Kants Idee des Ewigen Friedens. In Habermas, Juergen (ed), *Die Einbeziehung des Anderen.* Frankfurt aM: Suhrkamp.

Habermas, J. (1999b). *The Structural Transformation of the Public Sphere: An Inquiry into a Category of Bourgeois Society* (10. print). Cambridge, MA: MIT Press.

Habermas, J. (2015). *Strukturwandel der Öffentlichkeit: Untersuchungen zu einer Kategorie der bürgerlichen Gesellschaft ; mit einem Vorwort zur Neuauflage 1990* (14. Aufl). Frankfurt am Main: Suhrkamp.

Haddad, C. (2015). Services in the Self: Embodied Labor and the Global Bioeconomy. *Life Sciences, Society and Policy, 11*(1). doi:10.1186/s40504-015-0027-x.

Haddock, A., et al. (ed) (n.d.). *Social Epistemology.* Oxford and New York, NY: Oxford University Press.

Hallett, T. (2010). The Myth Incarnate: Recoupling Processes, Turmoil, and Inhabited Institutions in an Urban Elementary School. *American Sociological Review, 75*(1), 52–74.

Hallett, T. & Ventresca, M. J. Theor Soc (2006). Inhabited Institutions: Social Interactions and Organizational Forms in Gouldner's Patterns of Industrial Bureaucracy. *Theory and Society, 35*, 213–236.

Haraway, D. J. (1988). Situated Knowledges: The Science Question in Feminism and the Privilege of Partial Perspective. *Feminist Studies, 14*(3), 575–599.

Haraway, D. J. (2015). Anthropocene, Capitalocene, Plantationocene, Chthulucene: Making Kin. *Environmental Humanities, 6*(1), 159–165.

Haraway, D. J. (2016). *Staying with the Trouble: Making Kin in the Chthulucene.* Durham, NC: Duke University Press.

Harding, S. (2008). *Sciences from Below: Feminisms, Postcolonialities, and Modernities.* Durham, NC: Duke University Press.

Harding, S. (ed) (2011). *The Postcolonial Science and Technology Studies Reader.* Durham, NC: Duke University Press Books.

Harrington, A. (ed) (2000). *The Placebo Effect: An Interdisciplinary Exploration* (3. Print). Cambridge, MA: Harvard University Press.

Harrington, A. (2008). *The Cure Within: A History of Mind-Body Medicine* (1st ed.). New York, NY: W.W. Norton.

Harris, C. I. (1993). Whiteness As Property. *Harvard Law Review, 106*(8), 1707. doi:10.2307/1341787.

Haslanger, S. (1995). Ontology and Social Construction. *Philosophical Topics, 23*(2), 95–125.

Haslanger, S. (1999). What Knowledge Is and What It Ought to Be: Feminist Values and Normative Epistemology. *Noûs, 33*, 459–480.

Haslanger, S. (2015). Distinguished Lecture: Social Structure, Narrative and Explanation. *Canadian Journal of Philosophy, 45*(1), 1–15.

Heidegger, M. (1994) Creative Landscape: Why Do We Stay in the Provinces? In A. Kaes, M. Jay, & E. Dimendberg (eds), *The Weimar Republic Sourcebook* (pp. 426–428). Berkeley, California: University of California Press.

Hengehold, L. (2007). *The Body Problematic.* University Park, PA: Penn State University Press.

Henrich, J., Heine, S. J., & Norenzayan, A. (2010). The Weirdest People in the World? *Behavioral and Brain Sciences, 33*(2–3), 61–83. doi:10.1017/S0140525X0999152X.

Hester, H. (2018). *Xenofeminism.* Chichester: Wiley.

Hoenig, B. (2017). Competing for Status: Dynamics of Scientific Disciplines in the European Transnational Field. *Serendipities. Journal for the Sociology and History of the Social Sciences, 2*(1), 90–106; ISSN: 2521-0947. doi:10.25364/11.2:2017.1.6.

Hohle, R. (2015). *Race and the Origins of American Neoliberalism.* London: Routledge.

Holbraad, M. (2015). *Times of Security: Ethnographies of Fear, Protest and the future.* New York, NY: Routledge.

Holbraad, M. & Pedersen, M. A. (2013) *Times of Security: Ethnographies of Fear, Protest and the Future.* London: Routledge.

Hong, G. K. & Ferguson, R. A. (eds) (2011). *Strange Affinities: The Gender and Sexual Politics of Comparative Racialization.* Durham, NC: Duke University Press.

Hopkins, D. (2016). *How a TV Sitcom Triggered the Downfall of Western Civilization.* Retrieved from https://medium.com/s/story/how-a-tv-sitcom-triggered-the-downfall-of-western-civilization-336e8ccf7dd0 (last accessed April 21, 2019).

Hörl, E. (2013a). A Thousand Ecologies: The Process of Cyberneticization and General Ecology (Jeffrey Kirkwood, James Burton, and Maria Vlotides trans.). In D. Diederichsen & A. Franke (eds), *The Whole Earth. California and the Disappearance of the Outside* (pp. 121–130). Berlin: Sternberg Press.

Hörl, E. (2013b). The Artificial Intelligence of Sense: The History of Sense and Technology After Jean-Luc Nancy. *Parrhesia : A Journal of Critical Philosophy, 17*(9), 11–24.

Hörl, E. (2015). The Technological Condition. *Parrhesia, 22*, 1–15.

Howe, C., Lockrem, J., Appel, H., Hackett, E., Boyer, D., Hall, R., … Mody, C. (2016). Paradoxical Infrastructures: Ruins, Retrofit, and Risk. *Science, Technology, & Human Values, 41*(3), 547–565. doi:0.1177/0162243915620017.

Humphreys, S. & Otomo, Y. (2014). Theorising International Environmental Law. *LSE Law, Society and Economy Working Papers 9/2014 London School of Economics and Political Science Law Department.*

Iannucci, A. (2015). Politics Was Once About Beliefs and Society. Now It's a Worship of Money. *The Guardian*. Retrieved from www.theguardian.com/commentisfree/2015/mar/08/armando-iannucci-money-at-heart-of-politics-general-election-2015 (last accessed November 15, 2018).

Inglis, J. (2016). *Thomas Aquinas*. Retrieved from http://search.ebscohost.com/login.aspx?direct=true&scope=site&db=nlebk&db=nlabk&AN=1521014 (last accessed April 21, 2019).

Ingold, T. (2008). Anthropology is Not Ethnography Radcliffe-Brown Lecture in Social Anthropology held at the British Academy, *Proceedings of the British Academy*.

Ingold, T. (2017). Surface Visions. *Theory, Culture & Society*, *34*(7–8), 99–108. doi:10.1177/0263276417730601.

Innis, H. (1951). *The Bias of Communication*. Toronto, ON: University of Toronto Press.

Issac, B. H. (2006). *The Invention of Racism in Classical Antiquity* (4. print., and 1. paperback print). Princeton, NJ: Princeton University Press.

Issac, B. H. (2017). *Empire and Ideology in the Graeco-Roman World: Selected Papers*. Cambridge and New York, NY: Cambridge University Press.

Iyko, D. (2015). Being or Nothingness: Indigeneity, Antiblackness, and Settler Colonial Critique. *Critical Ethnic Studies*, *1*(2), 102. doi:10.5749/jcritethnstud.1.2.0102.

Jalbert, J. E. (2003). Time, Death, and History in Simmel and Heidegger. *Human Studies*, *26*(2), 259–283. doi:10.1023/A:1024024624911.

Jamme, C. (1999). *"Gott an hat ein Gewand": Grenzen und Perspektiven philosophischer Mythos-Theorien der Gegenwart ; mit einem Vorw. zur Taschenbuchausg* (1. Aufl). Frankfurt am Main: Suhrkamp.

Janara, L. (2004). Brothers and Others: Tocqueville and Beaumont, U.S. Genealogy, Democracy, and Racism. *Political Theory*, *32*(6), 773–800. doi:10.1177/0090591704265931.

Jasanoff, S. & Kim, S.-H. (2009). Containing the Atom: Sociotechnical Imaginaries and Nuclear Power in the United States and South Korea. *Minerva*, *47*(2), 119–146. doi:10.1007/s11024-009-9124-4.

Jones, J. P., Nast, H. J., & Roberts, S. M. (eds) (1997). *Thresholds in Feminist Geography: Difference, Methodology, and Representation*. Lanham, MD: Rowman & Littlefield Publishers.

Juris, J. S. & Khasnabish, A. (eds) (2013). *Insurgent Encounters: Transnational Activism, Ethnography, and the Political*. Durham, NC and London: Duke University Press.

Kamiyama, K. (2009b). *Eden of the East (東のエデン Higashi no Eden)*. Tokyo: Production I.G.

Karafyllis, N. (2003). *Biofakte: Versuch über den Menschen zwischen Artefakt und Lebewesen* (1., Aufl). Paderborn: mentis.

Kaul, N. (2008). *Imagining Economics Otherwise: Encounters with Identity Difference*. London: Routledge.

Kemple, T. (2012). The Spatial Sense of Empire: Encountering Strangers with Simmel, Tocqueville and Martineau. *Oñati Socio-Legal Series* [Online], *2*(7), 1–17.

Kemple, T. & Mawani, R. (2009). The Sociological Imagination and Its Imperial Shadows. *Theory, Culture & Society*, *26*(7–8), 228–249.

Kidd, I. J. & Carel, H. (2016). Epistemic Injustice and Illness. *Journal of Applied Philosophy*, 33, 172–190.

Kidman, J. & O'Malley, V. (Online First). Questioning the Canon: Colonial History, Counter-Memory and Youth Activism. *Memory Studies.*

Kinchy, A. J. (2012). *Seeds, Science, and Struggle: The Global Politics of Transgenic Crops.* Cambridge, MA: MIT Press.

Kirksey, E. (2015). *Emergent Ecologies.* Durham, NC: Duke University Press.

Knorr-Cetina, K. (1999). *Epistemic Cultures: How the Sciences Make Knowledge.* Cambridge, MA: Harvard University Press.

Kockelman, P. (2006). A Semiotic Ontology of the Commodity. *Journal of Linguistic Anthropology, 16*(1), 76–102.

Kockelman, P. (2016). *The Chicken and the Quetzal: Incommensurate Ontologies and Portable Values in Guatemala's Cloud Forest.* Durham, NC: Duke University Press.

Kohn, M. (2002). The Other America: Tocqueville and Beaumont on Race and Slavery. *Polity, 35*(2), 169–193. doi:10.1086/POLv35n2ms3235496.

Konkel, R. (2014). The Monetization of Global Poverty: The Concept of Poverty in World Bank History, 1944–1990. *Journal of Global History, 9*(02), 276–300. doi:10.1017/S1740022814000072.

Krause, S. S. (2017). Über Sklaverei, Rassismus und die Segmentierung der Gesellschaft. Tocquevilles Analyse sozialer Ausgrenzung in der Demokratie. *Berliner Journal für Soziologie, 27*(1), 151–167. doi:10.1007/s11609-017-0327-5.

Kuehn, M. (2001). *Kant: A Biography.* Cambridge: Cambridge University Press.

Kunze, K., Leipholz, N. S., Macfalda, L., Leuphana Universität Lüneburg, & Leuphana Universität Lüneburg (eds). (2018). *Eine Universität für das 21. Jahrhundert? 10 Jahre "Leuphana": eine kritische Festschrift* (1. Auflage). Lüneburg: AStA Universität Lüneburg.

Kuprecht, K. (2014). *Indigenous Peoples' Cultural Property Claims Repatriation and Beyond.* doi:10.1007/978-3-319-01655-9.

Lakoff, G. & Johnson, M. (2003). *Metaphors We Live By.* Chicago, IL: University of Chicago Press.

Lamla, J. (2013). *Verbraucherdemokratie: politische Soziologie der Konsumgesellschaft* (Erste Auflage). Berlin: Suhrkamp.

Landa, I. (2015a). Bataille: The Master, the Slave, and Consumption. *Critical Sociology, 41*(7–8), 1087–1102. doi:10.1177/0896920514526624.

Landa, I. (2015b). Bataille's Libidinal Economics: Capitalism as an Open Wound. *Critical Sociology, 41*(4–5), 581–596. doi:10.1177/0896920514526625.

Landecker, H. (2016). Antibiotic Resistance and the Biology of History. *Body & Society, 22*(4), 19–52. doi:10.1177/1357034X14561341.

Lange, K. W., Reichl, S., Lange, K. M., Tucha, L., & Tucha, O. (2010). The History of Attention Deficit Hyperactivity Disorder. *ADHD Attention Deficit and Hyperactivity Disorders, 2*(4), 241–255. doi:10.1007/s12402-010-0045-8.

Langer, E. (2009). *Counter Clockwise: Mindful Health and the Power of Possibility* (1st ed.). New York, NY: Ballantine Books.

Langman, L. & Smith, D. A. (eds) (2018). *Twenty-First Century Inequality & Capitalism: Piketty, Marx and Beyond.* Leiden and Boston, MA: Brill.

Larkin, B. (2013). The Politics and Poetics of Infrastructure. *Annual Review of Anthropology, 42*(1), 327–343. doi:10.1146/annurev-anthro-092412-155522.

Latour B (2004b) Why has critique run out of steam? From matters of fact to matters of concern. Critical Inquiry 30(2): 225–248.

Latour, B., Stengers, I., Tsing, A., & Bubandt, N. (2018). Anthropologists Are Talking – About Capitalism, Ecology, and Apocalypse. *Ethnos, 83*(3), 587–606. doi:10.1080/00141844.2018.1457703.

Latta, A. (2013). Locating Nature's Citizens: Latin American Ecologies of Political Space. *Citizenship Studies, 17*(5), 566–580. doi:10.1080/13621025.2013.818372.

Lawrence, B. & Dua, E. (2005). Decolonizing Antiracism. *Social Justice, 32*(4), 120–143 (120).

Lazzarato, M. (2012). *The Making of the Indebted Man: An Essay on the Neoliberal Condition.* Los Angeles, CA: Semiotext(e).

Lee, C. (2016). HBO's John Oliver Shows the Infuriating Truth About Credit Reporting Agencies. Retrieved from http://Fortune.Com/2016/04/11/Hbo-John-Oliver-Reveals-the-Awful-Business-behind-Credit-Background-Checks/ (last accessed November 15, 2018).

Lefebvre, A. (2005). *The Political Given: Decisionism in Schmitt's Concept of the Political. 132*(Telos), 83–98.

Lefebvre, H., Elden, S., Lebas, E., & Kofman, E. (2017). *Key Writings.* London: Bloomsbury Publishing.

Lemke, T. (2015). New Materialisms: Foucault and the 'Government of Things.' *Theory, Culture & Society, 32*(4), 3–25. doi:10.1177/0263276413519340.

Lemm, V. (2013). Nietzsche, *Einverleibung* and the Politics of Immunity. *International Journal of Philosophical Studies, 21*(1), 3–19. doi:10.1080/09672559.2012.746271.

Lemm, V. (2016). Is Nietzsche a Naturalist?: Or How to Become a Responsible Plant. *Journal of Nietzsche Studies, 47*(1), 61. doi:10.5325/jnietstud.47.1.0061.

Lenoir, T. (1989). *The Strategy of Life: Teleology and Mechanics in Nineteenth-Century German Biology.* Chicago, IL: University of Chicago Press.

Lepenies, W. (1985). *Die drei Kulturen.* Frankfurt am Main: Fischer.

Levina, M. & Silva, K. (2018). Cruel Intentions: Affect Theory in the Age of Trump. *Communication and Critical/Cultural Studies, 15*(1), 70–72.

Levins, R. & Lewontin, R. (1985). *The Dialectical Biologist.* Cambridge, MA: Harvard University Press.

Lévi-Strauss, C. (1973). *Tristes tropiques.* New York, NY: Jonathan Cape.

Levkoe, C. Z., Brem-Wilson, J., & Anderson, C. R. (2018). People, Power, Change: Three Pillars of a Food Sovereignty Research Praxis. *The Journal of Peasant Studies*, 1–24. doi:10.1080/03066150.2018.1512488.

Lewandowski, I. (2017). *Bioeconomy: Shaping the Transition to a Sustainable, Biobased Economy.* Dodrecht: Springer.

Lhuilier, G. (2013a). The Concept of 'Normative Space' – A Very Short Introduction. In Les Cahiers d'Ebisu. Occasional Papers No. 3: 67–70.

Lhuilier, G. (2013b). Three Views on Global Law and Global Legal Theory. In Les Cahiers d'Ebisu: Occasional Papers No. 3: 51–59.

Lhuilier, G. (2016). *Le droit transnational.* Paris: Dalloz.

Lindemann, H., Verkerk, M., & Walker, M. U. (eds) (2009). *Naturalized Bioethics: Toward Responsible Knowing and Practice.* Cambridge and New York, NY: Cambridge University Press.

Linke, U. (2006). Contact zones: Rethinking the sensual life of the state. *Anthropological Theory, 6*(2), 205–222.

Lockley, P. (2018). Open Initiatives for Decolonising the Curriculum. In G. K. Bhambra, D. Gebrial, & K. Nişancıoğlu (eds), *Decolonising the University* (pp. 145–170). Chicago, IL: Pluto Press.

Longino, H. (2013). *Studying Human Behavior: How Scientists Investigate Aggression and Sexuality.* Chicago, IL: University of Chicago Press.

Loraux, N. (2000). *Born of the Earth: Myth and Politics in Athens.* Ithaca, NY: Cornell University Press.

Lotz, C. (2014). *The Capitalist Schema: Time, Money, and the Culture of Abstraction.* Lanham, MD: Lexington Books.

Lowe, L. (2005). Insufficient Difference. *Ethnicities, 5*(3), 409–414. doi:10.1177/146879680500500308.

Lowe, L. (2015). *The Intimacies of Four Continents.* Durham, NC: Duke University Press.

Lübbe, H. (1965). *Säkularisierung.* Freiburg, München: Alber.Lugones, M. (1987). Playfulness, "World"-Travelling, and Loving Perception. *Hypatia, 2*(2), 3–19.

Lugones, M. (2008). The Coloniality of Gender. *Worlds & Knowledges Otherwise,* (Spring), 1–17.

Lugones, M. (2010). Toward a Decolonial Feminism. *Hypatia, 25*(4), 742–759.

Luper, S. (2004). Epistemic Relativism. *Philosophical Issues, 14,* 271–295.

Lynch, M. P. (2009). *Truth As One and Many.* Oxford and New York, NY: Clarendon Press.

Lynch, M. P. (2013). Epistemic Commitments, Epistemic Agency and Practical Reasons. *Philosophical Issues, 23*(1), 343–362.

Lyotard, J.-F. (1989). *Differend: Phrases in Dispute* (1st ed.). Minneapolis: University of Minnesota Press.

Lyotard, J.-F. (1993). *Political Writings.* Minneapolis: University of Minnesota Press.

Magubane, Z. (2016a). American Sociology's Racial Ontology: Remembering Slavery, Deconstructing Modernity, and Charting the Future of a Global Historical Sociology. *Cultural Sociology, 10*(3), 369–384.

Magubane, Z. (2016b). The Empire's Old Clothes: Debunking the "Newness" of the New Racism. Presented at the Mini-Conference at ESS Meeting 2016: Panel c. The Empire's Old Clothes: Debunking the "Newness" of the New Racism, Boston, MA.

Mah, A. (2017) Environmental Justice in the Age of Big Data: Challenging Toxic Blind Spots of Voice, Speed, and Expertise. *Environmental Sociology, 3*(2), 122–133. doi:10.1080/23251042.2016.1220849.

Maldonado-Torres, N. (2007). On the coloniality of Being. *Cultural Studies, 21*(Issue 2–3: Globalization and the De-Colonial Option), 240–270.

Mann, M. (1984). The Autonomous Power of the State: Its Origins, Mechanisms, and Results. *Archives européenes de sociologie, 25,* 185–213.

Manne, K. (2018). *Down Girl: The Logic of Misogyny.* New York, NY: Oxford University Press.

Manning, E. (2007). Relationscapes: How Contemporary Aboriginal Art Moves Beyond the Map. *Cultural Studies Review, 13*(2), 134–155.

Manning, E. (2009a). What If It Didn't All Begin and End With Containment? Toward a Leaky Sense of Self. *Body & Society, 15*(3), 33–45. doi:10.1177/1357034X09337785.

Manning, E. (2009b). *Relationscapes.* Cambridge, MA: MIT Press.

Mansfield, B. (2017). Folded Futurity: Epigenetic Plasticity, Temporality, and New Thresholds of Fetal Life. *Science as Culture, 26*(3), 355–379. doi:10.1080/09505431.2017.1294575.

Mansfield, B. & Guthman, J. (2015). Epigenetic Life: Biological Plasticity, Abnormality, and New Configurations of Race and Reproduction. *Cultural Geographies, 22*(1), 3–20. doi:10.1177/1474474014555659.

Marchart, O. (2008). *Post-Foundational Political Thought: Political Difference in Nancy, Lefort, Bafiou and Laclau* (Transferred to Digital Print). Edinburgh: Edinburgh University Press.

Marinetto, M. (2003). Governing Beyond the Centre: A Critique of the Anglo-Governance School. *Political Studies, 51*(3), 592–608. doi:10.1111/1467-9248.00443.

Markwardt, N. (2019). *Der Mythos der sozialen Marktwirschaft*. Retrieved from www.zeit.de/kultur/2019-05/soziale-marktwirtschaft-staat-wirtschaftspolitik-demokratie-sozialismus-neoliberalismus-ideengeschichte (last accessed April 21, 2019).

Massumi, B. (2009). National Enterprise Emergency: Steps Toward an Ecology of Powers. *Theory, Culture & Society, 26*(6), 153–185. doi:10.1177/0263276409347696.

Massumi, B. (2014). *What Animals Teach Us About Politics*. Durham, NC: Duke University Press.

Massumi, B. (2015). *Ontopower: War, Powers, and the State of Perception*. Durham, NC: Duke University Press.

Maus, I. (1994). *Zur Aufklärung der Demokratietheorie: rechts- und demokratietheoretische Überlegungen im Anschluß an Kant* (1. Aufl). Frankfurt am Main: Suhrkamp.

Mayer, E. (1987). Macro-Ethno-Economics. *Reviews in Anthropology, 14*(3), 250–256. doi:10.1080/00988157.1987.9977833.

Mayo, E. (1972). *The Psychology of Pierre Janet*. Westport, CT: Greenwood Press.

Mazzucato, V. (1997). Indigenous Economies: Bridging the Gap between Economics and Anthropology. *Indigenous Knowledge and Development Monitor*. Retrieved from http://library.wur.nl/WebQuery/wurpubs/39252 (last accessed April 21, 2019).

Marazzi, C. (2010). The Violence of Financial Capitalism, In A. Fumagalli & S. Mezzarda (eds), *Crisis in the Global Economy* (pp. 17–60). Los Angeles: Semiotexte.

Mbembe, A. (2015). Chille Mbembe on the State of South African Political Life. Retrieved from https://Africasacountry.Com/2015/09/Achille-Mbembe-on-the-State-of-South-African-Politics/ (last accessed December 15, 2018).

Mbembe, A. & Comaroff, J. (2010). Africa in Theory: A Conversation Between Jean Comaroff and Achille Mbembe. *Anthropological Quarterly, 83*(3): 653–678.

McCarthy, G. E. (2002). *Classical Horizons: The Origins of Sociology in Ancient Greece*. Albany: State University of New York Press.

McCarthy, G. E. (2009). *Dreams in Exile: Rediscovering Science and Ethics in Nineteenth-Century Social Theory*. Albany: State University of New York Press.

McComack, F. (2012). Fish Is My Daily Bread: Owning and Transacting in Maori Fisheries. *Anthropological Forum, 20*(1), 19–39.

McCormick, J. P. (1995). Dangers of Mythologizing Technology and Politics: Nietzsche, Schmitt and the Antichrist. *Philosophy & Social Criticism, 21*(4), 55–92. doi:10.1177/019145379502100404.

McCormick, J. P. (1997). *Carl Schmitt's Critique of Liberalism: Against Politics as Technology*. doi:10.1017/CBO9780511608988.

McDowell, D. M. (1989). The "Oikos" in Athenian law. *Classical Quarterly, 39*(i), 10–21.

McDowell, J. H. (1996). *Mind and World: With a New Introduction* (1st Paperback ed.). Cambridge, MA: Harvard University Press.

McLaughlin, N., Kowalchuk, L. & Turcotte, K. (2005). Why sociology does not need to be saved: Analytic reflections on public sociologiesThe American SociologistSeptember 2005, Volume 36, Issue 3–4, pp 133–151) 36: 133.

McWeeny, J. (2010). Liberating Anger, Embodying Knowledge: A Comparative Study of María Lugones and Zen Master Hakuin. *Hypatia, 25*(2), 295–315.

McWeeny, J. (2014). Topographies of Flesh: Women, Nonhuman Animals, and the Embodiment of Connection and Difference. *Hypatia, 29*(2), 269–286.

Merleau-Ponty, M. (2002). *Phenomenology of perception*. London: Routledge.

Merson, J. (2000). Bio-Prospecting or Bio-Piracy: Intellectual Property Rights and Biodiversity in a Colonial and Postcolonial Context. *Osiris, 15*, 282–296.

Merton, R. K. (1995). The Thomas Theorem and the Matthew Effect. *Social Forces, 74*(2), 379.

Metraux, G. (1999). Ancient Housing: "Oikos" and "Domus" in Greece and Rome. *Journal of the Society of Architectural Historians, 58*(3), 392–405. doi:10.2307/991533.

Metz, J. B. (1989). Anamnetische Vernunft. In A. Honneth, et al. (ed), *Zwischenbetrachtungen: im Prozeß d. Aufklärung; Jürgen Habermas zum 60. Geburtstag* (1. Aufl, pp. 733–738). Frankfurt am Main: Suhrkamp.

Mieg, H. A., Sundsboe, A. O., & Bieniok, M. (eds) (2011). *Georg Simmel und die aktuelle Stadtforschung*. Wiesbaden: VS Verlag.

Miéville, C. (2008). *The City and the City*. London: Macmillan.

Mignolo, W. D. (2009). Epistemic Disobedience, Independent Thought, and De-Colonial Freedom. *Theory, Culture, and Society, 26*(7–8), 159–182.

Mignolo, M. (2013). Geopolitics of sensing and knowing: On (de)coloniality, border thinking, and epistemic disobedience. *Confero*, 1 (1), 129 – 150.

Mignolo, W. D. (2011). *The Darker Side of Western Modernity: Global Futures, Decolonial Options*. Durham, NC: Duke University Press Books.

Mignolo, W. D. (2014). Spirit Out of Bounds Returns to the East: The Closing of the Social Sciences and the Opening of Independent Thoughts. *Current Sociology, 62*(4), 584–602. doi:10.1177/0011392114524513.

Mignolo, W. D. (2018). Decoloniality and Phenomenology: The Geopolitics of Knowing and Epistemic/Ontological Colonial Differences. *The Journal of Speculative Philosophy, 32*(3), 360–387.

Mignolo, W. D. (n.d.). Time to Bleed | The Chimurenga Chronic. Retrieved July 10, 2016, from http://chimurengachronic.co.za/time-to-bleed-an-extended-conversation-with-walter-mignolo/.

Mignolo, W.D. & Kaganof, A. (2016) *Time to Bleed*. Retrieved from http://chimurengachronic.co.za/time-to-bleed-an-extended-conversation-with-walter-mignolo/ (last accessed October 31, 2019).

Mignolo, W. D. & Tlostanova, M. (2006). Theorizing from the Borders Shifting to Geo- and Body-Politics of Knowledge. *European Journal of Social Theory, 9*(2), 205–221.

Mignolo, W. D. & Vázquez, R. (2013, July 15). Decolonial AestheSis: Colonial Wounds/Decolonial Healings. Retrieved December 24, 2014, from http://socialtextjournal.org/periscope_article/decolonial-aesthesis-colonial-woundsdecolonial-healings/.

Mignolo, W. D. & Walsh, C. (2018). *On Decoloniality*. Durham, NC: Duke University Press.

Milkov, N. (2002). Lotze's Concept of 'States of Affairs' and Its Critics. *Prima Philosophia*, 15 437–450.

Miller, D. A. (1999). *Crucible of Empire: The Spanish American War.* Alexandria, VA: PBS.

Miller, D. A. (2002). Turning Callon the Right Way Up. *Economy and Society, XXXI*, 218–233.

Miller, M. C. (2005). *Power and the Holy in the Time of the Investiture Conflict.* Boston, MA: Bedford/St.Martin's.

Miller, N. (2006). The Sexual Politics of Pain: Hannah Arendt Meets Shakespeare's Shrew. *Journal of Cultural and Religious Theory*, 7(2), 18–32.

Mills, C. W. (2002). *White Collar: The American Middle Classes* (50th Anniversary ed.). New York, NY: Oxford University Press.

Moore, J. W. (2011). Transcending the Metabolic Rift: A Theory of Crises in the Capitalist World-Ecology. *Journal of Peasant Studies, 38*(1), 1–46. doi:10.1080/03066150.2010.538579.

Morris, A. D. (2015). *The Scholar Denied: W.E.B. Du Bois and the Birth of Modern Sociology.* Oakland: University of California Press.

Motta, R. (2014). Social Disputes over GMOs: An Overview: Social Disputes over GMOs. *Sociology Compass, 8*(12), 1360–1376.

Mouffe, C. (2013). Agonistics: Thinking the World Politically (1st ed.). London and New York, NY: Verso.

Mukerji, C. (2010). The Territorial State as a Figured World of Power: Strategics, Logistics, and Impersonal Rule. *Sociological Theory, 28*, 402–424.

Murai, H. (2012). Hannah Arendt and Karl Jaspers: What Influence Did Jaspers Have on Arendt's Concept of Judgment? 『総合政策論叢』第2 3号 （2 0 1 2年3月）. Retrieved from http://hamada.u-shimane.ac.jp/research/32kiyou/10sogo/seisaku23.data/seisaku2303.pdf (last accessed April 21, 2019).

Murphy, M. (2017). *The Economization of Life.* Durham, NC and London: Duke University Press.

Myers, N. (2015). *Rendering Life Molecular: Models, Modelers, and Excitable Matter.* Durham, NC: Duke University Press. 2015.

Nafissi, M. (2005). *Ancient Athens & Modern Ideology. Value, Theory & Evidence in Historical Sciences: Max Weber, Karl Polanyi & Moses Finley.* London: University of London, Institute of Classical Studies.

Nancy, J.-L. (2013). Of Struction. *Parrhesia, 17*, 1–10.

Ndlovu-Gatsheni, S. (2015). *Empire, Global Coloniality and African Subjectivity.* Berghahn: New York and Oxford.

Ndlovu-Gatsheni, S. J. (2013). *Coloniality of Power in Postcolonial Africa. Myths of Decolonization.* Codesria.

Neckel, S. (2010). Refeudalisierung der Ökonomie Zum Strukturwandel kapitalistischer Wirtschaft. *MPIfG Working Paper MPIfG Working Paper 10/6.*

Neckel, S. (2013). 'Refeudalisierung' – Systematik und Aktualität eines Begriffs der Habermas'schen Gesellschaftsanalyse. *Leviathan*, 41(1), 39–56.

Neiman, S. (1997). *The Unity of Reason: Rereading Kant* (First issued as an Oxford University Press Paperback). New York, NY and Oxford: Oxford University Press.

Nelsestuen, G. A. (2017). Oikonomia as a Theory of Empire in the Political Thought of Xenophon and Aristotle. *Greek, Roman, and Byzantine Studies, 57*, 74–104.

Nelson, A. R. (2001). *Education and Democracy: The Meaning of Alexander Meiklejohn, 1872–1964.* Madison, WI: Wisconsin UP. Retrieved from http://public.eblib.com/choice/publicfullrecord.aspx?p=3444838.

Nelson, E. (2015). 6 Tracking the Hippocratic Woozle: Pseudepigrapha and the Formation of the Corpus. In L. Dean-Jones & R. M. Rosen (eds), *Ancient Concepts of the Hippocratic* (pp. 117–140). doi:10.1163/9789004307407_008.

Nelson, J. A. (1996). *Feminism, Objectivity and Economics*. London: Routledge.

Neptune, H. R. (2014). Savaging Civilization: Michel-Rolph Trouillot and the Anthropology of the West. *Cultural Dynamics*, *26*(2), 219–234.

Neumann, I. B. (2014). International Relations as a Social Science. *Millennium: Journal of International Studies*, *43*(1), 330–350. doi:10.1177/0305829814539860.

Nibert, D. A. (2002). *Animal Rights/Human Rights: Entanglements of Oppression and Liberation*. Lanham, MD: Rowman & Littlefield.

Nicholson, L. (ed) (1997). *The second wave: a reader in feminist theory*, New York: Routledge

Niemi, J. I. (2003). Habermas and Validity Claims. *International Journal of Philosophical Studies*, *13*(2), 227–244.

Noll, M. A. (2014). Tocqueville's *America*, Beaumont's *Slavery*, and the United States in 1831–32. *American Political Thought*, *3*(2), 273–302. doi:10.1086/677734.

Norris, C. (1998). Deconstruction, Postmodernism and Philosophy of Science: Some Epistemo-Critical Bearings. *Cultural Values*, *2*(1), 18–50.

Ober, J. (2002). *Political Dissent in Democratic Athens: Intellectual Critics of Popular Rule* (3. Print. and 1. Paperback Print). Princeton, NJ: Princeton University Press.

Ojakangas, M. (2016). *On the Greek Origins of Biopolitics: A Reinterpretation of the History of Biopower*. Retrieved from http://search.ebscohost.com/login.aspx?direct=true&scope=site&db=nlebk&db=nlabk&AN=1228705 (last accessed April 21, 2019).

Oliver, K. (2003). Subjectivity and Subject Position: The Double Meaning of Witnessing. *Studies in Practical Philosophy*, *3*(2), 132–143.

Oliver, K. (2004). Witnessing and Testimony. *Parallax*, *10*(1), 78–87.

O'Neill, O. (1989). *Constructions of Reason: Explorations of Kant's Practical Philosophy*. Cambridge and New York, NY: Cambridge University Press.

O'Neill, O. (2004). *Bounds of Justice*. Cambridge and New York, NY: Cambridge University Press.

Ong, A. (1999). *Flexible Citizenship*. Durham, NC: Duke University Press.

Ong, A. & Chen, N. N. (eds) (2010). *Asian Biotech: Ethics and Communities of Fate*. Durham, NC: Duke University Press.

Ophir, A., et al. (ed) (2009). *The Power of Inclusive Exclusion: Anatomy of Israeli Rule in the Occupied Palestinian Territories*. New York, NY and Cambridge, MA: Zone Books and Distributed by The MIT Press.

Opitz, S. & Tellmann, U. (2015). Europe as Infrastructure: Networking the Operative Community. *South Atlantic Quarterly*, *114*(1), 171–190. doi:10.1215/00382876-2831356.

Oreskes, N. & Conway, E. M. (2010). *Merchants of Doubt*. New York, NY: Bloomsbury.

Origgi, G. (2008). Trust, Authority and Epistemic Responsibility. *Theoria*, *23*(1), 35–44.

Osamu, N. (2006). Anthropos and Humanitas: Two Western Concept of the Human Being. In N. Sakai and J. Solomon (eds), *Traces 4: Translation, Biopolitics, Colonial Difference* (pp. 259–273). Hong Kong: Hong Kong University Press.

Otomo, Y. (2014). The Gentle Cannibal : The Rise and Fall of Lawful Milk. *Australian Feminist Law Journal, 40*(2), 215–228. doi:10.1080/13200968.2014.1008097.

Ovchinnikov, V. N. & Kolesnikov, Y. S. (2006). Ethnoeconomy as a Factor of Growth. *Studies on Russian Economic Development, 17*(1), 79–82. doi:10.1134/S1075700706010084.

Owens, P. (2015). *Economy of Force*. Cambridge, UK: Cambridge University Press.

Oyewumi, O. (1997). *The Invention of Women: Making an African Sense of Western Gender Discourses* (1st ed.). Minneapolis: University of Minnesota Press.

Papadopoulos, D. (2018). *Experimental Practice: Technoscience, Alterontologies, and More-Than-Social Movements*. Durham, NC: Duke University Press.

Parisi, L. (2007). Biotech: Life by Contagion. *Theory, Culture & Society, 24*(6), 29–52. doi:10.1177/0263276407078711.

Parrinder, P. (1990). Scientists in Science Fiction: Enlightenment and After. In R. Garnett and R.J. Ellis (eds), *Science Fiction Roots and Branches*. Basingstoke: Palgrave Macmillan.

Parsons, T. (1937). *The Structure of Social Action*. New York, NY: McGraw-Hill.

Parsons, T. (1963). On the Concept of Political Power. *Proceedings of the American Philosophical Society, 107*(3), 232–262.

Parsons, T. (1964). Evolutionary Universals in Society. *American Sociological Review*, 29, 3, 339–35.

Parsons, T. (1991) The Integration of Economic and Sociological Theory, *The Marshall Lectures, 61*(1), 10–59.

Parsons, T. & Smelser, N. J. (1956). *Economy and Society: A Study in the Integration of Economic and Social Theory* London: Routledge & Kegan Paul.

Passoth, J.-H. & Rowland, N. J. (2010). Actor-Network State: Integrating Actor-Network Theory and State Theory. *International Sociology, 25*(6), 818–841. doi:10.1177/0268580909351325.

Patel, G. (2006). Risky Subjects: Insurance, Sexuality, and Capital. *Social Text, 24*(4), 25–65. doi:10.1215/01642472-2006-010.

Paxson, H. (2014). *The Life of Cheese: Crafting Food and Value in America*. Berkeley: California University Press.

Pellegrini, P. (2015). The New Urban Question. A Conversation About Bernardo Secchi. Retrieved from www.Planum.Net/the-New-Urban-Question-Br-a-Conversation-on-the-Legacy-of-Bernardo-Secchi-with-Paola-Pellegrini-1 (last accessed April 21, 2019).

Pensky, M. (2004). Method and Time. In D. S. Ferris (ed), *The Cambridge Companion to Walter Benjamin* (pp. 177–198). doi:10.1017/CCOL0521793297.010.

Peters, B. (2007). *Der Sinn vn Öffentlchkeit*. Frankfurt am Main: Suhrkamp.

Pham, M.-H. T. (2016). Feeling Appropriately: On Fashion Copyright Talk and Copynorms. *Social Text*, 34(3 (128)): 51–74.

Pham, M.-H. T. (2017). Racial Plagiarism and Fashion. *QED: A Journal in GLBTQ Worldmaking, 4*(3), 67–80.

Pinzler, P. (2016). *Der Unfreihandel: die heimliche Herrschaft von Konzernen und Kanzleien*. Hamburg: Rowohlt Polaris Verlag.

Poe, M. R., McLain, R. J., Emery, M., & Hurley, P. T. (2013). Urban Forest Justice and the Rights to Wild Foods, Medicines, and Materials in the City. *Human Ecology, 41*(3), 409–422. doi:10.1007/s10745-013-9572-1.

Pommeranz, K. (2000). *The Great Divergence*. Princeton, NJ: Princeton University Press.

Povinelli, E. A. (2016). *Geontologies: A Requiem to Late Liberalism*. Durham, NC: Duke University Press.

Prasad, A. (2016). Discursive Contextures of Science: Euro/West-Centrism and Science and Technology Studies. *Engaging Science, Technology, and Society*, 2, 193.

Preciado, P. B. (n.d.). *Testojunkie*. New York, NY: Feminist Press (first published under B. P.)

Puig de la Bellacasa, M. (2017). *Matters of Care: Speculative Ethics in More Than Human Worlds*. Minneapolis: University of Minnesota Press.

Purcell, M. & Tyman, S. K. (2015). Cultivating Food As a Right to the City. *Local Environment, 20*(10), 1132–1147. doi:10.1080/13549839.2014.903236.

Quijano, A. (2000). Coloniality of Power, Eurocentrism, and Latin America. *Nepantla: Views from South, 1*(3), 533–580.

Raaflaub, K. A., Ober, J., & Wallace, R. W. (2008). *Origins of Democracy in Ancient Greece* (1st paperback print). Berkeley: University of California Press.

Rabinbach, A. (1992). *The Human Motor: Energy, Fatigue, and the Origins of Modernity* (1. paperback print). Berkeley: University of California Press.

Radkau, J. (2005). *Max Weber*. Muenchen: Hanser.

Radkau, J. (2011 [2009]). *Max Weber: A Biography* (English ed). Cambridge and Malden, MA: Polity.

Ramírez Barreto, A. C. (2010). Ontology and Anthropology of Interanimality: Merleau-Ponty from Tim Ingold's perspective. *AIBR. Revista de Antropología Iberoamericana, 05*(01), 32–57. doi:10.11156/aibr.050103e.

Rest, J. R., Narvaez, D., Thoma, S. J., & Bebeau, M. J. (2000). A Neo-Kohlbergian Approach to Morality Research. *Journal of Moral Education, 29*(4), 381–395. doi:10.1080/713679390.

Restivo, S. P. & Loughlin, J. (2000). The Invention of Science. *Cultural Dynamics, 12*(2), 135–149.

Rheinberger, H.-J. (2010). *An Epistemology of the Concrete: Twentieth-Century Histories of Life*. Durham, NC: Duke University Press.

Rhodes, R. A. W. (1994). The Hollowing Out of the State: The Changing Nature of the Public Service in Britain. *The Political Quarterly, 65*(2), 138–151. doi:10.1111/j.1467-923X.1994.tb00441.x.

Rhodes, R. A. W. (1997 [2010]). *Understanding Governance: Policy Networks, Governance, Reflexivity and Accountability* (Repr). Buckingham: Open University Press.

Richards, R. (2002). *The Romantic Conception of Life*. Chicago: Chicago University Press.

Richardson, R. D. (2007). *William James: In the Maelstrom of American Modernism; a Biography* (1. Mariner books ed.). Boston, MA: Houghton Mifflin.

Richter, D. (1996). *Nation als Form*. Wiesbaden: Springer VS.

Riehl A. (1897). *Friedrich Nietzsche*. Stuttgart: Frommann.

Ringguth, R. (1981). Rudolf Ringguth über Karl Jaspers: "Die großen Philosophen. Nachlaß". Retrieved from www.Spiegel.de/Spiegel/Print/d-14352968. Html (last accessed April 21, 2019).

Riskin, J. (2010). *Genesis Redux Essays in the History and Philosophy of Artificial Life*. Chicago, IL: University of Chicago Press.

Ritter, J. (2002). *Metaphysik und Politik*. Frankfurt am Main: Suhrkamp.

Robertson, E. (2014). Propaganda and 'Manufactured Hatred': A Reappraisal of the Ethics of First World War British and Australian atrocity propaganda. *Public Relations Inquiry, 3*(2), 245–266.

Roche, F. (n.d.). Fiction, Ecophysical Apparatus and Skizoid Machines: Animism, Vitalism and Machinism as a Way to Rearticulate the Need to Confront the Unknown in a Contradictory Manner. *Architectural Design*, *80*(6), 64–71.

Rodgers, D. T. (2000). *Atlantic Crossings: Social Politics in a Progressive Age*. Cambridge, MA and London: Belknap.

Rodríguez de Francisco, J. C. & Boelens, R. (2015). Payment for Environmental Services: Mobilising an Epistemic Community to Construct Dominant Policy. *Environmental Politics*, *24*(3), 481–500. doi:10.1080/09644016.2015.1014658.

Rohwetter, M. (2019). Retrieved from www.zeit.de/2019/20/strassenflaeche-verkehr-grossstadt-stadtentwicklung-verkehrspolitik/komplettansicht (last accessed April 21, 2019).

Rorty, R. (1981). Nineteenth-Century Idealism and Twentieth-Century Textualism: *Monist*, *64*(2), 155–174. The Hegeler Institute. doi:10.5840/monist198164211.

Rosa, H. (2010). *Alienation and Acceleration. Towards a Critical Theory of Late-Modern Temporality*. Malmö/Arhus: NSU Press.

Ross, S. T. (2002). Understanding Propaganda: The Epistemic Merit Model and Its Application to Art. *Journal of Aesthetic Education*, *36*, 16–20.

Rothkopf, D. (2012). *Power, Inc.: The Epic Rivalry Between Big Business and Government – and the Reckoning that lies ahead*. New York, NY: Farrar, Straus & Giroux.

Ruddick, S. (1995). *Maternal Thinking: Toward a Politics of Peace; With a New Preface*. Boston, MA: Beacon Press.

Russell, G. (2010). *Epistemic viciousness*. In D. Young and G. Priest (eds), *Martial arts and philosophy: beating and nothingness* (pp. 129–144). Chicago: Open Court.

Saguy, A. C. (2014). *What's Wrong With Fat?* S.l.: s.n.

Sakai, N. (2010). Theory and Asian Humanity: On the Question of Humanitas and Anthropos. *Postcolonial Studies: Culture, Politics, Economy*, *14*(4), 441–464.

Sakupap, T. C. (2018). Ecumenical Ecclesiology in the African Context: Towards a View of the Church as Ubuntu. *Scriptura*, *117*(1). doi:10.7833/117-1-1384.

Santos, B. de S. (2018). *The End of the Cognitive Empire: The Coming of Age of Epistemologies of the South*. Durham, NC: Duke University Press.

Sarasin, P. (2001). *Reizbare Maschinen: Eine Geschichte des Körpers 1765–1914*. Frankfurt am Main: Suhrkamp.

Sarasin, P. (2006). *Anthrax: Bioterror as Fact and Fantasy*. Cambridge, MA: Harvard University Press.

Sarasin, P. (2014). *Geschichtswissenschaft und Diskursanalyse* (Originalausgabe, 4. Auflage). Frankfurt am Main: Suhrkamp.

Sarasin, P. & Tanner, J. (eds) (1998a). *Physiologie und industrielle Gesellschaft: Studien zur Verwissenschaftlichung des Körpers im 19. und 20. Jahrhundert* (1. Aufl). Frankfurt am Main: Suhrkamp.

Sarasin, P. & Tanner, J. (eds) (1998b). *Physiologie und industrielle Gesellschaft: Studien zur Verwissenschaftlichung des Körpers im 19. und 20. Jahrhundert* (1. Aufl). Frankfurt am Main: Suhrkamp.

Sassen, S. (2007). *Territory, Authority, Rights*. Princeton, NJ: Princeton University Press.

Sassen, S. (2014). *Expulsions: Brutality and Complexity in the Global Economy*. Cambridge, MA: The Belknap Press of Harvard University Press.

Savransky, M. (2016). *The Adventure of Relevance: An Ethics of Social Inquiry (with a Foreword by Isabelle Stengers)*. London: Palgrave MacMillan.

Savransky, M. (n.d.). On Relevance (the Very Idea). *Somatosphere*. Retrieved from http://somatosphere.net/2014/10/on-relevance.html (last accessed June 20, 2018).

Sayle, M. (2000). Ethno-Economics in Japan. *Asian Perspective, 24*(4): 179–196.

Scaff, L. A. (2005). The Creation of the Sacred Text: Talcott Parsons Translates the Protestant Ethic and the Spirit of Capitalism. *Max Weber Studies, 5*(2), 205–228.

Scaff, L. A. (2017). *Max Weber in America*. doi:10.23943/princeton/9780691147796.001.0001.

Scharff, C. (2016). The Psychic Life of Neoliberalism: Mapping the Contours of Entrepreneurial Subjectivity. *Theory, Culture & Society, 33*(6), 107–122. doi:10.1177/0263276415590164.

Schaub, D. (1998). Perspectives on Slavery. *Legal Studies Forum, 22*, 607–626.

Schluchter, W. (1996). *Unversöhnte Moderne* (1. Aufl). Frankfurt am Main: Suhrkamp.

Schluchter, W. (2000). Handlungs- und Strukturtheorie nach Max Weber. *Berliner Journal Für Soziologie, 10*(1), S.125–136. ISSN 0863-1808.

Schoenfield, M. (2016). Moral Vagueness Is Ontic Vagueness. *Ethics, 126*(2), 257–282.

Schrader, A. (2010). Responding to Pfisteria piscida (the Fish Killer): Phantomatic Ontology, Indeterminacy, and Responsibility in Toxic Microbiology. *Social Studies of Science, 40*, 275–306.

Schuilenburg, M. (2015). *The Securitization of Society Crime, Risk, and Social Order.* New York: New York University Press.

Scott, J. C. (1999/2008). *Seeing Like a State: How Certain Schemes to Improve the Human Condition Have Failed* (Nachdr.). New Haven, CT: Yale University Press.

Selg, P. (2013). The Politics of Theory and the Constitution of Meaning. *Sociological Theory, 31*(1), 1–23.

Shenker-Osorio, A. (2013). Why Americans All Believe They Are "Middle Class". A Taxonomy of How We Talk About Class and Wealth in the United States Today. *The Atlantic*, www.theatlantic.com/politics/archive/2013/08/why-americans-all-believe-they-are-middle-class/278240/ (last accessed November 15, 2018).

Shiva, V. (2007). Bioprospecting as Sophisticated Biopiracy. *Signs: Journal of Women in Culture and Society, 32*(2), 307–313.

Siemens, H. W. (2013). Reassessing Radical Democratic Theory in the Light of Nietzsche's Ontology of Conflict. In K. Ansell-Pearson (ed), *Nietzsche and Political Thought. Bloomsbury Studies in Continental Philosophy* (pp. 83–106). London and New York, NY: Bloomsbury Academic.

Siles-Brügge, G. (2018). Bound by Gravity or Living in a 'Post Geography Trading World'? Expert Knowledge and Affective Spatial Imaginaries in the Construction of the UK's Post-Brexit Trade Policy. *New Political Economy, 24*(3), 1–18.

Silver, M. (2016). At the Base of Rome's Peculium Economy. *Fundamina, 22*(1), 67–93. doi:10.17159/2411-7870/2016/v22n1a5.

Simmel, G. (1892). *Einleitung in die Moralwissenschaft*, 2 vols. Stuttgart: Cotta.

Simmel, G. (2015). *The View of Life: Four Metaphysical Essays with Journal Aphorisms* (J. A. Y. Andrews & D. N. Levine, Trans.). Chicago, IL and London: The University of Chicago Press.

Simon, N. (n.d.). Retrieved from www.artofrelevance.org/ (last accessed April 21, 2019).

Simpson, L. (2014). Land as Pedagogy. *Decolonization: Indigeneity, Education & Society, 3*(3), 1–25.

Sirota, D. (n.d.). New Data Shows School "Reformers" Are Full of It. Retrieved from www.salon.com/2013/06/03/instead_of_a_war_on_teachers_how_about_one_on_poverty/ (last accessed April 21, 2019).

Skocpol, T. (n.d.). Keynote: Higher Education and the Challenge of Civic Engagement. Presented at the Eastern Sociological Society, Annual Meeting, Boston, MA.

Skocpol, T. & Williamson, V. (2016). *The Tea Party and the Remaking of Republican Conservatism*. New York, NY: Oxford University Press.

Slaughter, A.-M. (2004). *A New World Order*. Princeton, NJ: Princeton University Press.

Smith, L. T. (2012). *Decolonizing Methodologies: Research and Indigenous Peoples* (2nd ed.). London: Zed Books.

Smith, R. S. (1995). Giving Credit Where Credit Is Due: Dorothy Swaine Thomas and the "Thomas Theorem." *The American Sociologist, 26*(4), 9–28.

Soerenson, G. (2010). *The Transformation of the State*. New York, NY: Palgrave MacMillan.

Solms-Laubach, F. (2007). *Nietzsche and Early German and Austrian Sociology*. Berlin: de Gruyter.

Spillers, H. (1987). Mama's Baby, Papa's Maybe: An American Grammar Book. *Diacritics, 17*(2), Culture and Countermemory: The "American" Connection (Summer, 1987), 64–81.

Spinello, R. A. (2009). *A Defense of Intellectual Property Rights*. Cheltenham: Edward Elgar.

Staubmann, H. (ed). (2016). *Soziologie in Österreich – Internationale Verflechtungen*. doi:10.15203/3122-56-7.

Steinberg, H., et al. (2013). Johann Christian August Heinroth. *Psychiatria Danubina, 25*(1), 11–16.

Steinmetz, G. (ed) (2005). *The Politics of Method in the Human Sciences: Positivism and Its Epistemological Others*. Durham, NC: Duke University Press.

Steinmetz, G. (2013). *Sociology and Empire: The Imperial Entanglements of a Discipline*. Durham, NC: Duke University Press.

Stengers, I. (2013). Introductory Notes on an Ecology of Practices. *Cultural Studies Review, 11*(1), 183–196. doi:10.5130/csr.v11i1.3459.

Sterne, J. (2003). Bourdieu, Technique and Technology. *Cultural Studies, 17*(3–4), 367–389. doi:10.1080/0950238032000083863a.

Stichweh, R. (1984). *Zur Entstehung des modernen Systems wissenschaftlicher Disziplinen: Physik in Deutschland, 1740–1890* (1. Aufl). Frankfurt am Main: Suhrkamp.

Stichweh, R. (1991). *Der frühmoderne Staat und die europäische Universität: zur Interaktion von Politik und Erziehungssystem im Prozess ihrer Ausdifferenzierung (16.-18. Jahrhundert)* (1. Aufl). Frankfurt am Main: Suhrkamp.

Stichweh, R. (2000). *Die Weltgesellschaft: soziologische Analysen* (1. Aufl., Orig.-Ausg). Frankfurt am Main: Suhrkamp.

Stiegler, B. (2010). *Taking Care of Youth and the Generations*. Stanford, CA: Stanford University Press.

Stiegler, B. & Neyrat, F. (2012). Interview. *Parrhesia, 14*, 9–15.

Stingl, A. (2008). *Between Discursivity and Sensus Communis*. Erlangen: OPUS.

Stingl, A. (2009). *Aufklaerung als Flaschenpost oder Anthropologie der Gegenwart*. Saarbruecken: VDM.

Stingl, A. I. (2011). Truth, Knowledge, Narratives of Selves: An Account of the Volatility of Truth, the Power of Semantic Agency, and Time in

Narratives of the Self. *The American Sociologist, 42*(2–3), 207–219. doi:10.1007/s12108-011-9128-z.

Stingl, A. I. (2014a). Braining Your Life/Living Your Brain: Cyborg Gaze and Medical Images. In M. Grabowski (ed), *Neuroscience and Media: New Understandings and Representations.* London: Routledge.

Stingl, A. I. (2014b). Digital Fairground – The Virtualization of Health, Illness, and the Experience of "Becoming a Patient" As a Problem of Political Ontology and Social Justice. In H. F. Dahms (ed), *Current Perspectives in Social Theory* (Vol. 32, pp. 53–92). doi:10.1108/S0278-120420140000032004.

Stingl, A. I. (2015). The Rural Imaginary. In J. I. Bakker (ed), *The Methodology of Political Economy: Studying the Global Rural-Urban Matrix.* Lanham, MD: Lexington Books.

Stingl, A. I. (2016a). Digital Divide. In H. ten Have (ed), *Encyclopedia of Global Bioethics* (pp. 868–877). doi:10.1007/978-3-319-09483-0_246.

Stingl, A. I. (2016b). *The Digital Coloniality of Power: Epistemic Disobedience in the Social Sciences and the Legitimacy of the Digital Age.* Lanham, MD: Lexington Books.

Stingl, A. I. (2017). Exercising Epistemic Authority and the Deployment of Persuasive Technologies: The Case of Medical Images. In J. Vollmann, A. Nossek, J. Gather, & T. Henking (eds), *Beneficial Coercion in Medicine? Foundations, Areas of Conflict, Prevention* (publication Following BMBF Klausurwoche, Bochum 2016), Münster: Mentis.

Stingl, A. I. & Weiss, S. M. (2013). Beyond and Before the Label: The Ecologies and Agencies of ADHD. In M. Dellwing & M. Harbusch (eds), *Krankheitskonstruktionen und Krankheitstreiberei* (pp. 201–231). doi:10.1007/978-3-531-18784-6_9.

Stingl, A. I. & Weiss, S. M. (2014). Between Shell and Ghost: A Hauntology of Zombies in the Social Imaginary. Also cited as Stingl/Weiss (2015)

Stingl, A. I. & Weiss, S. M. (2015). Between Shell and Ghost: A Hauntology of Zombies in the Social Imaginary. In M. Dellwing & M. Harbusch (eds), *Vergemeinschaftung in Zeiten der Zombie-Apokalypse* (pp. 69–122). doi:10.1007/978-3-658-01722-4_4.

Stingl, A. I. & Weiss, S. M. (2016). Making Trouble: Mindfulness As a Care Ethic. In C. Brand (ed), *Dual-Process Theories in Moral Psychology* (pp. 315–341). doi:10.1007/978-3-658-12053-5_15.

Stoler, A. (2002). Colonial Archives and the Arts of Governance: On the Content in the Form. In C. Hamilton, V. Harris, J. Taylor, M. Pickover, G. Reid, & R. Saleh (eds). *Refiguring the Archive.* Dordrecht: Springer.

Stoler, A. (2009). *Along the Archival Grain.* Princeton, NJ: Princeton University Press.

Strathern, M. (ed) (2000). *Audit Cultures: Anthropological Studies in Accountability, Ethics and the Academy.* London: Routledge.

Strohl, M. P. (2011). Bradley's Benzedrine Studies on Children with Behavioral Disorders. *The Yale Journal of Biology and Medicine, 84*(1), 27–33.

Sunar, L. (2014). From Democracy to Despotism: Tocqueville on Slavery, Colonialism and "Other." *Ilahiyat Studies, 5*(2), 155–188. doi:10.12730/13091719.2014.52.107.

Supiot, A. (2013). The Public-Private Relation in the Context of Today's Refeudalization. *International Journal of Constitutional Law, 11*(1), 129–145. doi:10.1093/icon/mos050.

Swaine Thomas, D. (1929). Statistics in Social Research. *American Journal of Sociology, 35*(1), 1–17.

Swaine Thomas, D. (1952). *Experiences in Interdisciplinary Research.* In Presidential Address, American Sociological Association. Atlantic City, NJ.

Swaine Thomas, D. (n.d.). *The Methodology of Experimental Sociology. Introduction to Dorothy Swaine Thomas and Associates, Some New Techniques for Studying Social Behavior* (pp. 1–21). New York, NY: Teachers College, Columbia University. Retrieved from https://brocku.ca/MeadProject/Thomas/Dorothy/Thomas_1929.html (last accessed June 20, 2018).

Swyngedouw, E. (2006). Circulations and Metabolisms: (Hybrid) Natures and (Cyborg) Cities. *Science as Culture, 15*(2), 105–121.

Sykes. K. (ed) (2017). Interrogating Individuals: The Critique of Possessive Individualism in the Western Pacific. Special Issue *Anthropological Forum, 17*(3), 213–224.

Synek, E. M. (2005). '*Oikos*-Ecclesiology' and 'Church Order' in Eastern Christianity. In A. B. Mulder-Bakker & J. Wogan-Browne (eds), *Household, Women, and Christianities* (Vol. 14, pp. 37–70). doi:10.1484/M.MWTC-EB.3.3621.

Szakolczai, Á. (1998). *Max Weber and Michel Foucault: Parallel Life-Works.* London: Routledge.

Sze, J. (n.d.). *Sustainability: Approaches to Environmental Justice and Social Power.* New York: New York University Press.

Talay, Z. (2011). A Dialogue with Nietzsche: Blumenberg and Löwith on History and Progress. *History of European Ideas, 37*(3), 376–381. doi:10.1016/j.histeuroideas.2010.11.002.

Tamer, G. (2001). *Islamische Philosophie und die Krise der Moderne: Das Verhältnis von Leo Strauss zu Alfarabi, Avicenna und Averroes.* Leiden: Brill.

Taylor, C. (2004). *Modern Social Imaginaries.* Durham, NC: Duke University Press.

Taylor, D. E. (2009). *The Environment and the People in American Cities, 1600–1900s: Disorder, Inequality, and Social Change.* Durham, NC: Duke University Press.

Taylor, D. E. (2014). *Toxic Communities: Environmental Racism, Industrial Pollution, and Residential Mobility.* New York: New York University Press.

Taylor, D. E. (2016). *The Rise of the American Conservation Movement: Power, Privilege, and Environmental Protection.* Durham, NC: Duke University Press.

Taylor, D. H. (2017). *Take Us to Your Chief: And Other Stories: Classic Science-Fiction with a Contemporary First Nations Outlook.* Vancouver: Douglas & McIntyre.

Temin, P. (2006). The Economy of the Early Roman Empire. *Journal of Economic Perspectives, 20*(1), 133–151. doi:10.1257/089533006776526148.

Temin, P. (2013). *The Roman Market Economy.* Princeton, NJ: Princeton University Press.

Teoharova, G. (2005). *Karl Jaspers' Philosophie auf dem Weg zur Weltphilosophie.* Würzburg: Königshausen & Neumann.

Teubner, G. (2010). Fragmented Foundations: Societal Constitutionalism Beyond the Nation State. In Petra Dobner und Martin Loughlin (Hrsg.), *The Twilight of Constitutionalism?* Oxford: Oxford University Press.

Teubner, G. (2011). Constitutionalising Polycontexturality. *Social and Legal Studies*, 19, 209–252.

Teubner, G. (2012). *Verfassungsfragmente.* Berlin: Suhrkamp.

Thacker, E. (2011). Necrologies; or, the Death of the Body Politic. In P. T. Clough & C. Willse (eds), *Beyond Biopolitics: Essays on the Governance of Life and Death* (pp. 139–162). Durham, NC: Duke University Press.

Thévenot, L. (2001). Organized Complexity Conventions of Coordination and the Composition of Economic Arrangements. *European Journal of Social Theory, 4*(4), 405–425.

Thévenot, L. & Boltanski, L. (2006). *On Justification*. Princeton, NJ: Princeton University Press.

Thomas, R. H. (1986). *Nietzsche in German Politics and Society, 1890–1918*. La Salle, IL: Open Court.

Thornhill, C. (2018). On Misunderstanding States: The Transnational Constitution in the National Constitution. *International Journal of Constitutional Law, 16*(4), 1186–1198. doi:10.1093/icon/moy095.

Thornhill, C. (2019). The Sociology of Law and Global Sociology. *Zeitschrift Für Soziologie, 47*(6), 375–401. doi:10.1515/zfsoz-2018-0124.

Thrift, N. (2004). Remembering the Technological Unconscious by Foregrounding Knowledges of Position. *Environment and Planning D: Society and Space, 22*(1), 175–190. doi:10.1068/d321t.

Tlostanova, M. (2018). *Madina Tlostanova on Decolonizing the Equality Discourse*. Retrieved from www.youtube.com/watch?v=tVPVbpinrAI (last accessed April 21, 2019).

Torpey, J. (2009). The Problem of 'American Exceptionalism' Revisited. *Journal of Classical Sociology, 9*(1), 143–168. doi:10.1177/1468795X08098981.

Torri, M. C. (2011). Bioprospecting and Commercialisation of Biological Resources by Indigenous Communities in India: Moving Towards a New Paradigm? *Science, Technology and Society, 16*(2), 123–146.

Toscano, A. (2008). The Culture of Abstraction. *Theory, Culture & Society, 25*(4), 57–75. doi:10.1177/0263276408091983.

Trakman, L. (2012, January 26). Trakman, Leon, Investor State Arbitration or Local Courts: Will Australia Set a New Trend?. *Journal of World Trade, 46*(1), 83–120; UNSW Law Research Paper No. 2012-1. Retrieved from SSRN: https://ssrn.com/abstract=2000361 (last accessed April 21, 2019).

Tronto, J. C. (2013). *Caring Democracy: Markets, Equality, and Justice*. New York: New York University Press.

Trouillot, M.-R. (2001). The Anthropology of the State in the Age of Globalization: Close Encounters of the Deceptive Kind. *Current Anthropology, 42*(1), 125–138.

Trouillot, M.-R. (2002). North Atlantic Universals: Analytical Fictions 1792–1945. *South Atlantic Quarterly, 101*(4), 839–858.

Tsing, A. (2015). *The Mushroom at the End of the World*. Princeton, NJ: Princeton University Press.

Tsing, A. (2016). *The Mushroom at the End of the World*. Princeton, NJ: Princeton University Press.

Turner, B. S. (1982). Nietzsche, Weber and the Devaluation of Politics: The Problem of State Legitimacy. *The Sociological Review, 30*(3), 367–391. doi:10.1111/j.1467-954X.1982.tb00659.x.

Turner, B. S. (2011). Max Weber and the Spirit of Resentment: The Nietzsche Legacy. *Journal of Classical Sociology, 11*(1), 75–92. doi:10.1177/1468795X10391458.

Turner, C. (1993). Liberalism and the Limits of Science: Weber and Blumenberg. *History of the Human Sciences, 6*(4), 57–79. doi:10.1177/095269519300600404.

Turner, J. (2008). American Individualism and Structural Injustice: Tocqueville, Gender, and Race. *Polity, 40*(2), 197–215. doi:10.1057/palgrave.polity.2300088.

Tversky, B. (2004). Narratives of Space Time and Life. *Mind & Language, 19*(4), 380–392.

Uberoi, J. P. S. (1984). *The Other Mind of Europe: Goethe as a Scientist*. Delhi and New York, NY: Oxford University Press.

Valsiner, J. & Veer, R. van der. (2000). *The Social Mind: Construction of the Idea*. Cambridge and New York, NY: Cambridge University Press.

van den Scott, L.-J. (2016). The Extension of the Coloniality of Power into Digital Culture: Book Review. *Symbolic Interaction, 40*(1): 133–135.

Varela, M., Dhawan, N. (2015). *Postkoloniale Theorie*. Bielefeld: Transcript.

Varoufakis, Y. (2015). Yanis Varoufakis Full Transcript: Our Battle to Save Greece. Retrieved from www.Newstatesman.Com/World-Affairs/2015/07/Yanis-Varoufakis-Full-Transcript-Our-Battle-Save-Greece (last accessed April 21, 2019).

Verstraete, B. C. (1980). Slavery and the Social Dynamics of Male Homosexual Relations in Ancient Rome. *Journal of Homosexuality, 5*(3), 227–236. doi:10.1300/J082v05n03_06.

Vesselinov, E., Cazessus, M., & Falk, W. (2007). Gated Communities and Spatial Inequality. *Journal of Urban Affairs, 29*(2), 109–127. doi:10.1111/j.1467-9906.2007.00330.x.

Viglas, K. (2016). The Placement of Lucian's Novel True History in the Genre of Science Fiction. *Interlitteraia, 21*(1), 158–171.

Vitale, F. (2013). The Law of the Oikos. Jacques Derrida and the Deconstruction of the Dwelling. *SAJ, 5*(1), 59–73.

Viveiros de Casro, E. (2004). Perspectival Anthropology and the Method of Controlled Equivocation. *Tipití: Journal of the Society for the Anthropology of Lowland South America, 2*(1), 3–22.

Vlassopoulos, K. (2009). Slavery, Freedom and Citizenship in Classical Athens: Beyond a Legalistic Approach. *European Review of History: Revue Europeenne d'histoire, 16*(3), 347–363. doi:10.1080/13507480902916852.

von Kempski, J. (n.d.). *Schriften*, Vols. 1–3. Frankfurt am Main: Suhrkamp.

Wahl, D. (2005). "Zarte Empirie": Goethean Science as a Way of Knowing. *Janus Head, 8*(1), 58–76.

Wallerstein, I. (n.d.). *The Modern World System,* Vol. I–IV. Berkeley: California University Press.

Wark, M. (n.d.). *Molecular Red*. New York, NY: Verso.

Wark, M. (2015). Lazzarato and Pasoloni. Retrieved from www.Publicseminar.Org/2015/06/Lazzarato-and-Pasolini/ (last accessed April 21, 2019).

Wearne. B. (1989). The theory and scholarship of Talcott Parsons to 1951: A Critical Commentary. Cambriudge, UK: Cambridge University Press.

Weber, M., Baehr, P., & Wells, G. C. (2002). *The Protestant Ethic and the "Spirit" of Capitalism and Other Writings*. New York, NY: Penguin Books.

Weber, M., Winckelmann, J., Weber, M., & Weber, M. (1988). *Gesammelte Aufsätze zur Wissenschaftslehre* (7. Aufl., photomechanischer Nachdr. der 6. Aufl). Tübingen: Mohr.

Weheliye, A. G. (2014). *Habeas Viscus: Racializing Assemblages, Biopolitics, and Black Feminist Theories of the Human*. Durham, NC: Duke University Press.

Weigel, S. (1996). *Body-and Image-Space: Re-Reading Walter Benjamin*. London and New York, NY: Routledge.

Weiner, M. & Báez, A. C. (2018). *Smash the Pillars*. Lanham, MD: Lexington Books.

Weiss, S. M. (n.d.). Dynamic Being: A Digital/Analog Analysis of the Technosocial Anthropocology of Juvenile ADHD Care Decisions. 4S Conference, Copenhagen. Retrieved from http://citation.allacademic.com/meta/p_mla_apa_research_citation/5/8/0/8/8/p580885_index.html?phpsessid=pjnd7r4ftgo34v3ekqm3n0rpp0 (last accessed April 21, 2019).

Weiss, S. M., Restivo, S. P., & Stingl, A. I. (2014). *Worlds of ScienceCraft: New Horizons in Sociology, Philosophy, and Science Studies*. Farnham and Burlington, VT: Ashgate Pub.

Welch, C. (2007). Creating Concitoyens: Tocqueville on the Legacy of Slavery. In R. Geenens & A. De Dijn (eds), *Reading Tocqueville* (pp. 31–51). doi:10.1057/9780230599123_3.

Whitehead, A. N. (1929). *The Function of Reason*. Princeton, NJ: Princeton University Press.

Whitehead, A. N. (1997). *Science and the Modern World: Lowell Lectures, 1925*. New York, NY: The Free Press.

Whitehead, A. N. (2015). *The Concept of Nature: The Tarner Lectures Delivered in Trinity College, November 1919* (Cambridge philosophy classics edition). Cambridge: Cambridge University Press.

Whitehead, A. N. (n.d.). *Religion in the Making: Lowell Lectures 1926*. New York, NY: Fordham University Press.

Whitehead, A. N., Griffin, D. R., & Sherburne, D. W. (1985). *Process and Reality: An Essay in Cosmology: Gifford Lectures Delivered in the University of Edinburgh During the Session 1927–1928* (Corr. ed., first Free Press paperback edition). New York, NY: Free Press.

Williams, J. R. G. (2008). Ontic Vagueness and Metaphysical Indeterminacy. *Philosophy Compass*, *3*(4), 763–788.

Williamson, V., et al. (2011). The Tea Party and the Remaking of Republican Conservatism. *Perspectives on Politics*, *9*(01), 25–43.

Winthrop-Young, G. (2013). Cultural Techniques: Preliminary Remarks. *Theory, Culture & Society*, (online first). doi:10.1177/0263276413500828.

Winthrop-Young, G. (2014). The Kultur of Cultural Techniques: Conceptual Inertia and the Parasitic Materialities of Ontologization. *Cultural Politics an International Journal*, *10*(3), 376–388. doi:10.1215/17432197-2795741.

Wolmark, J. (1994). *Aliens and Others: Science Fiction, Feminism and Postmodernism*. Iowa: Iowa University Press.

Womack, Y. (2013). *Afrofuturism: The World of Black Sci-Fi and Fantasy Culture*. Chicago, IL: Chicago Review Press.

Wrong, D. H. (1961). The Oversocialized Conception of Man in Modern Sociology. *American Sociological Review*, *26*(2), 183. doi:10.2307/2089854.

Wu, T. (2018, forthcoming). Blind Spot: The Attention Economy and the Law. *Antitrust Law Journal*, 82. Retrieved from https://papers.ssrn.com/sol3/papers.cfm?abstract_id=2941094 (last accessed April 21, 2019).

Wulf, A. (2015). *The Invention of Nature: Alexander von Humboldt's New World* (1st American ed.). New York, NY: Vintage Books.

Wylie, S. A. (2018). *Fractivism: Corporate Bodies and Chemical Bonds*. Durham, NC: Duke University Press.

Wynter, S. (2000). The Re-Enchantment of Humanism: An Interview with Sylvia Wynter. *Small Axe, 8*, 119–207.

Wynter, S. (2003). Unsettling the Coloniality of Being/Power/Truth/Freedom: Towards the Human, After Man, Its Overrepresentation—An Argument. *CR: The New Centennial Review, 3*(3), 257–337.

Wynter, S. & McKittrick, K. (2014). *Sylvia Wynter: Being Human as Praxis*. Durham, NC: Duke University Press.

Ye, J., van der Ploeg, J. D., Schneider, S., & Shanin, T. (2019). The Incursions of Extractivism: Moving from Dispersed Places to Global Capitalism. *The Journal of Peasant Studies*, Online First, 1–29. doi:10.1080/03066150.2018.1559834.

Zamfir, K. (2013). *Men and Women in the Household of God: A Contextual Approach to Roles and Ministries in the Pastoral Epistles*. Goettingen: Vandenhoeck & Ruprecht.

Zerilli, L. M. G. (2005). "We Feel Our Freedom": Imagination and Judgment in the Thought of Hannah Arendt. *Political Theory, 33*(2), 158–188. doi:10.1177/0090591704272958.

Zerilli, L. M. G. (2016). *A Democratic Theory of Judgment*. Chicago, IL and London: The University of Chicago Press.

Žižek, S. (2014). *Trouble in Paradise: From the End of History to the End of Capitalism*. London: Allen Lane.

Zuern, M. (1998). *Regieren Jenseits des Nationalsaates*. Frankfurt am Main: Suhrkamp.

Index

For Product Safety Concerns and Information please contact our EU representative GPSR@taylorandfrancis.com
Taylor & Francis Verlag GmbH, Kaufingerstraße 24, 80331 München, Germany

www.ingramcontent.com/pod-product-compliance
Ingram Content Group UK Ltd.
Pitfield, Milton Keynes, MK11 3LW, UK
UKHW022004180926
475701UK00003B/193
9781032082714